A SURVEY OF CLASSICAL AND MODERN GEOMETRIES

WITH COMPUTER ACTIVITIES

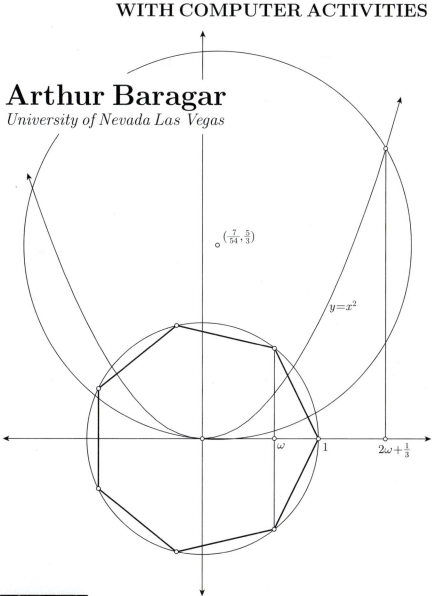

Arthur Baragar

University of Nevada Las Vegas

$\left(\frac{7}{54}, \frac{5}{3}\right)$

$y=x^2$

ω 1 $2\omega+\frac{1}{3}$

Prentice Hall

PRENTICE HALL
Upper Saddle River, New Jersey 07458

Library of Congress Cataloging in Publication Data

Baragar, Arthur.
    A survey of classical and modern geometries: with computer activities / Arthur Baragar.
        p. cm.
    Includes bibliographical references and index.
    ISBN 0-13-014318-9
    1. Geometry. I. Title.

QA445 .B318 2001
516–dc21   00-050201

Acquisition Editor: George Lobell
Editor in Chief: Sally Yagan
Vice President/Director of Production and Manufacturing: David W. Riccardi
Executive Managing Editor: Kathleen Schiaparelli
Senior Managing Editor: Linda Mihatov Behrens
Production Editor: Barbara Mack
Manufacturing Buyer: Alan Fischer
Manufacturing Manager: Trudy Pisciotti
Marketing Manager: Angela Battle
Marketing Assistant: Vince Jansen
Director of Marketing: John Tweeddale
Editorial Assistant: Gale Epps
Art Director: Jayne Conte
Cover Design: Bruce Kenselaar
Cover Photo: Museum of Fruit, Yamanashi. Mitsumasa Fujitsuka/Itsuko Hasegawa
    Atelier

Prentice Hall   ©2001 by Prentice-Hall, Inc.
                Upper Saddle River, New Jersey 07458

Printed in the United States of America
10   9   8   7   6   5   4   3   2   1

ISBN 0-13-014318-9

Prentice-Hall International (UK) Limited, London
Prentice-Hall of Australia Pty. Limited, Sydney
Prentice-Hall Canada Inc., Toronto
Prentice-Hall Hispanoamericana, S.A., Mexico
Prentice-Hall of India Private Limited, New Delhi
Prentice-Hall of Japan, Inc., Tokyo
Pearson Education Asia Pte. Ltd.
Editora Prentice-Hall do Brasil, Ltda., Rio de Janeiro

For Meg and Timothy

Contents

Preface for the Instructor and Reader

I never intended to write a textbook and certainly not one in geometry. It was not until I taught a course to future high school teachers that I discovered that I have a view of the subject which is not very well represented by the current textbooks. The dominant trend in American college geometry courses is to use geometry as a medium to teach the logic of axiomatic systems. Though geometry lends itself very well to such an endeavor, I feel that treating it that way takes a lot of excitement out of the subject. In this text, I try to capture the joy that I have for the topic. Geometry is a fun and exciting subject that should be studied for its own sake.

Though the primary target audience for this text is the future high school teacher, this text is also suitable for math majors, both because of the challenging problems throughout the text, and because of the quantity of material. In particular, I think this would make an excellent text for an undergraduate course in hyperbolic geometry.

To the Student

In the *Republic*, Plato (ca. 427 – 347 B.C.) wrote that his ideal State should be ruled by philosophers educated first in mathematics. He believed that the value of mathematics is how it trains the mind, and that its practical utility is of minor importance. This philosophy is as valid now as it was then. A modern education might include vocational or technical training (such as engineering, medicine, or law), but at its core, there are the English and mathematics courses which make up a liberal education. Though mathematics has rather surprising utility, for many students, the most important lesson to be learned in their math classes is how to think analytically, creatively, and rigorously.

Keep this in mind as you read this book. Recognize that the exercises are a fundamental and integral part of the text. This is where the most important lessons are learned. You will not solve them all, perhaps not even most, but I hope that the exercises you do solve will leave you with a feeling of satisfaction.

Recommended Courses

For a college geometry course for future high school teachers, the basic course outline that I recommend and usually teach is shown in Table 1.

Sections	Comments
1.1 – 1.12	Light on Sections 1.3 and 1.4.
1.13 – 1.15	Optional.
3.1 – 3.7	Section 3.7 is optional.
4.1 – 4.4	Integrate with Chapter 3.
5.1 – 5.5	Section 5.3 is optional.
6.1 – 6.2, 6.4 – 6.6	Cover quickly and sparingly.
7.1 – 7.4, 7.6 – 7.13	
8.1 – 8.2, 8.4 – 8.5	Use an overhead.

Table 1. A recommended course outline for future high school teachers.

Chapter 2 on Greek astronomy provides some interesting material which can be mixed in with Chapter 1, or used on 'optional' days, such as the Wednesday before Thanksgiving. I usually begin integrating Sketchpad (Chapter 4) after I have completed the first few sections on constructions (Chapter 3). A laptop and computer projector come in handy. Polyhedra (Chapter 5) might be considered optional, but I think it can be very valuable for a future high school teacher. In particular, Exercise 5.14 should not be missed, both as a class project and again as an exercise. These are lessons which can be easily brought into the high school classroom and have the potential to be memorable. I usually skip most of Chapter 6, and only introduce the 'crutch,' the concepts of parallel and ultraparallel lines, and the concept of asymptotic triangles. The beginning of Chapter 7 poses a bit of a dilemma. Most of my students are not familiar enough with path integrals and differentials to understand the arguments of Sections 7.2 and 7.3. I could not see a way of introducing the Poincaré upper half plane model that avoids these arguments or something as difficult. I usually ask those students to accept these results and not worry too much if they do not understand the proofs. If I reach Chapter 8, it is usually covered during the last week of classes. I think of it as a cushion which allows the students a little extra time to absorb the difficult material of Chapter 7 before their final.

One of the constraints I face when I teach this course is the weak background of some of our students. Education students who have chosen mathematics as their second teaching field are required to take our geometry course. Outside of this course, the most sophisticated course they are currently required to take is the first semester of calculus. We are in the process of changing this, so that these students must also take a course in linear algebra. I think a rather nice alternative for a class of these students would be to omit Chapters 6 and 7, and instead introduce the pseudosphere

(Chapter 12) as the model of hyperbolic geometry, after covering spherical geometry (Sections 10.1 – 10.5). With such a course, I would not overly emphasize the axioms of geometry. I would instead emphasize the relations between these geometries through the similar results, most notably in the different trigonometries. Such a plan would require a little more thought on the part of the instructor, since Chapter 12 was not written with this organization in mind. Nevertheless, a good instructor thoroughly familiar with the contents of Chapter 7 should be able to pull it off.

Special Notes

There are many places where the treatment of this subject could have been done differently. I would like to take a moment to explain some of my choices, as well as draw attention to and justify some of the unusual placements of material. Instructors may wish to occasionally return to this section as they teach.

In Chapter 1, I never do define the measure of an angle. Though I use degrees earlier, there is no real need to talk about the measure of an angle until the Law of Cosines is introduced. Before that, for example in the Star Trek lemma, we only need a notion of congruent angles, which is defined via isometries. Since I already assume knowledge of trigonometry when I introduce the Law of Cosines, I do not see the point of formally defining the measure of angles. The student is eventually asked to formally define the measure of angles in Exercise 9.21. In Chapter 1, when we do use the measure of angles, we use degrees, which is the measure most commonly used in high schools. Later, when we introduce hyperbolic geometry, we switch to radians.

There is a nice proof of Ceva's theorem (see Exercise 1.120) which does not use Menelaus' theorem. This can be used by an instructor who wishes to skip Menelaus' theorem. One advantage of the proof of Ceva's theorem using Menelaus' theorem is that it also works in both spherical and hyperbolic geometry.

There is a very nice proof that $\cos(2\pi/5)$ is constructible (see Exercise 3.18). The advantage of the algebraic proof given in the text is that similar arguments are required in the proof that $\cos(2\pi/7)$ and similar quantities are not constructible.

There are a number of programs similar to Geometer's Sketchpad (like Cabri and Cinderella), but I believe Sketchpad currently dominates the market, particularly in the high schools. This is why I chose to learn and write about Sketchpad.

I have grown to appreciate the value of Geometer's Sketchpad and encourage instructors and readers to not just shrug off Chapter 4. It can be very useful for weak students and can be very valuable for future high school teachers. It can also be very fascinating and instructive for talented students. There are a lot of questions about constructions that I would never have considered had I not been familiar with Sketchpad. For example, which tilings of the Poincaré plane can be drawn using only a straightedge and compass? How can we construct a regular 7-gon using a straightedge, compass, and something else (see Exercise 3.39)? Some theorems, for example Feuerbach's theorem, are also a little more satisfying when played with using dynamic software (see Exercise 4.22).

Results in hyperbolic trigonometry are included in Section 7.16. It is appropriate to first read about spherical trigonometry, which appears later in Sections 10.2 and 10.3. I chose to introduce hyperbolic trigonometry first only because I wanted to keep it together with the rest of Chapter 7. This could have been avoided by introducing spherical geometry first, but because we introduce new geometries via a change in Euclid's axioms, hyperbolic geometry naturally comes first.

Tilings are first introduced in the exercises of Chapter 5 together with the regular and semiregular polyhedra. They are introduced again in Chapter 8, together with tilings of hyperbolic geometry.

Chapter 9 is an unusual treatment of the foundations of geometry. It is intended for students who have already taken a course in analysis and assumes an axiomatic development of the real line.

When compared to contemporary textbooks, the placement of Chapter 9 might also seem unusual, but it is not so unusual when compared with history. A sound axiomatic system for geometry was not developed until the late nineteenth century, well after the development of models for hyperbolic geometry. Though the logical order of geometry begins with the axioms, I do not believe that it should be taught that way. A strong intuitive understanding of geometry is necessary for anyone to understand the subtleties of the axiomatic foundation.

As mentioned earlier, the placement of Chapter 10 is a matter of taste. If the instructor wishes to introduce spherical geometry earlier, there is no problem. The only prerequisites for Sections 10.1 – 10.5, other than Chapter 1, are trigonometry and some vector geometry (dot products and cross products). Parts of Chapter 5 should be done before Section 10.6, and Chapter 9 is a prerequisite for Sections 10.7 and 10.8.

If the instructor really wishes to emphasize axiomatic systems, I encourage them to look closely at Chapter 13. In this chapter, the finite affine and projective planes are first introduced as algebraic objects. We then define them as incidence geometries together with Desargues' theorem and eventually show that the two definitions are equivalent. This beautiful result due to Hilbert really emphasizes the relationship between algebra and geometry.

Chapter Dependence

Though most of this book is meant to be read in order, there are only a few chapters which have a heavy dependence on earlier chapters. These

Sections	Prerequisites and co-requisites
Chapter 1	Trigonometry and mathematical sophistication.
Chapter 2	None, not even Chapter 1.
Chapter 3	None (except Chapter 1).
Chapter 4	Co-requisite: Chapter 3.
Chapter 5	None, except that Section 5.6 requires some integration.
Chapter 6	None.
Chapter 7	Chapter 6 is helpful, but not required. Familiarity with 2×2 matrix algebra is required. Some sections require additional background, as described below.
7.1 – 7.3	Line integrals and multivariable differentials.
7.13	Double integrals.
7.16 – 7.17	Recommended co-requisite: Sections 10.2 and 10.3.
Chapter 8	Chapter 5 and most of Chapter 7 are recommended.
Chapter 9	Some real analysis is recommended (Cauchy sequences and the development of the real line.)
Chapter 10	Vector geometry (dot and cross products), Section 1.15, and nothing more, except as noted below.
10.6	Chapter 5.
10.7	Chapter 9.
Chapter 11	Familiarity with 3×3 matrix algebra.
Chapter 12	Familiarity with 3×3 matrix algebra, dot and cross products, Section 1.15, and Chapter 10. Section 7.14 is required for Section 12.4.
Chapter 13	Modular arithmetic and Chapter 11. A course in abstract algebra will help.
Chapter 14	Chapter 3 and at least linear algebra, though a course in abstract algebra would be helpful.
Chapter 15	Sections 15.2, 15.3, and 15.4 should be read in order. Section 15.5 can be read separately but is related to Section 15.3. Section 15.6 requires Section 5.3. Section 15.7 assumes Chapter 7. Section 15.8 does not require any other background. Section 15.9 references finite geometries, the subject of Chapter 13.
Chapter 16	None.

Table 2

dependencies are outlined in Table 2. Depending on course objectives, several chapters can be safely skipped, and in particular, spherical geometry (Chapter 10) can immediately follow Chapter 1. I expect that the reader has at least a decent high school education, including trigonometry, and that they have some mathematical sophistication. I also expect that all readers cover the bulk of Chapter 1 (say, Sections 1.1 – 1.11) before moving on. More background is required for some of the text, as outlined in Table 2.

Errata and Web Support

Supporting material for this textbook will be made available at

http://www.nevada.edu/~baragar/geometry.html

I anticipate that this page will include further exercises, perhaps solutions, links to related sites, and an errata sheet. Comments and reports of errors are sincerely appreciated and can be sent to

baragar@nevada.edu

Acknowledgments

This text evolved from a course I taught several times at the University of Nevada Las Vegas. I would like to thank all the students who took this course, and in particular, I would like to thank the class of fall '97. They showed a great deal of character by embracing this subject with nothing but classroom notes and a text we never used. Their enthusiasm was inspirational and helped motivate the creation of this text. I would like to extend special thanks to Robin Fulmer and Brenda Walker, who both lent me their notes from the fall class.

I would like to thank my advisor, Joseph Silverman, and my editor, George Lobell, whose encouragement helped transform those classroom notes into a textbook. I would like to thank Peter Shiue and Dorette Pronk, who both provided feedback after they used versions of this text in courses they taught. I would like to thank Jeff Johannes, who also carefully read the text and who participated in frequent conversations about geometry and the history of mathematics. I would like to thank the reviewers too for their input. I would like to thank my production and copy editors, Barbara Mack and Martha Williams, who taught me a little about grammar.

I would like to thank John Scherk, from whom I took my first undergraduate course in geometry at the University of Alberta.

I would like to thank the members and coaches of the '98 and '99 Canadian IMO (International Mathematical Olympiad) teams. Some of the more sophisticated gems in this book are due to my association with these teams. It was also through my association with the IMO that I was exposed to Kiran Kedlaya's beautiful book [Ke]. I highly recommend this text to anyone with a serious interest in competition mathematics.

I would like to thank Hanns-Heinrich Langmann of Germany, Åke H. Samuelsson of Sweden, and Bogdan Enescu of Romania, for graciously allowing me to use the IMO logos from '89, '91, and '99, the years their respective countries hosted the International Mathematical Olympiad.

Finally, I would like to thank my wife Meg, and my son Timothy, whose support and tolerance made writing this text smoother and more enjoyable.

Arthur Baragar

Introduction

We live in a geometrical world, and at least since the time of the ancient Babylonians and Egyptians, we have tried to model our world. One of the ancient Greeks, Euclid (ca. 300 B.C.), proposed five postulates[1] that essentially define the geometry that bears his name – Euclidean geometry. This version of geometry is the most widely known and accepted of all geometries. In fact, for many centuries, many people attempted to prove that it is the only geometry. More precisely, if we accept the first four of Euclid's postulates, is it possible to prove the fifth postulate – the parallel postulate? This last postulate asserts that given any line and any point not on that line, there exists a unique line through that point which never intersects the original line – that is, the fifth postulate asserts the existence of a unique parallel. It is now known that the fifth postulate does not follow from the first four, and if we instead assume that through any point there exist at least two lines which do not intersect a given line not through that point, then we get a geometry known as *hyperbolic* geometry. Because it was widely believed that the fifth postulate should follow from the first four, centuries passed before hyperbolic geometry was discovered, and after it was discovered, it took years for it to be accepted, even by the original discoverers. For a very intriguing and expansive history of the discovery of hyperbolic geometry, consult Greenberg's book *Euclidean and Non-Euclidean Geometries*[Gre].

The first four of Euclid's postulates may also be modified. There exist valid geometries – for example spherical geometry – for which these postulates are not all true. In this text, we will investigate several geometries, including Euclidean, hyperbolic, and spherical geometry.

0.1   The Geometry of Our World

Since we regularly experience the geometry of our world, we already have notions of lengths, angles, and lines. We typically define a line segment

---

[1] *Postulate*: Something assumed without proof as being self-evident or generally accepted, especially when used as a basis for an argument. An axiom. Mathematicians sometimes make a subtle distinction between axioms and postulates, but in this text, we will consider them to be synonyms.

between two points to be the shortest path between them. Thus, a string, when pulled taut, describes a straight line segment. For example, carpenters use chalk lines to draw straight lines. For longer distances, though, the force required to pull a string taut enough gets excessive, so we typically use our line of sight (or lasers) as a means of describing a straight line. Sight, too, is limited, so we can only construct line segments. The notion of a line of indefinite length is therefore *theoretical* – we can visualize them, but cannot construct any.

Euclidean, spherical, and hyperbolic geometry are distinguished by their differing parallel postulates. In Euclidean geometry, for any line l and any point P not on the line l, there exists exactly one line through P which does not intersect l. In hyperbolic geometry, there exists an infinite number of lines through P which do not intersect l. In spherical geometry, any two lines intersect – that is, there are no parallel lines.[2]

An intriguing question to ask is, *Which geometry do we live in?* If we pull out a piece of paper and on it draw a line l and point P, we might be able to convince ourselves that there exists a unique parallel line through that point. But let us now lay that piece of paper on the ground and orient the line so that it points north and south. Now imagine ourselves as living on the equator of a perfectly spherical world. How does our line extend? It is a line of longitude which travels through the North and South Poles. How should we construct a line parallel to it? We draw a line segment l_2 through P and perpendicular to l. We then draw the line through P and perpendicular to l_2. It should be our parallel, but if we continue it along the surface of the Earth, it too goes through both the North and South poles. Thus, it intersects l. That is, there are no parallel lines.

The Earth is not our entire universe either, and a ray of light does not travel along a line of longitude, so perhaps this model is fanciful. But it points out that our view of the world is limited. Our view is *local*.[3] It is like the drawing on that piece of paper. It may look like Euclidean geometry, but the features which distinguish Euclidean geometry from spherical and hyperbolic geometry are features which lie beyond the piece of paper. Those features are *global*. We cannot tell, from our limited experience, what the geometry of our universe really is.

The current popular theory among physicists is that we live in a geometry which is much more complicated than any of the geometries we will study. This geometry, as you might imagine, is also locally indistinguishable from Euclidean geometry.

---

[2]Euclid's first four axioms must also be modified to generate spherical geometry. No such modification is necessary for hyperbolic geometry.

[3]In mathematics, local means in a sufficiently small neighborhood. In reality, we should probably distinguish local from *microscopic*. For example, light travels locally in a straight line, but at a microscopic (submicroscopic?) level, we might notice that light travels like a wave. In this text, we will never distinguish the local and microscopic levels.

0.2 A Review of Terminology

Right angle: An angle which measures 90°. We will formally define right angles later in this text.

Obtuse angle: An angle whose angular measure is greater than 90°.

Acute angle: An angle whose angular measure is less than 90°.

Circle: A circle with *center O* and *radius r* is the set of points a distance r away from O.

Chord: A chord of a circle is a line segment with endpoints on the circle.

Diameter: A diameter of a circle is a chord which passes through the center. The diameter may also refer to the length of a diameter.

Radius: A radius is a line segment from the center of a circle to the circle. The radius may also refer to the length of a radius.

Arc: A portion of a circle. The angular measure of an arc is the measure of the angle created by the radii which go to the ends of the arc.

Triangle: A set of three points (vertices) together with the three line segments joining them.

Degenerate triangle: A triangle whose three vertices are collinear.

Isosceles triangle: A triangle with two sides of equal length.

Scalene triangle: A triangle for which no two sides are equal.

Right angle triangle: A triangle with a right angle.

Quadrilateral: A set of four points joined by four nonintersecting segments.

Square: A quadrilateral whose sides have equal length and whose angles are equal.

Rhombus: A quadrilateral whose sides are equal.

Parallelogram: A quadrilateral whose opposite sides are parallel.

Supplementary angles: Two angles which together form a straight line.

Adjacent angles: Two angles which share a common ray.

Complementary angles: Two angles which together form a right angle.

Vertical angles: Two opposite angles at the intersection of two lines.

0.3 Notes on Notation

There are several notations used in this text for which there are other, possibly more standard, notations. Below is a list of notations we use, together with some common alternatives.

We denote the length of the segment AB with $|AB|$. The notation AB is common.

We denote the segment AB with 'the segment AB.' A common alternative is $\overline{AB}$.

We denote the line AB with 'the line AB.' A common alternative is $\overleftrightarrow{AB}$

We denote the ray AB with 'the ray AB.' A common alternative is $\overrightarrow{AB}$.

We denote the arc AB with 'the arc AB.' A common alternative is $\overparen{AB}$.

We denote the area of $\triangle ABC$ with $|\triangle ABC|$. A common alternative is $[ABC]$.

We denote the measure of the angle $\angle ABC$ with '$\angle ABC$' and use the same notation for the angle $\angle ABC$. We expect the distinction between the two to be obvious from the context. The usual alternative is to denote the angle $\angle ABC$ with $\angle ABC$, and the measure of the angle $\angle ABC$ with $m\angle ABC$. A usual exception to this notation is the abbreviation A for $\angle BAC$ in a triangle $\triangle ABC$. This abbreviation is used, for example, in the Law of Sines and Law of Cosines (pages 42 and 38). In these cases, it is clear that A represents an angle since it is used as the argument of a trigonometric function.

0.4   Notes on the Exercises

The level of difficulty of the exercises is indicated with zero or more asterisks. Exercises with no asterisk are considered normal and may range in difficulty from trivial to moderately challenging. An exercise with one asterisk considered to be either hard, or very long. An exercise with two asterisks is harder, possibly at the level of a national or international contest problem. In general, the average level of difficulty for exercises increases as the chapter or book progresses.

Some exercises are further developments in the theory and should not be missed. Such exercises are denoted with a dagger (†).

Most exercises are placed at the end of the section, but some are not. These are often questions which naturally arise from the current discussion. They are usually no more important than those at the end of the section, but often either the question or the result is illuminating and relevant to the current discussion. These exercises should therefore be read as part of the text, even if there is no intent to solve them.

Some exercises are followed by one or more of the symbols [H], [A], or [S]. These symbols mean that there is, respectively, a hint, answer, or solution in Appendix B.

Chapter 1

Euclidean Geometry

The geometry which we will first study is known as Euclidean geometry, named after the Greek mathematician Euclid (ca. 300 B.C.), who proposed a set of postulates which (essentially) define the geometry. It is the geometry taught in schools in America, and we tend to think of it as the geometry we live in.

We will take the point of view that geometry is entirely theoretical and is meant to interpret the world we live in. We will construct it from scratch – making several assumptions which are inspired by our experiences, and from them we will derive the truths of our geometry. Since our view of the world is local, and our artificial geometry is supposed to mimic reality, the assumptions we make which are least likely to be debated are those that are local in nature. Our global assumptions will be more contentious, and by varying these assumptions, we will construct several different geometries. Each geometry will be equally valid. That is, they will all model our perception of reality in a satisfactory way (so locally, they will all look alike) and they will all be *coherent* (that is, in each geometry, there are no statements which can be proved to be both true and false).

Since locally each geometry looks alike, one might wonder why we should care to develop the different geometries. After all, our experience is merely local. The reason is, of course, curiosity. We have an insatiable desire to ask and answer questions. Some questions, though, can only be answered by making some global assumptions. Sometimes, those assumptions seem so obvious that we do not even notice we are making them. The assumption that we live in a world that globally follows the laws of Euclidean geometry is an example of one of those assumptions. For example, consider the following question asked by many through the ages: Imagine yourself out in the nighttime desert, lying under the stars. You fix your gaze on one star in particular and wonder just how far away it is. Such questions have occurred to many people in the past, and a rather simple experiment can be devised to answer it.

Measure the angle between the sun, you, and the star. Half a year later, make the same measurement. The two angles should sum to less than 180°. We now have a triangle for which we know all three angles, and the length of one side. We should therefore be able to compute the length of the other two sides.

Exercise 1.1. Suppose that at the spring solstice the angle between a star, the Earth, and the sun measures 79.1°, and at the fall solstice the angle measures 100.8°. How far away is the star from the Earth? (The Earth is 93 million miles from the sun.)

Solution. The star, Earth during the fall, and Earth in springtime form a very long and narrow triangle (see Figure 1.1).

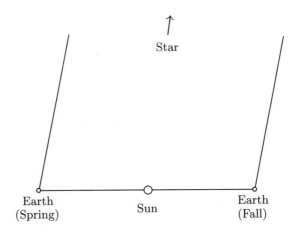

Figure 1.1

We know two angles and the side between in this triangle, so by ASA, we should be able to solve for the other two sides. We do this using the Law of Sines (proved in Theorem 1.11.1 in Section 1.11), which says, for a triangle $\triangle ABC$ with side a opposite angle A, b opposite angle B, and c opposite angle C, we have

$$\frac{a}{\sin A} = \frac{b}{\sin B} = \frac{c}{\sin C}.$$

Let A be the angle at the star. Since the angles in a triangle sum to 180°, we know $A = .1°$. We can now solve for b and c, and find that, up to two significant digits (which is the degree of accuracy implied by the measurement from the sun to the Earth), the two are equal:

$$b \cong c \cong 100,000,000,000 \text{ miles}$$

Note that, as one would expect, the difference in the distance between the Earth and the star during the two seasons is insignificant.   □

In the above calculation, we made a very strong assumption. We assumed that our universe is Euclidean. This assumption was made first when we assumed that the angles in a triangle sum to $180°$, and again when we used the Law of Sines. If the universe is spherical,[1] then the star is in fact closer. If the universe is hyperbolic, then the star is in fact further away. To answer the question truthfully, we must decide which geometry most closely models our universe. Notice that we are not *a priori*[2] considering the star-Earth-sun triangle to be local – which is why to answer this question we must use some of the global assumptions of our geometry.

To calculate the actual distance in a spherical or hyperbolic world, we would also need to know the *curvature* of the universe. In a spherical universe, this quantity is positive and is the inverse of the square of the radius of the sphere. In a hyperbolic universe, it is negative. The curvature of Euclidean space is zero.

Suppose we could measure the distance between the star and the Earth by some other concrete means. If this distance is greater or less than the calculated distance, we will know which model is best, and we will know the curvature of the universe. However, if the actual and calculated distances differ by less than the errors in our measurements, then we will be exactly where we started – we will still not know.

This example also illustrates a need and desire to come up with experiments which can decide which model best describes reality. As of this date, no definitive experiment has been performed. However, if we do assume our world is either spherical, hyperbolic, or Euclidean, then by observing distant stars and galaxies, we can find bounds on the curvature. These bounds are so close to zero that, for near stars, we might as well assume our universe is Euclidean. Keep in mind, though, that these three geometries are not the only geometries that exist, and our universe may be different from all three.

Exercise 1.2. The numbers in Exercise 1.1 were cooked up. From your general knowledge, are these numbers reasonable? Explain.

Trivia. The speed of light is $186,000$ miles/sec.

Exercise 1.3. What order of magnitude is the difference between $180°$ and the sum of the angles for the summer and winter observations of Alpha Centauri? (See Figure 1.2.)

Trivia. The number found in Exercise 1.3 is very small – so small that it is unlikely that an experiment as described above could actually be successfully implemented. The measurements are instead done by taking pictures of the star and its background stars both in summer and winter. The

---

[1]In this text, we will only study two-dimensional versions of spherical and hyperbolic geometry. The universe, of course, is (at least) three-dimensional. There are three-dimensional versions of both spherical and hyperbolic geometry.

[2]*a priori* – at first. That is, without any prior assumptions.

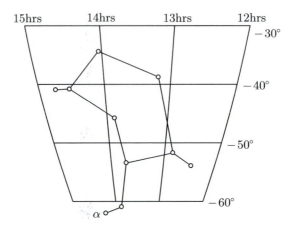

Figure 1.2. The nearest star (other than the sun) is Alpha Centauri, also known as Rigil Kentaurus, which is 4.3 light years away. This star is the α or brightest star in the Centaurus constellation, which is just south of Libra. It is visible from June to December in the nighttime sky of the southern hemisphere. It briefly appears above the southern horizon in parts of Florida and southern Texas.

astronomer then compares how much Alpha Centauri moves with respect to the background stars (this assumes the background stars are much further away – a reasonable assumption given that they move very little with respect to each other when compared with the amount Alpha Centauri moves).

Exercise 1.4*. In Figure 1.2, how much does Alpha Centauri move (use units of length) between the summer and winter observations?

1.1 The Pythagorean Theorem

Let us begin with a familiar theorem.

Theorem 1.1.1 (The Pythagorean Theorem). *Suppose a right angle triangle* $\triangle ABC$ *has a right angle at* C, *hypotenuse* c, *and sides* a *and* b. *Then*

$$c^2 = a^2 + b^2.$$

Proof. On the side AB of $\triangle ABC$, construct a square of side c. Draw congruent triangles on each of the other three sides of this square, as in Figure 1.3.

Since the angles at A and B sum to $90°$, the angle CBC' is $180°$. That is, it is a straight line. Thus, the resulting figure is a square. The area of the

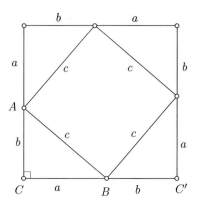

Figure 1.3

larger square can be calculated in two different ways – first by considering it as a square with side $a + b$, and second by summing the area of the four triangles with the area of the interior square. Equating the two, we get

$$(a + b)^2 = 4\left(\frac{1}{2}ab\right) + c^2$$
$$a^2 + 2ab + b^2 = 2ab + c^2$$
$$a^2 + b^2 = c^2$$

as desired. □

In this proof, we accepted several assumptions as fact. We assumed that squares exist, that the area of a square of side s is s^2, and that the interior angles in a triangle sum to $180°$. All of these assumptions are valid in Euclidean geometry, but not in either spherical or hyperbolic geometry. In fact, the Pythagorean theorem is a theorem in Euclidean geometry, and not in either of the other two.

I hear you thinking to yourself, *Isn't this evidence that our world exists in a Euclidean geometry?* Think of the diagram in Figure 1.3 drawn on a piece of paper and laid flat on the surface of the Earth. Our proof still looks valid – we have drawn a figure that looks like a square, and the sum of the angles in these triangles look like they sum up to $180°$. But in fact, as a figure in spherical geometry, where the sphere is the surface of the Earth, the angles in the 'square' are a little more than $90°$ and the angles in the triangles sum up to a little more than $180°$ – it only looks right because the error is so small as to be undetectable.

This is analogous to the differences between Newtonian and Einstein mechanics. We now know that Einstein's model of the universe is more accurate than Newton's model. But for objects traveling at slow speeds compared to the speed of light (the analogue of a small triangle in spherical or hyperbolic geometry), Newton's model is perfectly adequate. That is, the two models are indistinguishable when describing slow moving objects.

An important consequence of the Pythagorean theorem (in a geometry where the Pythagorean theorem is valid) is its converse:

Theorem 1.1.2 (The Converse of the Pythagorean Theorem). *Suppose we are in a geometry where the Pythagorean theorem is valid. Suppose in triangle $\triangle ABC$ we have*

$$a^2 + b^2 = c^2.$$

Then the angle at C is a right angle.

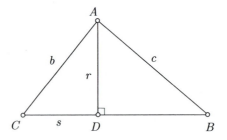

Figure 1.4

Proof. Let the perpendicular at A intersect the line BC at D, (see Figure 1.4). Let $r = |AD|$ and $s = |DC|$. Then, by the Pythagorean theorem, $r^2 + s^2 = b^2$. Also, by the Pythagorean theorem, $r^2 + (a \pm s)^2 = c^2$ (the choice of $\pm$ depends on whether the angle C is acute or obtuse). Combining these two equations, we get

$$a^2 \pm 2sa + b^2 = c^2,$$

and hence, since $c^2 = a^2 + b^2$, we get $2sa = 0$. Thus, $s = 0$ and $D = C$. That is, C is a right angle. $\square$

Trivia. The converse of the Pythagorean theorem is frequently used by carpenters. When framing a house, a carpenter will build each wall on the ground first. This frame is a quadrilateral so is not stable. Before erecting the wall, the carpenter will measure points three feet and four feet from a corner in different directions. If the wall is square, then these points should be five feet apart (since $3^2 + 4^2 = 5^2$). If they are not, the carpenter will nudge the appropriate corner until it is square. He will then nail a temporary stud diagonally across the wall so that it will not go out of square when he erects it.

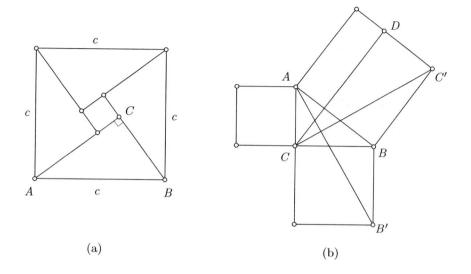

(a)   (b)

Figure 1.5. See Exercises 1.5 and 1.6.

Exercise 1.5. The diagram in Figure 1.5(a) suggests a different proof of the Pythagorean theorem. Fill in the details.

Exercise 1.6. The diagram in Figure 1.5(b) suggests another proof of the Pythagorean theorem. Fill in the details.   [H]

Exercise 1.7 (Pappus' Variation on the Pythagorean Theorem). Let $\triangle ABC$ be a triangle (not necessarily right). Let $ACDE$ and $BCFG$ be parallelograms whose sides DE and FG intersect at H (see Figure 1.6). Let $ABIJ$ be a parallelogram with sides AJ and BI parallel to and with

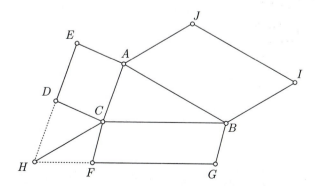

Figure 1.6. See Exercise 1.7

the same length as CH. Prove that the area of $ABIJ$ is equal to the sum of the areas of the other two parallelograms.

1.2   The Axioms of Euclidean Geometry

After our proof of the Pythagorean theorem, we questioned the validity of some of the assumptions we made. It is clear that we cannot continue in this fashion. If we are to successfully prove results in geometry, we must do so without making questionable assumptions. We must therefore first agree on which facts we should accept as absolute or unquestionable. Such facts are called axioms or postulates. Euclid proposed that the geometry which bears his name be defined by the following five postulates:

1. We can draw a unique[3] line segment between any two points.

2. Any line segment can be continued indefinitely.

3. A circle of any radius and any center can be drawn.

4. Any two right angles are congruent.

5. Given a line l and a point P not on l, there exists a unique line l_2 through P which does not intersect l.[4]

These axioms already presuppose several notions. They first of all presuppose the existence of a set of points which we call the *Euclidean plane*. In this set, there exists a notion of length, of lines, of circles, of angular measure, and of congruence. Euclid also presupposed that the plane is two-dimensional. This can be expressed axiomatically, as we do with the 'separation axiom' in Chapter 9, but in this chapter, we will assume an intuitive understanding of two dimensions.

Definition 1. *Distance.* Let $d(P,Q)$ be a function which assigns a positive real number to any pair of points in the plane. We say d is a *distance function* (or *metric*) if it satisfies the following three properties for any three points P, Q and R in the plane:

$d(P,Q) = d(Q,P)$

$d(P,Q) \geq 0$ With equality if and only if $P = Q$,

$d(P,R) \leq d(P,Q) + d(Q,R)$.   (The triangle inequality.)

We say the *distance* between P and Q is $d(P,Q)$ and usually denote this distance by $|PQ|$.

---

[3]Euclid's version did not include the word *unique*. Proclus (410 – 485 A.D.) notes that Euclid must have intended that segments be unique, since he uses this later in his text.

[4]This is actually Playfair's version of the fifth postulate. We will investigate Euclid's version in Section 1.4 and show that his statement is equivalent to the one given here.

We will think of a *line segment* as the shortest path between two points. A *line* is an indefinite continuation of a line segment.

Exercise 1.8. How should we measure the length of a path?

The *circle* $\mathcal{C}_P(r)$ centered at P with radius r is the set

$$\mathcal{C}_P(r) = \{Q : |PQ| = r\}.$$

We introduce the notion of *congruence* axiomatically via a notion of *isometries*:

Definition 2. *Isometry.* An *isometry* of the plane is a map from the plane to itself which preserves distances. That is, f is an isometry if for any P and Q in the plane, we have

$$d(f(P), f(Q)) = d(P, Q).$$

We are familiar with several isometries – namely the *translations*, *rotations* and *reflections*. We will formally define these isometries at the end of Section 1.3.

Definition 3. *Congruence.* Two sets of points (defining a triangle, angle, or some other figure) are *congruent* if there exists an isometry which maps one set to the other.

In particular, two angles are congruent if there exists an isometry which sends one angle to the other. We will write $\triangle ABC \equiv \triangle A'B'C'$ if there exists an isometry f such that $f(A) = A'$, $f(B) = B'$, and $f(C) = C'$. For angles, we will write $\angle BAC = \angle B'A'C'$. This notation is a bit unusual but is consistent with the notion that congruent angles have equal measures.

Our notion of congruence is completed by the following axioms which guarantee the existence of the isometries we desire:

6. Given any points P and Q, there exists an isometry f so that $f(P) = Q$ (e.g., translations).

7. Given a point P and any two points Q and R which are equidistant from P, there exists an isometry which fixes P and sends Q to R (e.g., rotations, but also reflections).

8. Given any line l, there exists an isometry which fixes every point in l but fixes no other points in the plane (the reflection through l).

Note that Axiom 6 does not explicitly guarantee the existence of translations. In fact (depending on how translations are defined), translations do not exist in spherical geometry, yet Axiom 6 is valid in spherical geometry.

We can now define right angles:

Definition 4. *Right Angle.* Two lines l_1 and l_2 intersect at *right angles* if any two adjacent angles at the point of intersection are congruent. That is, they intersect at right angles if there exists an isometry which sends an angle to one of its adjacent angles.

There is a final notion, which was not considered by Euclid, that should be introduced in the axioms. These are the axioms of *completeness*. We will not deal with these axioms here either. For a careful treatment, see Chapter 9.

There are also some issues concerning the *independence* and *coherence* of these axioms which we will not deal with. For example, how do we know that there does not exist a statement that can be proved from these axioms to be both true and false? If no such statement exists, then we call the set of statements *coherent* (if such a statement does exist, we call our system *incoherent*, contradictory, or just plain garbage). We say the set of statements is *independent* if none of these statements can be derived from the others. This set is in fact not independent, since Euclid's fourth axiom can be derived from the definition of right angle and Axioms 6 through 8. Can we further pare the list?

Exercise 1.9. The Cartesian plane $\mathbb{R}^2$ is a model of Euclidean geometry. In this model, explicitly describe an isometry which has no fixed points and is not a translation.

Exercise 1.10. Which of Axioms 1 – 8 are local in nature, and which are global? It may help to ask yourself if any of these properties are true on a sphere.

Exercise 1.11. The triangle inequality states

$$|PQ| + |QR| \geq |PR|.$$

Show that we can have equality if and only if Q is a point on the line segment PR.

Exercise 1.12. It is sometimes possible to 'define away' an axiom by choosing a suitable definition. For example, with our definition of a circle, Euclid's third axiom is vacuous – all circles exist by definition, though some may be empty sets. Euclid's first axiom can also be defined away by making the following definition for a line segment:

$$PQ = \{R : |PQ| = |PR| + |RQ|\}.$$

With this definition, between any two points P and Q there exists a unique line segment PQ. Find an example of a familiar geometry where this definition for a line segment does not correspond with our intuitive notion of what a line segment should be.

1.3   SSS, SAS, and ASA

You will probably recall from high school geometry that two triangles are congruent if their three sides are equal (SSS or side-side-side); or if two sides and the angle between them are equal (SAS or side-angle-side); or if two angles and the side between are equal (ASA or angle-side-angle). In this section, we will prove SSS and leave the proofs of SAS and ASA as exercises. None of these proofs require the parallel postulate. This means that the proofs are equally valid for hyperbolic geometry.

Theorem 1.3.1 (SSS). *If the corresponding sides of two triangles $\triangle ABC$ and $\triangle A'B'C'$ have equal lengths, then the two triangles are congruent.*

To prove this, we will need a very believable lemma with a rather difficult proof. This result is proved in Chapter 9:

Lemma 1.3.2. *Two distinct circles intersect in zero, one, or two points. If there is exactly one point of intersection, then that point lies on the line joining the two centers.*

Proof of SSS. Let us assume these triangles are not degenerate. That is, let us assume that C does not lie on the line AB, and that C' does not lie on $A'B'$. We leave this degenerate case to the reader.

To show that two figures are congruent, we must show that there exists an isometry which sends one to the other. By Axiom 6, there exists an isometry f_1 such that $f_1(A) = A'$. Since f_1 is an isometry, and $|AB| = |A'B'|$, we have that

$$|A'f_1(B)| = |f_1(A)f_1(B)| = |AB| = |A'B'|,$$

so by Axiom 7, there exists an isometry f_2 such that $f_2(A') = A'$ and $f_2(f_1(B)) = B'$. If $f_2(f_1(C)) = C'$, then there is nothing more to do.

If $f_2(f_1(C)) \neq C'$, then consider the circle centered at A' with radius $|AC|$ and the circle centered at B' with radius $|BC|$. These two circles intersect in at most two points (by the previous lemma), one of which is C' and the other of which must be $f_2(f_1(C))$. By Axiom 8, there exists an isometry f_3 which fixes every point on $A'B'$ but fixes no other points. Since this map is an isometry, and since C' is not on $A'B'$, the point C' must get mapped to $f_2(f_1(C))$, and vice versa. Composing these three isometries, we get

$$f_3(f_2(f_1(A))) = A'$$
$$f_3(f_2(f_1(B))) = B'$$
$$f_3(f_2(f_1(C))) = C',$$

and hence the two triangles are congruent.   □

By definition, two triangles are congruent if there exists an isometry which sends one to the other. We will write

$$\triangle ABC \equiv \triangle A'B'C'$$

if there exists an isometry f such that $f(A) = A'$, $f(B) = B'$, and $f(C) = C'$. Note that our notation is more informative than the statement that the two triangles are congruent – it also says which pairs of vertices correspond to each other.

Remark. The rigidity of a triangle is guaranteed by SSS. This is why the triangle is used so often in engineering, for example in trusses. A quadrilateral is not a rigid structure. That is, two quadrilaterals with equal-length sides might not have equal angles. Four timbers bolted together at the corners can easily collapse. On the other hand, three timbers bolted together at the corners will not collapse unless something breaks.

Exercise 1.13. Prove SAS.

Exercise 1.14. Prove ASA.

Let us now formally categorize the isometries. We begin by defining the *orientation* of a triangle. We say a nondegenerate triangle $\triangle ABC$ is oriented clockwise if the path from A to B to C is oriented clockwise. If a nondegenerate triangle is not oriented clockwise, then we say the it is oriented counterclockwise. We are relying on an intuitive understanding of clockwise and counterclockwise orientation. The concept of same orientation can be made exact in Euclidean, spherical, and hyperbolic geometry (see Exercise 9.11), and even in projective geometry, though in projective geometry, all triangles have the same orientation.

Definition 5. *Direct Isometry.* We call an isometry a *direct isometry* or *proper isometry* if the image of every clockwise oriented triangle is oriented clockwise. We call an isometry which is not direct an *improper isometry*. Another common terminology is to classify isometries as orientation *preserving* or *reversing* isometries.

Definition 6. *Translation.* We call an isometry f a *translation* if f is a direct isometry and either f is the identity or f has no fixed points.

Definition 7. *Rotation.* We call an isometry f a *rotation* if f is a direct isometry and if either f is the identity or there exists exactly one point P such that $f(P) = P$. We call P the *center of rotation* for f.

Definition 8. *Reflection.* We call an isometry f a *reflection* through the line l if $f(P) = P$ for every point P on l and if $f(P) \neq P$ for every point P not on l.

Exercise 1.15. Suppose f is an isometry and suppose there exist two distinct points P and Q such that $f(P) = P$ and $f(Q) = Q$. Show that f is either the identity or a reflection.

Exercise 1.16. Suppose f is a reflection. Prove that f is not a direct isometry.

Exercise 1.17. Prove that if a line $l_1 \neq l$ is sent to itself under a reflection through l, then l_1 and l intersect at right angles.

Exercise 1.18. Suppose that f is an isometry for which there exists exactly one point P such that $f(P) = P$. Prove that f is a rotation. That is, prove that f is a direct isometry.

Exercise 1.19†. Suppose f and g are two isometries such that $f(A) = g(A)$, $f(B) = g(B)$, and $f(C) = g(C)$ for some nondegenerate triangle $\triangle ABC$. Show that $f = g$. That is, show that $f(P) = g(P)$ for any point P.

Exercise 1.20*. Suppose P and Q are two distinct points. Prove that there exists exactly one translation which sends P to Q. [H]

1.4 Parallel Lines

Euclid stated his fifth postulate in this form:

> 5. Suppose a line meets two other lines so that the sum of the angles on one side is less than two right angles. Then the two other lines meet at a point on that side.

This is equivalent to our Axiom 5, but on occasion, this statement will be more useful. So let us begin by showing that Euclid's version follows from our version.

Theorem 1.4.1. *Let P be a point not on l, and let Q lie on l so that PQ is perpendicular to l. Let l_2 be the line through P which is parallel to l (as guaranteed by our Axiom 5). Then l_2 intersects PQ at $90°$.*

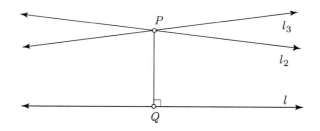

Figure 1.7

Proof. Suppose l_2 does not intersect PQ at $90°$. Then the reflection l_3 of l_2 through the line PQ is not the same as l_2 (see Figure 1.7). On the other hand, the image of l under this reflection is itself. Thus, l_3 cannot intersect l, for if it did, then the reflection of that point would be the intersection of l_2 and l, which cannot exist since we chose l_2 to be a line parallel to l. Thus, l_3 is also parallel to l, and we have two distinct lines through P which are parallel to l, a contradiction of Axiom 5. Thus, l_2 must intersect PQ at $90°$.   □

The point Q can always be found:

Lemma 1.4.2. *Let l be a line and P a point not on l. Then there exists a point Q on l so that PQ is perpendicular to l.*

Proof. Let P' be the image of P under reflection through l. Since P does not lie on l, we know $P' \neq P$. Let Q be the intersection of PP' and l. Let R be any point on l not equal to Q. Then $\angle RQP = \angle RQP'$, since one is the image of the other under the reflection. But, these two angles are adjacent, so they are right angles, as desired.   □

The converse of Theorem 1.4.1 is also true:

Theorem 1.4.3. *Suppose l intersects two distinct lines l_1 and l_2 perpendicularly. Then l_1 and l_2 are parallel.*

Proof. Suppose l_1 and l_2 are not parallel. Then they must intersect at some point R. Consider the reflection R' of R through l. Since l_1 and l_2 are perpendicular to l, their images under this reflection must be themselves. Thus l_1 and l_2 must intersect at both R and R'. Since l_1 and l_2 are distinct, R and R' must be the same point, so must lie on l. But this too would imply that l_1 and l_2 are not distinct, a contradiction. Hence, R cannot exist, so l_1 and l_2 are parallel.   □

Corollary 1.4.4. *Suppose a line l intersects two other lines l_1 and l_2 so that the opposite interior angles are equal. Then l_1 and l_2 are parallel.*

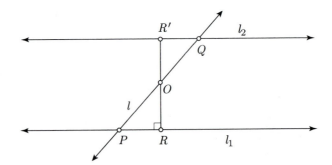

Figure 1.8

Proof. Let l intersect l_1 and l_2 at P and Q, respectively, as in Figure 1.8. Let O be the midpoint of PQ. Find the point R on l_1 so that OR is perpendicular to l_1. Now, consider the rotation centered at O which sends P to Q. Let the image of R be R', which *a priori* does not lie on l_2. Note that $\triangle ORP$ is congruent to $\triangle OR'Q$, since one is the image of the other under an isometry. Thus, $\angle OPR = \angle OQR'$, and since the opposite interior angles are equal, we now know R' lies on l_2. But then $\angle PRO = \angle QR'O = 90°$ and by Theorem 1.4.1, l_1 and l_2 are parallel. □

Corollary 1.4.5 (Euclid's Axiom 5). *Suppose a line l meets two other lines l_1 and l_2 so that the sum of the angles on one side is less than $180°$. Then the lines l_1 and l_2 meet at a point on that side.*

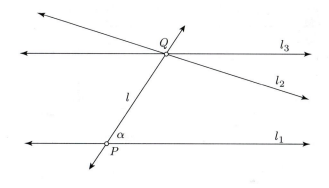

Figure 1.9

Proof. Let l intersect l_1 and l_2 at P and Q respectively, as in Figure 1.9. Let α be the angle l makes with l_1. There exists a line l_3 through Q which makes an angle of $180° - \alpha$ with l. Then, by Corollary 1.4.4, l_3 is parallel to l_1, and hence l_2 is not parallel to l_1, so intersects l at some point R. If R is to the left of l, then l_3 enters $\triangle PQR$, so it must exit the side PR. That is, it must intersect l_1, which is a contradiction. Thus, the point R must be to the right of l. □

In this proof, we stated that a line which enters a triangle through a vertex must exit that triangle through the opposite side. This may seem obvious from pictures, but how does it follow from our axioms? This is the sort of question which eventually led mathematicians to the conclusion that Euclid's axioms are not sufficient to define Euclidean geometry. We will return to this question in Chapter 9.

Theorem 1.4.6. *The three angles in a triangle sum up to $180°$.*

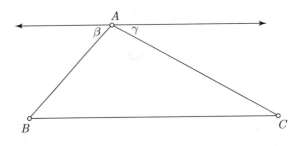

Figure 1.10

Proof. Let $\triangle ABC$ be a triangle. Find the line l through A and parallel to BC. By Euclid's version of Axiom 5, the angle labeled β in Figure 1.10 is equal to $\angle ABC$, and the angle γ is equal to $\angle ACB$. But

$$\beta + \angle BAC + \gamma = 180°,$$

so

$$\angle ABC + \angle BAC + \angle ACB = 180°,$$

as desired. $\square$

Definition 9. *Exterior angle.* The *exterior angle* at A in $\triangle ABC$ is one of the angles adjacent to $\angle BAC$ at the intersection of the lines AB and AC.

Corollary 1.4.7. *The exterior angle at A is equal to the sum of the other two interior angles.*

Exercise 1.21*. Show that Euclid's version of Axiom 5 implies our version of Axiom 5.

Exercise 1.22. Prove that the angles in a quadrilateral sum up to 360°. Generalize this result to an n-sided polygon.

Exercise 1.23. What is the sum of the exterior angles of a triangle? What is the sum of the exterior angles of a quadrilateral? What is the sum of the exterior angles of an n-gon?

1.5   Pons Asinorum

In this section, we prove a very obvious theorem which has many applications. The theorem is known as *pons asinorum* or the *ass's bridge*, probably because of the drawing used in Euclid's proof and because of the dimness of anyone who cannot grasp the theorem [C]. Our proof does not include the same picture (and is shorter) because we have chosen a different foundation for Euclidean geometry.

Theorem 1.5.1 (Pons Asinorum). *The base angles of an isosceles triangle are equal.*

Proof. By SSS, $\triangle ABC$ is congruent to $\triangle ACB$. That is, there exists an isometry which fixes A and sends B to C (said isometry is, of course, reflection through the angle bisector at A). But isometries preserve angles, so the angle at B must be equal to the angle at C. □

Exercise 1.24†. Prove the converse of *pons asinorum*. That is, show that if in $\triangle ABC$ we have $\angle ABC = \angle ACB$, then $|AB| = |AC|$.

Exercise 1.25†. Prove that if a diameter of a circle bisects a chord which itself is not a diameter, then the diameter is perpendicular to the chord. Also, prove that the perpendicular bisector of a chord goes through the center of the circle. And finally, prove that if a diameter is perpendicular to a chord, then the diameter bisects the chord.   [S]

1.6   The Star Trek Lemma

One of the most important theorems of circle geometry is a consequence of *pons asinorum*. In the spirit of Euclid, we will refer to this theorem as the Star Trek lemma because of the figure associated with the statement of the theorem. In this proof, we will need to know that the sum of angles in a triangle is 180°, so this is a statement in Euclidean geometry only. We will no longer emphasize this point until we begin the chapter on hyperbolic geometry.

Let A, B, and C be three points on a circle centered at O. We call angle $\angle BAC$ an *inscribed angle*, since it is inscribed in a circle. The angular measure of the *arc BC* is the measure of the angle $\angle BOC$ where this angle is measured on the same side of O as the arc. We say angle $\angle BAC$ *subtends* the arc BC.

Theorem 1.6.1 (Star Trek Lemma). *The measure of an inscribed angle is half of the angular measure of the arc it subtends.*

Before Star Trek, as far as I know, this theorem had no name, though some might call it Euclid III.20, which is its proposition number in Euclid's *Elements* (Book III, Proposition 20). Otherwise, it is often just quoted: 'The measure of an inscribed angle is half of the angular measure of the arc it subtends.'

Proof. We will prove this only in the case when $\angle BAC$ is acute and the center O is within this angle, as in Figure 1.11 (which resembles the Star Trek insignia). Note that OA, OB, and OC are radii, so we have several isosceles triangles. We have continued the segment OA to intersect the

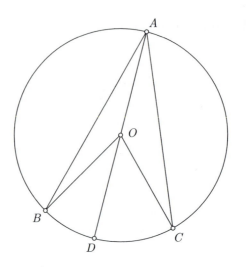

Figure 1.11

circle at D. Since ΔAOB is isosceles, $\angle BAO = \angle OBA$. Since the sum of angles in a triangle is $180°$,

$$\angle BOD = \angle OBA + \angle BAO = 2\angle BAO.$$

Similarly,
$$\angle DOC = 2\angle OAC.$$

Adding these two equations, we get

$$\angle BOC = 2\angle BAC. \qquad \square$$

We just proved the Star Trek lemma in one case. The other three cases are left as exercises. The proofs in two of these cases (Exercises 1.26 and 1.27) are almost identical to the case dealt with above. We will often offer a 'proof' which is only a proof for one case and leave the other cases to the reader. The core ideas of the proof for the other cases are usually the same as those for the case that is considered.

Exercise 1.26. Prove the Star Trek lemma for an acute angle for which the center O is outside the angle.

Exercise 1.27. Prove the Star Trek lemma for an obtuse angle.

Exercise 1.28†. Suppose $\angle ABC$ is a right angle inscribed in a circle. Prove that AC is a diameter.   [S]

Exercise 1.29† (Bow Tie Lemma). Let A, A', B, and C lie on a circle, and suppose $\angle BAC$ and $\angle BA'C$ subtend the same arc (as in Figure 1.12(a)). Show that

$$\angle BAC = \angle BA'C.$$

Again, because of the diagram, this lemma is sometimes known as the *Bow Tie* lemma. We say 'The angles at A and A' are equal since they subtend the same arc.'

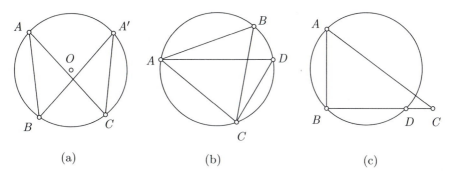

(a)   (b)   (c)

Figure 1.12. See Exercises 1.29, 1.30, and 1.31.

Exercise 1.30. If $|AB| = |AC| = |BC|$, what is the angle at D? (See Figure 1.12(b).)   [A]

Exercise 1.31. If $|AB| = 12$, $|BD| = 9$, $|BC| = 16$, and $|AC| = 20$, then what is the length of the diameter? (See Figure 1.12(c).)   [A]

Exercise 1.32. If $|AB| = |AC| = |BC|$ and AD is perpendicular to BC, then what is $\angle BCD$? (See Figure 1.13(a).)

Exercise 1.33† (**The Tangential Case of the Star Trek Lemma**). Suppose AT is a line segment that is tangent to a circle. Prove that $\angle ATB$ is half the measure of the arc TB which it subtends. Do this by picking a point C on the circle such that $\angle TCB$ subtends the arc TB (as in

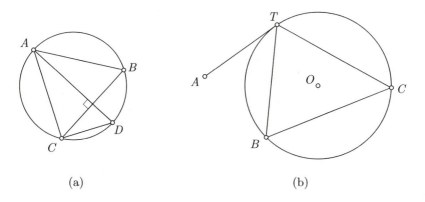

(a)   (b)

Figure 1.13. See Exercises 1.32 and 1.33.

Figure 1.13(b)). Show that

$$\angle ATB = \angle TCB.$$

Exercise 1.34†. Suppose two lines intersect at P inside a circle and meet the circle at A and A' and at B and B', as shown in Figure 1.14(a). Let α and β be the measures of the arcs $A'B'$ and AB respectively. Prove that

$$\angle APB = \frac{\alpha + \beta}{2}.$$

Exercise 1.35†. Suppose an angle α is defined by two rays which intersect a circle at four points, as in Figure 1.14(b). Suppose the angular measure of the outside arc it subtends is β, and the angular measure of the inside arc it subtends is γ. (So, in Figure 1.14(b), $\angle AOB = \beta$ and $\angle A'OB' = \gamma$.) Show that

$$\alpha = \frac{\beta - \gamma}{2}.$$ [S]

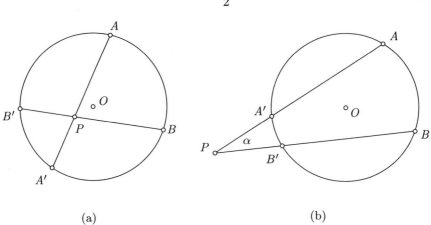

(a)   (b)

Figure 1.14. See Exercises 1.34 and 1.35.

Exercise 1.36. Prove that the opposite angles in a convex quadrangle inscribed in a circle sum to 180°. Conversely, prove that if the opposite angles in a convex quadrilateral sum to 180°, then the quadrilateral can be inscribed in a circle. Such a quadrilateral is called a *cyclic quadrilateral*.

Exercise 1.37†. Let two circles Γ and Γ' intersect at A and B, as in Figure 1.15. Let CD be a chord on Γ. Let AC and BD intersect Γ' again at E and F. Prove that CD and EF are parallel. [S]

Exercise 1.38†. Let $ABCD$ be *nonconvex cyclic quadrilateral*. That is, $ABCD$ is inscribed in a circle and two of its opposite sides intersect (as in Figure 1.16(a)). Prove that $\angle ABC = \angle CDA$ and $\angle DAB = \angle BCD$. Conversely, suppose $\angle ABC = \angle DAB$ and $\angle BCD = \angle DAB$ in a quadrilateral with intersecting opposite sides. Show that $ABCD$ is cyclic.

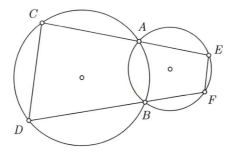

Figure 1.15. See Exercise 1.37.

Exercise 1.39. Suppose $ABCDEF$ is a hexagon inscribed in a circle. Show that

$$\angle ABC + \angle CDE + \angle EFA = 360°.$$

Prove that the converse is not true. That is, find an example of a hexagon $ABCDEF$ whose angles B, D, and F sum to $360°$ but which cannot be inscribed in a circle.

Exercise 1.40. Let E be a point inside a square $ABCD$ such that $\triangle BCE$ is an equilateral triangle, as in Figure 1.16(b). Show that $\angle EAD = 15°$.

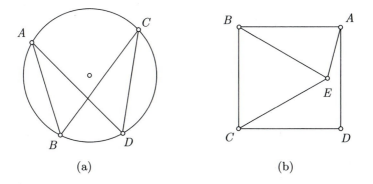

(a)   (b)

Figure 1.16. See Exercises 1.38 and 1.40.

1.7 Similar Triangles

The following result is of fundamental importance in Euclidean geometry:

Theorem 1.7.1. *Let B' and C' be on the respective sides AB and AC of a triangle $\triangle ABC$. Then $B'C'$ is parallel to BC if and only if*

$$\frac{|AB'|}{|AB|} = \frac{|AC'|}{|AC|}.$$

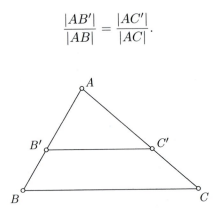

Figure 1.17

We will postpone the proof of this result until Chapter 9, where we prove the following piece:

Theorem 1.7.2 (Theorem 9.9.1). *Let $\triangle ABC$ be a triangle. Let B' be on AB, and let the line through B' and parallel to BC intersect AC at C'. Then*

$$\frac{|AB'|}{|AB|} = \frac{|AC'|}{|AC|}.$$

From this, we can prove the converse:

Theorem 1.7.3. *Let B' and C' be points on the sides AB and AC of a triangle $\triangle ABC$. Suppose*

$$\frac{|AB'|}{|AB|} = \frac{|AC'|}{|AC|}.$$

Then $B'C'$ is parallel to BC.

Proof. Let the line through B' and parallel to BC intersect AC at C''. Then, by Theorem 1.7.2,

$$\frac{|AB'|}{|AB|} = \frac{|AC''|}{|AC|},$$

so $C'' = C'$.   □

Note that the angles in triangles $\triangle ABC$ and $\triangle AB'C'$ (in Figure 1.17) are equal, since $B'C'$ is parallel to BC. We call these two triangles *similar*.

Definition 10.   *Similar Triangles.* We say two triangles $\triangle ABC$ and $\triangle A'B'C'$ are *similar* if their angles are congruent. We write

$$\triangle ABC \sim \triangle A'B'C'.$$

In this text, we will further adopt the convention that $\triangle ABC \sim \triangle A'B'C'$ means the angles $\angle A = \angle A'$, $\angle B = \angle B'$, and $\angle C = \angle C'$. That is, $\triangle ABC \sim \triangle A'B'C'$ if their corresponding angles are congruent.

Corollary 1.7.4. *If $\triangle ABC \sim \triangle A'B'C'$, then*

$$\frac{|A'B'|}{|AB|} = \frac{|A'C'|}{|AC|} = \frac{|B'C'|}{|BC|}.$$

Proof. Since $\angle BAC = \angle B'A'C'$, there exists an isometry which sends A' to A and sends B' and C' to points on the lines AB and AC, respectively. Since $\angle ABC = \angle AB'C'$, the line $B'C'$ is parallel to BC, so by Theorem 1.7.2,

$$\frac{|A'B'|}{|AB|} = \frac{|A'C'|}{|AC|}.$$

Similarly, by sending B' to B, we get

$$\frac{|B'A'|}{|BA|} = \frac{|B'C'|}{|BC|}.$$

Combining, we get the desired result.   □

One of the conveniences of the notational convention that the statement

$$\triangle ABC \sim \triangle A'B'C'$$

means that the corresponding angles are congruent is that we can decide which sides correspond to which. For example, AB corresponds to $A'B'$, since they share the same spots in their respective notations.

Exercise 1.41. Suppose $\triangle ABC$ is similar to $\triangle A'B'C'$. Show that

$$|\triangle A'B'C'| = \left(\frac{|A'B'|}{|AB|}\right)^2 |\triangle ABC|.$$

That is, the ratio of the areas of the two triangles is the same as the square of the ratios of the sides. Here, we have used the notation $|\triangle ABC|$ to represent the area of $\triangle ABC$.

The above exercise gives an amusing proof of the Pythagorean theorem. Let $\triangle ABC$ be a right angle triangle, as in Figure 1.18. Let the altitude at C intersect AB at D. Then triangles $\triangle ABC$, $\triangle ACD$, and $\triangle CDB$ are similar, and hence

$$\frac{|\triangle ABC|}{|\triangle ACD|} = \frac{|AB|^2}{|AC|^2} = \frac{c^2}{b^2}$$

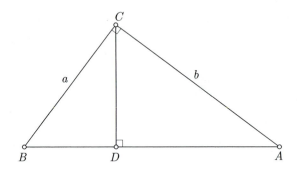

Figure 1.18

and

$$\frac{|\triangle ABC|}{|\triangle CDB|} = \frac{|AB|^2}{|BC|^2} = \frac{c^2}{a^2}.$$

But, clearly

$$|\triangle ACD| + |\triangle CDB| = |\triangle ABC|,$$

so

$$\frac{b^2}{c^2}|\triangle ABC| + \frac{a^2}{c^2}|\triangle ABC| = |\triangle ABC|$$

$$b^2 + a^2 = c^2.$$

Exercise 1.42. Let $\triangle ABC$ be an arbitrary triangle. Let A', B', and C' be the midpoints of the opposite sides. Draw lines through A', B', and C' that make an angle of $60°$ with each side, as in Figure 1.19(a). These lines intersect at A'', B'', and C'' as shown. Prove that $\triangle A''B''C''$ is similar to $\triangle ABC$.

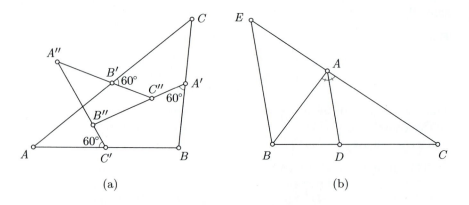

Figure 1.19. See Exercises 1.42 and 1.43.

Exercise 1.43† (**The Angle Bisector Theorem**). In an arbitrary triangle $\triangle ABC$, let the interior angle bisector at A intersect the side BC at D. Show that

$$\frac{|BD|}{|DC|} = \frac{|AB|}{|AC|}.$$

Hint: Construct the line parallel to AD and through B, as shown in Figure 1.19(b). Let this intersect AC at E. Show $|AB| = |AE|$.

Exercise 1.44. The pentagon in Figure 1.20(a) is regular and each side has length one. Show that

$$\frac{|AF|}{|FD|} = \frac{-1 + \sqrt{5}}{2}.$$

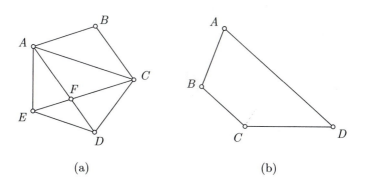

(a)   (b)

Figure 1.20. See Exercises 1.44 and 1.45.

Exercise 1.45. In the quadrilateral $ABCD$ in Figure 1.20(b), AD is parallel to BC, $\angle C = 2\angle A$, $|CD| = 3$, and $|BC| = 2$. What is $|AD|$?   [A]

Exercise 1.46†. Let $\triangle ABC$ and $\triangle A'B'C'$ be two triangles such that

$$\angle BAC = \angle B'A'C' \qquad \text{and} \qquad \frac{|A'B'|}{|AB|} = \frac{|A'C'|}{|AC|}.$$

Prove that $\triangle ABC \sim \triangle A'B'C'$.

Exercise 1.47†. Suppose $\triangle ABC$ and $\triangle A'B'C'$ are two similar but non-congruent triangles such that AB is parallel to $A'B'$, AC is parallel to $A'C'$, and BC is parallel to $B'C'$, as in Figure 1.21. Prove that the lines AA', BB', and CC' are concurrent.

Exercise 1.48. Let Γ be a circle. Suppose Γ' is another circle whose center lies on Γ. Let these two circles intersect at A and B, as in Figure 1.22(a). Let P be a point on Γ, and let PB intersect Γ' again at Q. Show that $\triangle PQA$ is isosceles.   [H]

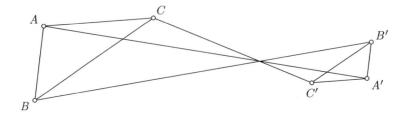

Figure 1.21. See Exercise 1.47.

Exercise 1.49. Let two circles Γ and Γ' intersect at A and B (see Figure 1.22(b)). Let P be a point on the circle Γ. Let PA intersect Γ' again at C, and let PB intersect Γ' again at D. Show that the length $|CD|$ is independent of the location of P. (Note that there are two cases to consider.) [H]

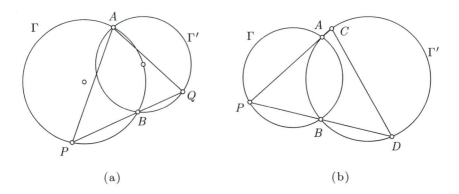

(a)   (b)

Figure 1.22. See Exercises 1.48 and 1.49.

1.8   Power of the Point

The definition of the *power of the point* P with respect to a circle Γ is inspired by the following theorem which we also call *power of the point.*

Theorem 1.8.1 (Power of the Point). *Let P be a point inside a circle Γ (see Figure 1.23). Let QQ' and RR' be two chords which intersect at P. Then*

$$|PQ||PQ'| = |PR||PR'|.$$

Proof. By the Star Trek lemma, $\angle RR'Q = \angle RQ'Q$ and $\angle Q'RR' = \angle Q'QR$. Thus, the triangles $\triangle RQ'P$ and $\triangle QR'P$ are similar, so

$$\frac{|PR|}{|PQ|} = \frac{|PQ'|}{|PR'|}.$$

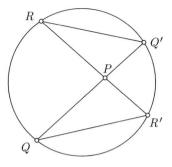

Figure 1.23

Cross multiplying gives us the desired result.   □

This result is also true for P outside the circle (see Exercise 1.51). For any chord QQ' through P, we call the product $\Pi(P) = \pm|PQ||PQ'|$ the *power of the point P with respect to the circle* Γ. We choose the sign to be positive if P is outside the circle, and negative if P is inside the circle. If Γ has radius r and center O, then the power of the point P, together with the correct sign, is given by

$$\Pi(P) = |OP|^2 - r^2.$$

To see this, choose RR' so that RR' goes through P and is a diameter of Γ.

Exercise 1.50. If in Figure 1.24(a), $|AP| = 2$, $|AB| = 6$, and $|PC| = 3$, then what is $|PD|$?   [A]

Exercise 1.51†. Theorem 1.8.1 is also true if the point P lies outside the circle, as in Figure 1.25(a). Prove it for this case.

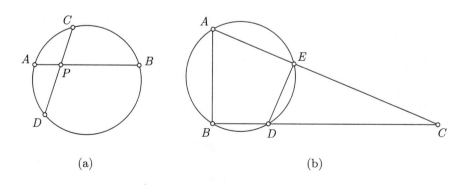

(a)   (b)

Figure 1.24. See Exercises 1.50 and 1.52.

Exercise 1.52. If in Figure 1.24(b), $|AB| = 5$, $|BC| = 12$, $|AC| = 13$, and $|BD| = 3$, then what is $|DE|$? [A]

Exercise 1.53† (**The Tangential Version of Power of the Point**). In Figure 1.25(a), suppose we let R and R' move toward each other until they meet, so that the 'chord' RR' is really the tangent at R, as in Figure 1.25(b). Then we get yet another version of the power of the point. Suppose a chord QQ' intersects a tangent at R at a point P. Show that

$$|PQ||PQ'| = |PR|^2.$$

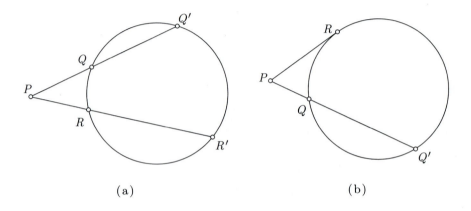

(a)   (b)

Figure 1.25. See Exercises 1.51 and 1.53.

Exercise 1.54. Use the tangential version of power of the point (Exercise 1.53) to come up with yet another proof of the Pythagorean theorem. [H][S]

Exercise 1.55. In Figure 1.26(a), if BC is tangent to the circle centered at A, $|BC| = 3$, and $|CD| = 1$, then what is the radius of the circle? [A]

Exercise 1.56. In Figure 1.26(a) as described in Exercise 1.55, what is the area of $\triangle ABD$? [A]

Exercise 1.57. Let $\triangle ABC$ be an isosceles triangle with $|AB| = |AC|$ (see Figure 1.26(b)). Let A' and B' be the midpoints of sides BC and AC, respectively. Let Γ be the circle that goes through B, A', and B'. Let the extended side AB intersect Γ at D. Suppose $|AD| = 5$ and $|AB'| = 4$. What are the lengths of the sides of the triangle?

Exercise 1.58. In Figure 1.27, the point C is on the diameter AB of a half circle. The two smaller half circles have diameters AC and CB. The region bounded by the curved edges of the half circles was called an *arbelos* or *butcher's knife* by Archimedes. The perpendicular at C intersects the larger circle at D. Prove that the circle with diameter D has the same area as the arbelos.

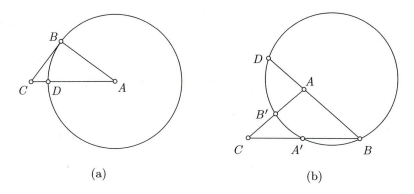

Figure 1.26. See Exercises 1.55, 1.56, and 1.57.

Exercise 1.59. Show that the two inscribed circles in the arbelos in Figure 1.27 have the same radii.   [H]

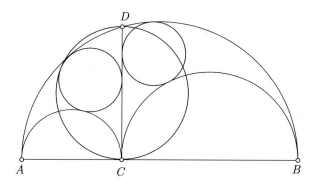

Figure 1.27. See Exercises 1.58 and 1.59.

The *radical axis* of two circles Γ and Γ' is the set of points P with the property that the powers of the point P with respect to both Γ and Γ' are equal.

Exercise 1.60. Let two circles Γ and Γ' have distinct centers O and O'. Prove that the radical axis of Γ and Γ' is a line perpendicular to OO'.

Exercise 1.61. Suppose two circles intersect at A and B. Prove that the radical axis of the two circles is the line AB.

Exercise 1.62. Let the circle Γ intersect the circle Γ' at A and B, and the circle Γ'' at C and D. Let P be the point of intersection of AB and CD. Prove that P lies on the radical axis for Γ' and Γ''.

Exercise 1.63 (The Radical Center). Let Γ, Γ', and Γ'' be three circles. Show that the three radical axes of these circles intersect at a common point. This point is called the *radical center* of the three circles.

Exercise 1.64. Let P be the radical center of three circles Γ, Γ', and Γ''. Suppose P is outside circle Γ. Prove that P is also outside circles Γ' and Γ''.

Exercise 1.65. Let P be the radical center of three circles Γ, Γ', and Γ'', and suppose P lies outside of Γ. Show that P is the center of a circle which intersects Γ, Γ', and Γ'' perpendicularly. Furthermore, show that the radius of this circle is $\sqrt{\Pi(P)}$.

Exercise 1.66. Let Γ_1, Γ_2, Γ_3, and Γ_4 be four circles. Let Γ_1 and Γ_2 intersect at P_1 and P_2. Let Γ_3 and Γ_4 intersect at P_3 and P_4. Let Γ_1 and Γ_4 intersect at Q_1 and Q_2, and let Γ_2 and Γ_3 intersect at Q_3 and Q_4. Show that P_1, P_2, P_3, and P_4 are collinear or cyclic if and only if Q_1, Q_2, Q_3, and Q_4 are collinear or cyclic (not necessarily respectively).

1.9   The Medians and Centroid

In a triangle $\triangle ABC$, let A', B', and C' be the midpoints of BC, AC, and AB, respectively, as in Figure 1.28(a). The line segments AA', BB', and CC' are called the *medians* of $\triangle ABC$.

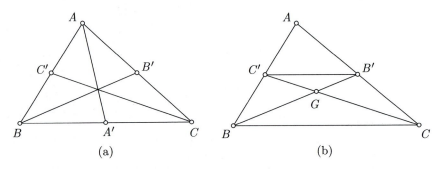

(a)   (b)

Figure 1.28

Theorem 1.9.1. *The three medians of a triangle $\triangle ABC$ intersect at a common point G. Furthermore, $|AG| = 2|A'G|$, $|BG| = 2|B'G|$, and $|CG| = 2|C'G|$.*

Proof. Let G be the intersection of the medians BB' and CC', as in Figure 1.28(b). Note that

$$\frac{|AC'|}{|AB|} = \frac{|AB'|}{|AC|} = \frac{1}{2},$$

since B' and C' are midpoints. Thus, $B'C'$ is parallel to BC, and furthermore, $|B'C'| = \frac{1}{2}|BC|$. Since $\angle C'B'B$ and $\angle B'BC$ are opposite interior angles of parallel lines, they are equal. Similarly, $\angle B'C'C = \angle C'CB$, so $\triangle B'C'G$ is similar to $\triangle BCG$. In particular, since the ratio of $|B'C'|$ to $|BC|$ is 1/2, we get $|B'G| = \frac{1}{2}|GB|$ and $|C'G| = \frac{1}{2}|BC|$.

We now apply the same argument to the medians AA' and BB'. Let these two intersect at G'. We again conclude that $|B'G'| = \frac{1}{2}|G'B|$, so $G' = G$. Thus, the three medians intersect at a common point. $\square$

Definition 11. *Centroid.* We call G the *centroid* or *center of mass* of $\triangle ABC$.

Exercise 1.67. In triangle $\triangle ABC$ pictured in Figure 1.29, let A' and B' be the midpoints of the sides BC and AC, respectively. Let AA' and BB' intersect at G. Suppose $|AG| = 2$ and suppose also that the circle through A, A', and B bisects the segment $B'G$. What is the length $|BG|$?   [A]

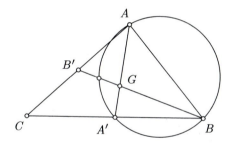

Figure 1.29. See Exercise 1.67.

Exercise 1.68. Cut a triangle out of card stock. Locate its centroid G. Now take a needle and hang the triangle by thread through G. The triangle should be balanced, which is why we call G the center of mass or centroid of the triangle.

Exercise 1.69. Suppose the medians AA' and BB' of $\triangle ABC$ intersect at right angles, and suppose $a = 3$ and $b = 4$. What is c?   [A]

Exercise 1.70. Prove that the sum of the medians of a triangle lies between $\frac{3}{4}p$ and p, where p is the length of the perimeter of the triangle.

Exercise 1.71. Prove that the diagonals of a parallelogram bisect each other.

1.10 The Incircle, Excircles, and the Law of Cosines

In the last section, we defined the centroid, a type of center of a triangle. In this section, we describe a different type of center – the *incenter*. By the end of this chapter, we will have described four types of centers for a triangle.

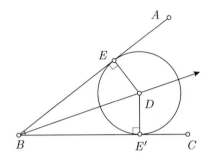

Figure 1.30

Consider an angle $\angle ABC$ and let D be any point on its angle bisector. Let E and E' be the points on BA and BC, respectively, so that $\angle BED$ and $\angle BE'D$ are right angles (see Figure 1.30). Then, $\triangle BED$ and $\triangle BE'D$ are congruent, by ASA. Hence, $|DE| = |DE'|$, and the circle centered at D with radius $|DE|$ is tangent to both BA and BC. Thus, we can prove:

Theorem 1.10.1. *The angle bisectors of a triangle intersect at a common point I called the incenter, which is the center of the unique circle inscribed in the triangle (called the incircle).*

Proof. Let I be the intersection of the angle bisector of $\angle ABC$ and $\angle ACB$. The perpendiculars from I to AB and BC are equal since I lies on the angle bisector of $\angle ABC$, and similarly, the perpendiculars from I to AC and BC are equal, since I lies on the angle bisector of $\angle ACB$. Thus, the perpendiculars from I to AB and AC are equal, so I lies on the angle bisector of $\angle BAC$, so the three angle bisectors intersect at a common point. $\square$

In a triangle $\triangle ABC$, we can define both the *interior angle bisector* at A, and the *exterior angle bisector* at A. The incenter I is defined via the interior bisectors, but we can also define the *excenters* I_a, I_b, and I_c. The excenter I_a is the point of intersection of the interior angle bisector at A, and the exterior angle bisectors at B and C. It is the center of the circle which is tangent to the side BC, is tangent to the extended sides AB and AC, and lies outside $\triangle ABC$ (see Figure 1.31). This circle is called an *excircle*.

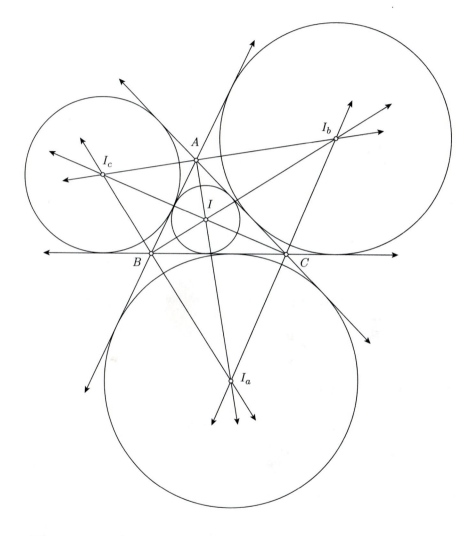

Figure 1.31. A triangle together with its incircle and three excircles.

Let the *inradius* r be the radius of the incircle, and let r_a, r_b, and r_c be the *exradii*. Let $s = \frac{1}{2}(a + b + c)$ be the semiperimeter of $\triangle ABC$.

Theorem 1.10.2. *Let r be the inradius of $\triangle ABC$, and let s be the semiperimeter of $\triangle ABC$. Then*

$$|\triangle ABC| = rs.$$

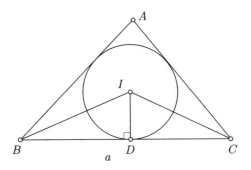

Figure 1.32

Proof. Consider $\triangle BCI$ in Figure 1.32, which has base a and height r, so has area $|\triangle BCI| = \frac{1}{2}ar$. Thus,

$$|\triangle ABC| = \frac{1}{2}(ar + br + cr) = rs. \qquad \square$$

Theorem 1.10.3 (The Law of Cosines). *For any triangle $\triangle ABC$, we have*

$$c^2 = a^2 + b^2 - 2ab\cos C.$$

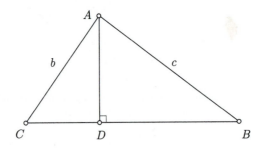

Figure 1.33

Proof. Consider a triangle $\triangle ABC$ with altitude AD, as in Figure 1.33. By the Pythagorean Theorem,

$$c^2 = |AD|^2 + |DB|^2.$$

But

$$|AD| = b\sin C$$
$$|DB| = |a - b\cos C|. \tag{1.1}$$

Thus

$$c^2 = b^2\sin^2 C + a^2 - 2ab\cos C + b^2\cos^2 C$$
$$c^2 = a^2 + b^2 - 2ab\cos C. \qquad \square$$

Theorem 1.10.4 (Heron's Formula). *For any triangle $\triangle ABC$,*

$$|\triangle ABC| = \sqrt{s(s-a)(s-b)(s-c)}.$$

Proof. Note that

$$|\triangle ABC| = \frac{1}{2}ab\sin C.$$

By the Law of Cosines,

$$\cos C = \frac{a^2 + b^2 - c^2}{2ab}.$$

Thus,

$$
\begin{aligned}
|\triangle ABC| &= \frac{1}{2}ab\sqrt{1 - \cos^2 C} \\
&= \frac{1}{2}ab\frac{\sqrt{4a^2b^2 - (a^2 + b^2 - c^2)^2}}{2ab} \\
&= \frac{1}{4}\sqrt{4a^2b^2 - (a^2 + b^2 - c^2)^2} \\
&= \frac{1}{4}\sqrt{(2ab + a^2 + b^2 - c^2)(2ab - a^2 - b^2 + c^2)} \\
&= \frac{1}{4}\sqrt{((a+b)^2 - c^2)(c^2 - (a-b)^2)} \\
&= \frac{1}{4}\sqrt{(a+b+c)(a+b-c)(c+a-b)(c-a+b)} \\
&= \sqrt{\frac{(a+b+c)}{2}\frac{(a+b-c)}{2}\frac{(a-b+c)}{2}\frac{(-a+b+c)}{2}} \\
&= \sqrt{s(s-c)(s-b)(s-a)}. \qquad \square
\end{aligned}
$$

Heron's formula is named after Heron of Alexandria, who most probably lived in the third century A.D., but possibly as early as 100 B.C. The formula, though, dates back to Archimedes (ca. 250 B.C.).

Exercise 1.72. In a triangle $\triangle ABC$, suppose $\angle ABC = 60°$, $a = 5$, and $c = 8$. What is b?

Exercise 1.73. In a quadrilateral $ABCD$, $|AB| = 3$, $|BC| = 11$, $|CD| = 8$, $|DA| = 4$, and $\angle BAD$ is a right angle. What is the area of the quadrilateral? [A]

Exercise 1.74. What is the area of $\triangle ABC$ if $a = 3$, $b = 5$, and $c = 6$?

Exercise 1.75. What is the area of the incircle of $\triangle ABC$ if $a = 5$, $b = 6$, and $c = 7$? [A]

Exercise 1.76. In Figure 1.34(a), suppose $|AC'| = |C'B| = |CE| = 2$, $|CD| = 3$, and $|BF| = 1$. What is the area of $\triangle ABC$? [A]

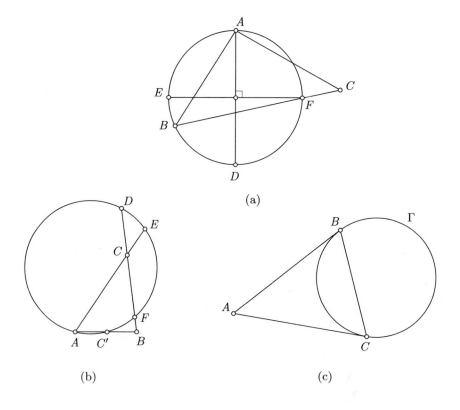

Figure 1.34. See Exercises 1.76, 1.77, and 1.79.

Exercise 1.77. In Figure 1.34(b), suppose AD and EF are perpendiculars which intersect at the center of the circle. Suppose also that $|AB| = 2$ and $|BC| = 3$. Then what is $|AC|$? [A]

Exercise 1.78. Explain why there is an absolute value around $a - b\cos C$ in Equation 1.1. [H]

Exercise 1.79. For a circle Γ and a point A outside the circle, let B and C be the points on Γ such that AB and AC are tangent to the circle Γ, as in Figure 1.34(c). Prove that the incenter I and the excenter I_a of $\triangle ABC$ both lie on Γ.   [S]

Exercise 1.80†. Prove that

$$|\triangle ABC| = (s - a)r_a.$$

Exercise 1.81†. Show that the distance from the vertex A to the tangent on the side AB of

a. the incircle is $s - a$;

b. the excircle centered at I_a is s;

c. the excircle centered at I_c is $s - b$.

Exercise 1.82. Show that

$$rr_a r_b r_c = |\triangle ABC|^2.$$ [H][S]

Exercise 1.83. Show that

$$\frac{1}{r_a} + \frac{1}{r_b} + \frac{1}{r_c} = \frac{1}{r}.$$

Exercise 1.84 (Theorem of Apollonius). In a triangle $\triangle ABC$, let A' be the midpoint of BC and let $l = |AA'|$. Prove that

$$b^2 + c^2 = 2l^2 + \frac{a^2}{2}.$$

Exercise 1.85. In this exercise, we are asked to find a different proof of the Law of Cosines. In $\triangle ABC$, suppose without loss of generality that $a \geq b$. Draw the circle centered at C that goes through B. Extend the sides AC and AB. Find the power of the point A with respect to this circle. Conclude the Law of Cosines.

Exercise 1.86*. Before we started using trigonometric functions, there was no need to define the measure of an angle. Come up with a suitable definition of the measure of an angle.

1.11   The Circumcircle and the Law of Sines

The third center we define is the *circumcenter O*, which is the center of the circle that circumscribes the triangle $\triangle ABC$.

Notice that the sides AB, AC, and BC are chords of the circumcircle, so their perpendicular bisectors all go through the center O. In particular, they all meet at a common point. The radius R of the circumcircle is called the *circumradius*.

Theorem 1.11.1 (The Extended Law of Sines).

$$\frac{a}{\sin A} = \frac{b}{\sin B} = \frac{c}{\sin C} = 2R.$$

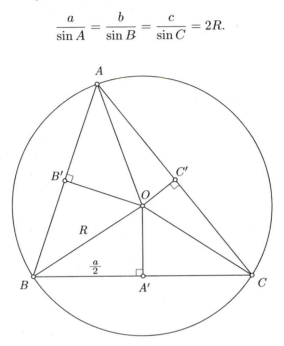

Figure 1.35

Proof. In $\triangle ABC$, let $A'O$ be the perpendicular bisector of BC, as in Figure 1.35. Note that $\triangle BOC$ is isosceles, that $\angle BOA' = \angle COA'$, and that $|BA'| = |A'C| = \frac{a}{2}$. Note also that, by the Star Trek lemma, $\angle BOC = 2A$, so $\angle BOA' = A$. Thus,

$$R \sin A = \frac{a}{2}$$

or

$$2R = \frac{a}{\sin A}.$$

But in a similar fashion,

$$2R = \frac{b}{\sin B} = \frac{c}{\sin C}. \qquad \square$$

Exercise 1.87. Though we have defined O to be the circumcenter, we have not shown that the circumcircle always exists. Prove that the perpendicular bisectors of two sides of a nondegenerate triangle $\triangle ABC$ intersect, and that this point is equidistant from all three vertices. Conclude that $\triangle ABC$ has a circumcircle and that the perpendicular bisectors are coincident.

Exercise 1.88. What are the areas of the incircle, circumcircle, and excircles of an equilateral triangle with sides of length 1?

Exercise 1.89. The Law of Sines (not the extended Law of Sines) merely states

$$\frac{a}{\sin A} = \frac{b}{\sin B} = \frac{c}{\sin C}.$$

There is a simpler proof of this result. Find it.   [H]

Exercise 1.90†. Let D be the base of the altitude at A of the triangle $\triangle ABC$. Let O be the circumcenter for $\triangle ABC$, and let AO intersect the circumcircle at E, as in Figure 1.36(a). As before, let R be the circumradius. Show that

$$|AD| = \frac{bc}{2R}.$$

Conclude that

$$|\triangle ABC| = \frac{abc}{4R}.$$

Hint: Show that $\triangle ABD$ is similar to $\triangle AEC$. Or, express $|AD|$ in terms of $\sin B$ and use the extended Law of Sines.   [S]

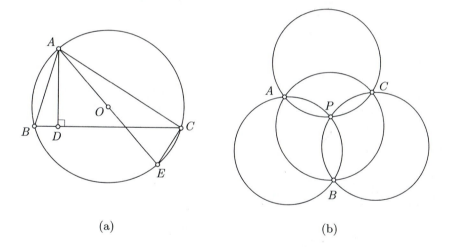

(a)   (b)

Figure 1.36. See Exercises 1.90 and 1.93.

Exercise 1.91. Prove that

$$4R = r_a + r_b + r_c - r.$$

Exercise 1.92. Where is the circumcenter of a right angle triangle, and why?

Exercise 1.93 (The Four Coin Problem). Suppose three congruent circles meet at a common point P and meet in pairs at the points A, B, and C, as in Figure 1.36(b). Show that the circumcircle of $\triangle ABC$ has the same radius as the original three circles. Use a coin to draw three such circles, and note that the coin can be used as the circumcircle for the resulting triangle $\triangle ABC$.

Exercise 1.94. In a triangle $\triangle ABC$, let O be the circumcenter, and let I be the incenter. Let the line AI intersect the circumcircle at D. Prove that $\triangle BID$ is isosceles.

Exercise 1.95*. Prove that a triangle is a right angle triangle if and only if

$$r + 2R = s.$$

Figure 1.37. The four coin problem was used as the logo for the 40th International Mathematical Olympiad held in Bucharest, Romania, 1999.

1.12 The Euler Line

The circumcenter O is the intersection of the perpendicular bisectors. The centroid G is the intersection of the medians. If these two points are concurrent, then the medians are the perpendicular bisectors, so the triangle is equilateral. Let us suppose $\triangle ABC$ is not equilateral, so that O and G are distinct.

Theorem 1.12.1. *In an arbitrary triangle $\triangle ABC$, the three altitudes AD, BE, and CF intersect at a common point.*

Proof. We prove this in a roundabout way. If $\triangle ABC$ is equilateral, then the altitudes are perpendicular bisectors and medians, so the altitudes all meet at $G = O$. So let us assume $\triangle ABC$ is not equilateral, so that O and G are distinct.

Let H be the point on the line OG so that $|GH| = 2|OG|$ and the points O, G, and H are in that order, as in Figure 1.38. We will show that H

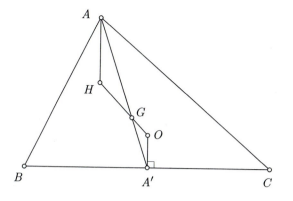

Figure 1.38

lies on the altitude at A. Let A' be the midpoint of BC, so G lies on the median AA' and OA' is the perpendicular bisector of BC. Consider the triangles $\triangle GOA'$ and $\triangle GHA$. The angles $\angle A'GO$ and $\angle AGH$ are vertical angles, so are congruent. Recall that the centroid divides the medians in thirds, so $|AG| = 2|GA'|$. And by construction, $|GH| = 2|OG|$, so these two triangles are similar. Hence, AH is parallel to OA', and therefore, if we continue AH until it intersects BC at D, then AD is perpendicular to BC. That is, AD is the altitude at A, and H lies on the altitude. But there was nothing special about starting with the side BC, so by the same argument, H lies on the altitudes at B and at C. □

The point of intersection of the altitudes is called the *orthocenter* and is usually denoted with an H. There are easier proofs of Theorem 1.12.1, as we will see in Exercise 1.96. However, in the above proof, we have also shown the following result:

Theorem 1.12.2 (The Euler Line). *The circumcenter O, the centroid G, and the orthocenter H are collinear. Furthermore, G is between O and H, and*

$$\frac{|OG|}{|GH|} = \frac{1}{2}.$$

The line which contains O, G, and H is called the *Euler line*, named after the man who discovered it, Leonhard Euler (1707 – 1783). It is surprising that such a basic theorem in geometry was discovered so recently.

Exercise 1.96. Draw an arbitrary triangle $\triangle ABC$. Through each vertex, draw a line parallel to the opposite side. These three lines form a new triangle. Where are the perpendicular bisectors of this new triangle? Use this to give a different proof that the altitudes of $\triangle ABC$ intersect at a common point.

Exercise 1.97. Suppose we are told that a triangle $\triangle ABC$ is isosceles with $|AB| = |AC|$ but are given only the length of the median BB' and the length of the altitude BE. Use this information to construct $\triangle ABC$.

Exercise 1.98. Let AD be the altitude of an arbitrary triangle $\triangle ABC$. Prove that
$$|AD| = 2R \sin B \sin C.$$
Conclude that
$$|\triangle ABC| = 4R^2 \sin A \sin B \sin C.$$

Exercise 1.99. Suppose the Euler line of a triangle $\triangle ABC$ passes through the vertex A. Prove that $\triangle ABC$ is either a right angle triangle or an isosceles triangle (or both).

Exercise 1.100. Let $\triangle ABC$ be a triangle with orthocenter H and circumcenter O. Let A' be the midpoint of BC and let D be the base of the altitude from A. Suppose that $HOA'D$ is a rectangle with sides $|HO| = 11$ and $|OA'| = 5$. What is the length $|BC|$? This question (without Figure 1.39) was question A-1 on the 1997 Putnam Mathematical Competition. [H][A]

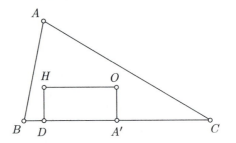

Figure 1.39. See Exercise 1.100.

1.13   The Nine Point Circle

Most would agree that a Mercedes is a pretty good car. This is because of its engine, its chassis, its comfortable interior, and its engineering. But most would also agree that it is not complete without a hood ornament. If we were to make an analogy between Euclidean geometry and an automobile, the nine point circle would have to be called a hood ornament. It is not a useful theorem – it just looks nice and embodies everything we have learned so far.

As usual, let A', B', and C' be the midpoints of the sides of a triangle $\triangle ABC$, as in Figure 1.40. Let D, E, and F be the bases of the altitudes. Let H be the orthocenter, and let A'', B'', and C'' be the midpoints of AH, BH, and CH, respectively.

Theorem 1.13.1 (The Nine Point Circle). *The nine points A', B', C',*
A'', B'', C'', D, E, and F all lie on a circle.

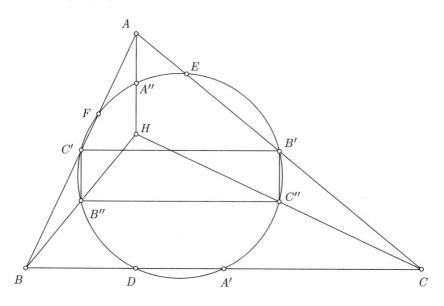

Figure 1.40

Proof. Since B' and C' are midpoints, $B'C'$ is parallel to BC. Since C'
and B'' are midpoints of two sides of $\triangle AHB$, we also have $B''C'$ is parallel
to AH, which is perpendicular to BC. Similarly, $B'C''$ is parallel to AH.
Thus, $B'C'B''C''$ is a rectangle. Construct the circle with diameter $C'C''$.
Since $\angle C'B'C''$ and $\angle C'B''C''$ are right angles, we know B' and B'' lie on
the circle. Furthermore, since $|B'B''| = |C'C''|$, $B'B''$ is a diameter. Since
CF is an altitude, $\angle C''FC'$ is a right angle, so F also lies on the circle.
Since $B'B''$ is a diameter and $\angle B'EB''$ is a right angle, E lies on the circle.
Finally, by a similar argument, $C'A''C''A'$ is a rectangle, so A' and A'' lie
on the circle, and $A'A''$ is a diameter, so D lies on the circle. $\qquad\square$

Exercise 1.101** (Feuerbach's Theorem). Prove that the nine point
circle of $\triangle ABC$ is tangent to the incircle and excircles of $\triangle ABC$. The nine
point circle is sometimes known as the *Feuerbach circle*.

Exercise 1.102**. Prove that the center of the nine point circle lies on
the Euler line.

Exercise 1.103**. Prove that the tangent to the nine point circle at the
point C' is parallel to the line DE.

1.14   Pedal Triangles and the Simson Line

The Simson line is another interesting result which showcases the techniques we have learned so far. Before we define it, let us first introduce the notions of cyclic quadrilaterals and pedal triangles.

A *cyclic quadrilateral* is a quadrilateral that can be inscribed in a circle. Cyclic quadrilaterals were introduced in Exercise 1.36, where we were asked to prove the following:

Theorem 1.14.1. *A convex quadrilateral ABCD is a cyclic quadrilateral if and only if* $\angle ABC + \angle CDA = 180°$.

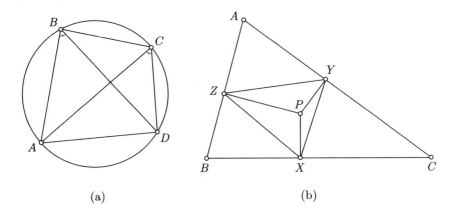

(a)   (b)

Figure 1.41

In a cyclic quadrilateral $ABCD$, the angles $\angle ABD$ and $\angle ACD$ are equal, since they subtend the same arc (see Figure 1.41(a)). This simple observation is quite useful, as we will see in the following lemma.

Let P be an arbitrary point either inside or outside a triangle $\triangle ABC$. Let X be the foot of the perpendicular to the extended side BC and through P. Similarly, define Y and Z on the extended sides AC and AB respectively, as in Figure 1.41(b). The triangle $\triangle XYZ$ is called the *pedal triangle* with respect to the point P and the triangle $\triangle ABC$.

Lemma 1.14.2. *Let P be a point inside triangle $\triangle ABC$, and let $\triangle XYZ$ be the pedal triangle with respect to P. Then*

$$\angle APB = \angle ACB + \angle XZY.$$

Proof. Let CP intersect AB at C' (see Figure 1.42). Let us write

$$\angle APB = \angle APC' + \angle C'PB.$$

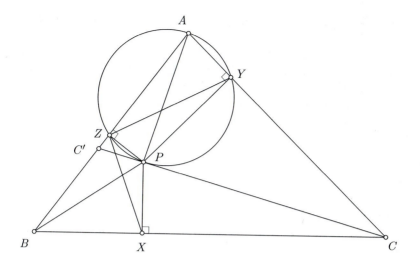

Figure 1.42

Since $\angle APC'$ is an exterior angle of $\triangle APC$, we get

$$\angle APC' = \angle PAC + \angle ACP.$$

Note that the quadrilateral $AYPZ$ is cyclic, since $\angle PZA = \angle AYP = 90°$, and so their sum is $180°$. Thus,

$$\angle PAC = \angle PAY = \angle PZY,$$

and hence,

$$\angle APC' = \angle PZY + \angle ACP.$$

Similarly,

$$\angle C'PB = \angle XZP + \angle PCB.$$

Summing, we get

$$\begin{aligned}
\angle APB &= \angle APC' + \angle C'PB \\
&= (\angle PZY + \angle XZP) + (\angle ACP + \angle PCB) \\
&= \angle XZY + \angle ACB,
\end{aligned}$$

as desired. □

Though this result is stated only for P inside $\triangle ABC$, similar results exist for arbitrary P. For example, note that in Figure 1.42, the points A, B, and C are oriented counterclockwise, as are the points X, Y, and Z. If we move P far enough across the line AC so that the points X, Y, and Z become oriented in a clockwise fashion, then the result becomes

$$\angle APB = \angle ACB - \angle XZY.$$

Such variety can be dealt with in a systematic way by introducing the concept of an *oriented angle*. When using oriented angles, the notation $\angle ABC$ means the measure of the angle from the ray BA to the ray BC, measured counterclockwise. This angle may be larger than $180°$. Consequently, an angle sum like $\angle ACB + \angle XZY$ may be larger than $360°$, so in equations involving oriented angles, we say two angle sums are equal if they differ by a multiple of $360°$. So, for example,

$$\angle ABC = 360° - \angle CBA = -\angle CBA.$$

Using this convention, Lemma 1.14.2 is true for any point P, and the given proof works for any diagram. We therefore get, as a corollary, the following theorem:

Theorem 1.14.3 (The Simson Line). *Let Γ be the circumcircle for $\triangle ABC$. Let P be a point on Γ, and let $\triangle XYZ$ be the pedal triangle with respect to P. Then $\triangle XYZ$ is a degenerate triangle. That is, the points X, Y, and Z are collinear. This line is called the Simson line.*

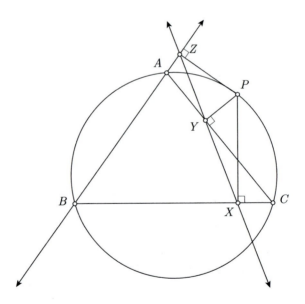

Figure 1.43

Proof. Without loss of generality, we may assume P is on the arc AC, as in Figure 1.43. Then,

$$\angle APB = \angle ACB$$

since they subtend the same arc. Hence, by Lemma 1.14.2,

$$\angle XZY = 0.$$

That is, $\triangle XYZ$ is degenerate, or equivalently, X, Y, and Z are collinear.

$\square$

Exercise 1.104. The proof of Lemma 1.14.2 works for any diagram. Prove Simson's result directly by adapting the proof of Lemma 1.14.2 for a point P on the circumcircle.

Exercise 1.105. Let D be the center of the square constructed on the hypotenuse of a right angle triangle $\triangle ABC$ with right angle at C. Prove that $\angle ACD = 45°$.

[H]

Exercise 1.106. Let H' be the reflection of the orthocenter H of $\triangle ABC$ through the line BC. Prove that $ABCH'$ is a cyclic quadrilateral.

Exercise 1.107. The Star Trek lemma is worded in terms of arcs instead of interior angles since we normally consider an angle to measure no more than $180°$. Reword the Star Trek lemma using oriented interior angles. Check that the proof given in Section 1.6 works for any diagram if we understand the angles to be oriented.

Exercise 1.108. Reword the results in Exercises 1.34 and 1.35 using oriented angles and in such a way that the wording for each is identical.

Exercise 1.109† (Ptolemy's Theorem). Suppose that $ABCD$ is a cyclic quadrilateral. Prove that

$$|AC||BD| = |AB||CD| + |BC||DA|.$$

Hint: Draw AE so that E is on BD and $\angle BAE = \angle CAD$, as in Figure 1.44(a).

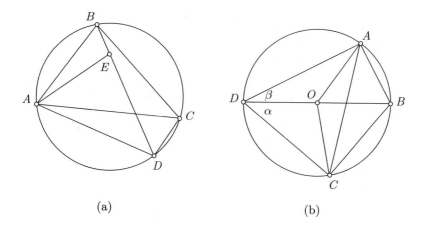

(a)   (b)

Figure 1.44. See Exercises 1.109 and 1.111.

Exercise 1.110. Suppose triangle $\triangle ABC$ has sides of length $a = 6$, $b = 5$, and circumradius $R = 4$. What is the length c?   [A]

Exercise 1.111. Use Ptolemy's theorem and Figure 1.44(b) to prove the angle sum formula for sines (for acute angles α and β.) That is, show

$$\sin(\alpha + \beta) = \sin\alpha\cos\beta + \cos\alpha\sin\beta.$$

Exercise 1.112. Extend the result in Exercise 1.111 to any angles. Make the substitution $\alpha' = \frac{\pi}{2} - \alpha$ and $\beta' = -\beta$ to derive the angle sum formula for cosines.

Exercise 1.113. Prove the converse of Exercise 1.111. That is, use Figure 1.44 and the angle sum formula for sines to prove Ptolemy's theorem.

Exercise 1.114. There are other proofs of the angle sum formulas. Use Figure 1.45 to come up with one of them. Though it is not necessary to notice any cyclic quadrilaterals, the argument might be made neater by doing so. For which angles is this proof valid?   [S]

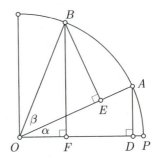

Figure 1.45. See Exercise 1.114.

Exercise 1.115 (Brahmagupta's Formula). Heron's formula generalizes to cyclic quadrilaterals in the following way. Let a, b, c, and d be the lengths of the four sides of the cyclic quadrilateral. Let $s = \frac{1}{2}(a + b + c + d)$. Show that the area of the quadrilateral is

$$\sqrt{(s - a)(s - b)(s - c)(s - d)}.$$

Exercise 1.116†. Let $\triangle XYZ$ be the pedal triangle with respect to P and a triangle $\triangle ABC$. Show that

$$|YZ| = |AP|\sin A. \qquad\qquad \text{[H]}$$

Exercise 1.117. Let $\triangle ABC$ be an equilateral triangle. Suppose D is a point outside the triangle such that $\angle ADB = 120°$. Show that

$$|CD| = |AD| + |BD|.$$

Exercise 1.118**. Let $ABCDEF$ be a convex hexagon with $|AB| = |BC| = |CD|$, $|DE| = |EF| = |FA|$, and $\angle BCD = \angle EFA = 60°$. Let G and H be two points in the interior of the hexagon such that $\angle AGB = \angle DHE = 120°$. Prove that

$$|AG| + |GB| + |DH| + |HE| + |GH| \geq |CF|.$$

This was Question 5 on the 1995 International Mathematical Olympiad exam.

1.15 Menelaus and Ceva

Suppose D, E, and F are three points on the three (extended) sides BC, AC, and AB of a triangle $\triangle ABC$. Menelaus (ca. 100 A.D.) showed that if these points are collinear, then they satisfy a certain condition. To state this condition, we must first define the notion of a signed ratio of lengths. If P, Q, and R are collinear, then the *signed ratio of lengths* is the ratio $\dfrac{|PQ|}{|QR|}$ together with a positive sign if Q is between P and R, and a negative sign otherwise. In this section, every ratio of lengths should be considered to be a signed ratio of lengths, unless otherwise stated. The converse of Menelaus' theorem is also true, as stated in the following. Though Menelaus did not prove the converse, he assumed that it was true.

Theorem 1.15.1 (Menelaus' Theorem). *Let D, E, and F be three points on, respectively, the extended sides BC, CA, and AB of $\triangle ABC$. Then the points D, E, and F are collinear if and only if*

$$\frac{|AF|}{|FB|} \frac{|BD|}{|DC|} \frac{|CE|}{|EA|} = -1.$$

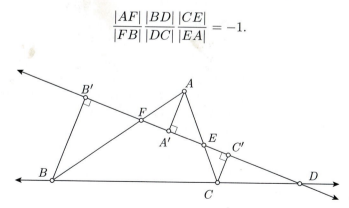

Figure 1.46

Proof. Let us suppose a line intersects the (extended) sides of $\triangle ABC$ at D, E, and F. Note that the line either intersects two of the proper sides (that is, between two vertices), or intersects none of the proper sides. In either case, the product $\dfrac{|AF|}{|FB|}\dfrac{|BD|}{|DC|}\dfrac{|CE|}{|EA|}$ is negative. We will therefore only concern ourselves with the magnitude of this quantity.

Drop the perpendiculars to the line from each of the vertices. Let these perpendiculars intersect the line at A', B', and C', as in Figure 1.46. Note that $\angle A'AF = \angle B'BF$, since they are opposite interior angles of a transversal to two parallel lines. Thus, $\triangle AA'F \sim \triangle BB'F$. Hence (with no signed ratios)

$$\frac{|AF|}{|BF|} = \frac{|AA'|}{|BB'|}.$$

Similarly, $\triangle AA'E \sim \triangle CC'E$ and $\triangle BB'D \sim \triangle CC'D$, giving

$$\frac{|AE|}{|CE|} = \frac{|AA'|}{|CC'|} \qquad \text{and} \qquad \frac{|BD|}{|CD|} = \frac{|BB'|}{|CC'|}.$$

Hence, ignoring the convention concerning the sign, we get

$$\left| \frac{|AF|}{|BF|}\frac{|BD|}{|CD|}\frac{|CE|}{|AE|} \right| = \left| \frac{|AA'|}{|BB'|}\frac{|BB'|}{|CC'|}\frac{|CC'|}{|AA'|} \right| = 1.$$

To prove the converse, let us suppose $\dfrac{|AF|}{|FB|}\dfrac{|BD|}{|DC|}\dfrac{|CE|}{|EA|} = -1$, but that D, E, and F are not collinear. Because of the sign, we know either two points are on proper sides, or none of the points are on proper sides. Let us suppose, without loss of generality, that the line DE intersects the third side at F' where F' is not between A and B. Then, by the first part of this proof,

$$\frac{|AF'|}{|F'B|}\frac{|BD|}{|DC|}\frac{|CE|}{|EA|} = -1.$$

Hence,

$$\frac{|AF'|}{|F'B|} = \frac{|AF|}{|FB|}.$$

Let $|AF'| = |AF| \pm |FF'|$. Then, $|F'B| = |FB| \pm |FF'|$. The choice of sign depends on whether F is between A and F' or not. Note that the sign is the same for both expressions, since F' does not lie between A and B. Thus,

$$(|AF| \pm |FF'|)|FB| = |AF|(|FB| \pm |FF'|)$$
$$\pm |FF'|(|FB| - |AF|) = 0.$$

Since F is not between A and B, we know $|FB| \neq |AF|$. Hence, we must have $|FF'| = 0$. That is, $F = F'$. $\square$

The following result, due to Giovanni Ceva (1678), has a flavor similar to the result above. It is surprising that 1600 years passed before it was discovered.

Theorem 1.15.2 (Ceva's Theorem). *Let D, E, and F be three points on, respectively, the sides BC, CA, and AB of $\triangle ABC$. Then the lines AD, BE, and CF are concurrent if and only if*

$$\frac{|AF|}{|FB|}\frac{|BD|}{|DC|}\frac{|CE|}{|EA|} = 1.$$

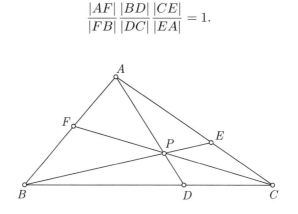

Figure 1.47

Proof. Suppose AD, BE, and CF intersect at a common point P. Let us consider $\triangle ABE$. Note that F, P, and C are collinear (see Figure 1.47), so by Menelaus' theorem, we have (ignoring the sign)

$$\left|\frac{|AF||BP||EC|}{|FB||PE||CA|}\right| = 1.$$

Similarly, from $\triangle BCE$, we have

$$\left|\frac{|BD||CA||EP|}{|DC||AE||PB|}\right| = 1.$$

Multiplying, we get

$$1 = \frac{|AF||BP||EC|}{|FB||PE||CA|}\frac{|BD||CA||EP|}{|DC||AE||PB|} = \frac{|AF||EC||BD|}{|FB||DC||AE|},$$

as desired. The sign must, of course, be positive.

For the other direction, we proceed as in the previous proof. Let us assume

$$\frac{|AF|}{|FB|}\frac{|BD|}{|DC|}\frac{|CE|}{|EA|} = 1,$$

but that AD, BE and CF are not coincident. Let AD and BE intersect at P, and let CP intersect AB at F'. Then, by the first part of this proof, we have

$$\frac{|AF'|}{|F'B|}\frac{|BD|}{|DC|}\frac{|CE|}{|EA|} = 1.$$

Hence,

$$\frac{|AF'|}{|F'B|} = \frac{|AF|}{|FB|}.$$

Again, suppose $|AF'| = |AF| + x$. Then, $|F'B| = |FB| - x$, so

$$\frac{|AF| + x}{|FB| - x} = \frac{|AF|}{|FB|}$$

$$x(|AF| + |FB|) = 0.$$

Hence, $x = 0$, and $F = F'$, as desired. □

If D is a point on the side BC of a triangle $\triangle ABC$, then AD is called a *Cevian*.

There is also a trigonometric version of Ceva's theorem:

Theorem 1.15.3 (Trig Ceva). *Let D, E, and F be points on the extended sides BC, CA, and AB, respectively. Let $\alpha = \angle DAB$ and $\alpha' = \angle CAD$, and define β, β', γ, and γ' similarly, as in Figure 1.48. Then AD, BE, and CF are coincident if and only if*

$$\sin \alpha \sin \beta \sin \gamma = \sin \alpha' \sin \beta' \sin \gamma'.$$

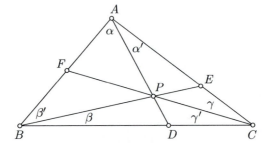

Figure 1.48

Exercise 1.119. Prove the trigonometric version of Ceva's theorem.

Exercise 1.120. In Figure 1.47, show that

$$\frac{|\triangle APB|}{|\triangle APC|} = \frac{|BD|}{|DC|}.$$

Use this to come up with a proof of Ceva's theorem which does not use Menelaus' theorem.

Exercise 1.121. What is the analogue of Menelaus' theorem for the case when D (in Figure 1.46) is at infinity?

Exercise 1.122. Use Ceva's theorem to show that the medians intersect at a common point.

Exercise 1.123. Use Ceva's theorem to show that the altitudes intersect at a common point.

Exercise 1.124. Use Ceva's theorem to show that the angle bisectors intersect at a common point.   [H]

Exercise 1.125 (The Gergonne Point). Let the incircle of a triangle $\triangle ABC$ be tangent to the sides BC, AC, and AB at, respectively, the points D, E, and F. Show that AD, BE, and CF are coincident. This point is known as the *Gergonne point*.

Exercise 1.126 (The Nagel Point). Let the excircle opposite A be tangent to BC at D. Define E and F similarly. Show that AD, BE, and CF are coincident. This point is known as the *Nagel point*.

Exercise 1.127. Let $\triangle ABC$ be a triangle. Suppose D, E, and F are points on the sides BC, AC, and AB, respectively, and such that AD, BE, and CF are coincident. Let the reflection of the ray AD through the angle bisector of $\angle BAC$ intersect BC at D'. Similarly define E' and F', as in Figure 1.49. Show that AD', BE', and CF' are coincident.

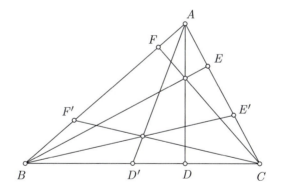

Figure 1.49. See Exercise 1.127.

Exercise 1.128. Let AD be an altitude to $\triangle ABC$. Prove

$$|AB|^2 - |BD|^2 = |AC|^2 - |CD|^2.$$

Exercise 1.129. Let X, Y, and Z be points on the sides BC, AC, and AB, respectively, of a triangle $\triangle ABC$. Prove that the perpendiculars to the sides of $\triangle ABC$ through the points X, Y, and Z are concurrent if and only if

$$|AZ|^2 - |ZB|^2 + |BX|^2 - |XC|^2 + |CY|^2 - |YA|^2 = 0. \qquad \text{[H]}$$

Exercise 1.130. Generalize the previous exercise: Let X, Y, and Z be any points in the plane. Prove that the perpendiculars to the sides BC, AC, and AB through the points X, Y, and Z, respectively, are concurrent if and only if

$$|AZ|^2 - |ZB|^2 + |BX|^2 - |XC|^2 + |CY|^2 - |YA|^2 = 0.$$

Chapter 2

Geometry in Greek Astronomy

The heavenly bodies have been a source of wonderment since at least the dawn of civilization. The most notable records of mankind's contemplation of the heavens in ancient Western civilization are those of the Babylonians and the Greeks.

The ancient Babylonians (2000 – 600 B.C.) charted the stars and named the constellations. They identified the twelve signs of the *zodiac* – those stars which lie in the plane described by the Earth's revolution about the sun. They further partitioned each sign into 30 degrees, so the zodiac circle has 360 degrees. Their system of angular measure – degrees, minutes, and seconds – is still in use today.

The Babylonians also tracked the movements of other celestial bodies, namely the sun, the moon, and at least three of the five planets visible to the naked eye: Venus, Mars, and Jupiter (Mercury and Saturn are the other two planets visible to the naked eye). They kept accurate records of celestial phenomena and were able in particular to predict solar and lunar eclipses.

If the Babylonians ever theorized about the meaning of these observations, such theories have been lost to us. Such theories were the domain of the ancient Greeks.

At the end of this chapter, we present a time line of significant evolutions in the theories of astronomy, but for now, a few remarks seem appropriate. The belief that the Earth is spherical dates back to the time of Pythagoras (ca. 550 B.C.). Anaxagoras (ca. 475 B.C.) believed that the moon is illuminated by the sun and that it is of an earthy nature (with plains, mountains and valleys). Plato (ca. 400 B.C.) wrote that the Earth is at the center of the universe but is said to have regretted that concept in his later years. Heraclides of Pontus (ca. 360 B.C.) believed that the Earth rotates on its axis, rather than that the sun and stars rotate daily about the Earth. He also believed that Mercury and Venus revolve about the sun.

The most significant theory is due to Aristarchus of Samos (310 – 230 B.C.), who is said to have proposed the following (and we quote from Archimedes)[H2]:

> You are aware that "universe" is the name given by most astronomers to the sphere the center of which is the center of the Earth, while its radius is equal to the straight line between the center of the sun and the center of the Earth. This is the common account, as you have heard from astronomers. But Aristarchus brought out a book consisting of certain hypotheses, wherein it appears, as a consequence of the assumptions made, that the universe is many times greater than the "universe" just mentioned. His hypotheses are that the fixed stars and sun remain unmoved, that the Earth revolves about the sun in the circumference of a circle, the sun lying in the middle of the orbit, and that the sphere of the fixed stars, situated about the same center as the sun, is so great that the circle in which he supposes the Earth to revolve bears such a proportion to the distance of the fixed stars as the center of the sphere bears to its surface.

Aristarchus proposed the heliocentric theory of the universe almost 2000 years before Copernicus! Unfortunately, his theory was heretic in his time, and geocentric theories of the universe prevailed.

Of celestial measurements, there is significant work of a few. Aristarchus measured the relative sizes of the moon, Earth, and sun, and the relative sizes of their orbits. The size of the Earth was measured by Eratosthenes of Cyrene (ca. 275 – 195 B.C.), and by Posidonius (135 – 51 B.C.). Their measurements are the subject of this chapter.

2.1   The Relative Size of the Moon and Sun

Aristarchus (310 – 230 B.C.) made several observations and with them calculated the relative diameters of the Earth, moon, and sun, and the relative radii of their orbits. Though his measurements were not always accurate, his method is sound.

The moon and the sun, when seen from the Earth, are about the same size. We know this from observation and also because total solar eclipses cast a shadow on only a small portion of the Earth. The angle they subtend is approximately .5 degrees.[1] Thus, we get

$$\frac{D_M}{R_M} = \frac{D_S}{R_S} = \frac{2\pi}{720},$$

---

[1] Aristarchus used the figure of $2°$ in his manuscript, but is said to have later used $.5°$. Hipparchus (161 – 126 B.C.) gave the measurements of $31'$ when the moon is furthest away from the Earth, and $35'$ when it is closest. Recall, $60' = 1°$ (Read 'sixty minutes equals one degree.')

where D_M and D_S are the diameters of, respectively, the moon and sun, and R_M and R_S are the distances from the Earth to, respectively, the moon and sun.

To compare the relative diameters of the moon and sun, Aristarchus made the following observation: At half moon, when the lighted portion of the moon exactly bisects the visible portion of the moon, the triangle made by the sun, moon, and Earth is a right angle triangle with the right angle at the moon (see Figure 2.1).

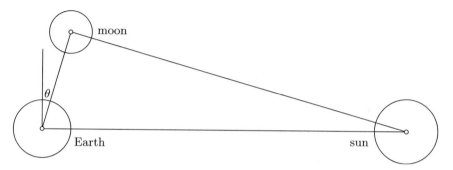

Figure 2.1

By measuring the angle the moon makes with the sun at this time, one can find $\dfrac{R_M}{R_S}$ and $\dfrac{D_M}{D_S}$. Aristarchus found this angle to be 87 degrees (in his words, "less than a quadrant by one-thirtieth of a quadrant.") That is, according to Aristarchus, the angle θ in Figure 2.1 is 3°.

We therefore get

$$\sin \theta = \frac{R_M}{R_S}.$$

Exercise 2.1. Express R_M, R_S, and D_S in terms of θ and D_M.

To compare the diameter of the moon D_M with the diameter of the Earth D_E, Aristarchus measured the duration of a lunar eclipse. From this observation, he concluded that the shadow of the Earth is twice as wide as the moon. In the following calculation, we will use the more accurate measurement made by Hipparchus (161 – 126 B.C.). He found that the shadow of the Earth is 2.5 times the diameter of the moon.

Figure 2.2 illustrates the shadow of the Earth cast on the moon. In this diagram, we have two similar triangles, which give

$$\frac{R_M}{R_S} = \frac{D_E - 2.5 D_M}{D_S - D_E}$$
$$R_M D_S - R_M D_E = R_S D_E - 2.5 R_S D_M.$$

In the above, we have simplified our argument by using the approximation $D_E - 2.5 D_M$. We remark that Aristarchus was more precise.

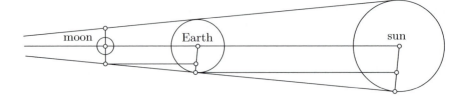

Figure 2.2

Recall that

$$\frac{R_M}{R_S} = \frac{D_M}{D_S},$$

so $R_M D_S = R_S D_M$. Thus, we have

$$3.5 R_S D_M = (R_S + R_M) D_E$$
$$3.5 D_M = \left(1 + \frac{R_M}{R_S}\right) D_E$$
$$= (1 + \sin\theta) D_E.$$

Exercise 2.2. Express R_S, R_M, D_S, and D_M in terms of θ and D_E.

Exercise 2.3. Explain why the figure $D_E - 2.5 D_M$ is an approximation. That is, what facts have we trivialized?

Exercise 2.4. It is fairly easy to measure the angle the moon subtends. On a clear night, use a stopwatch to time the moonrise. That is, measure the amount of time between when a sliver of the moon is first visible on the horizon and when the moon is completely above the horizon. Explain how this gives the angular measure of the moon. We know that the sun has approximately the same angular measure since solar eclipses occur and when they do, the shadow of the moon is so small on the Earth.

2.2 The Diameter of the Earth

James Smart, in his text *Modern Geometry*, gives a romantic version of the measurement of the Earth made by Eratosthenes of Cyrene (ca. 275 – 195 B.C.)[Sm]. Eratosthenes knew, according to Smart, that at high noon on the first day of summer, the sun shone directly into a well at Syene (now called Aswan). At a well in Alexandria, 787 km north, the sun did not shine all the way down but instead shone at an angle of 7.2° from the vertical. By assuming the sun is at a point at infinity, Eratosthenes was able to calculate the radius of the Earth.

Exercise 2.5. Use the above measurements and assumptions to find the circumference of the Earth.

Trivia. The meter was originally defined (by the French in 1799) so that the distance from the North Pole to the equator (through Paris) measures 10,000 km. Thus, the (polar) circumference of the Earth is 40,000 km.

Exercise 2.6. Was Eratosthenes' assumption that the sun is a point at infinity a safe assumption? Find an upper bound on the error in the above calculation of the circumference of the Earth, given that the sun is 150,000,000 km from the Earth.   [S]

Exercise 2.7. Note that Eratosthenes' measurements are by no means a testimony that the Earth is round. This time, assume the Earth is flat, and use Eratosthenes' measurements to calculate the distance from the Earth to the sun.   [S]

Of Eratosthenes' actual measurements, we have the description of Cleomedes (probably ca. 50 B.C., but possibly as late as 200 A.D.), in which Eratosthenes is said to have used sundials (not wells), the distance from Syene to Alexandria is said to be 5000 *stadia*, and the angle of the shadow at Alexandria is said to be one-fiftieth of a circle. The circumference of the Earth is therefore, according to Eratosthenes, 250,000 stadia. He later revised this to 252,000 stadia, but it is not known why he made this revision.

The accuracy of Eratosthenes' measurements have been the subject of some debate. The accepted length of the *stade*[2] in the time of Eratosthenes is not known, but the "most probable assumption" puts Eratosthenes' figure of 252,000 stadia at about 39,800 km [H2]. The accuracy seems to be a matter of luck more than anything else, since his measurements (5000 stadia and 1/50th of a circle) do not imply a significant degree of accuracy.

Exercise 2.8. Alexandria is not due north of Syene (Aswan). Consult a map of Egypt and measure the distance between Alexandria and Aswan, and also the distance between the lines of latitude through Aswan and through Alexandria. Use both measurements, together with the figure 7.2°, to calculate the circumference of the Earth. How much of an error is introduced by assuming that Alexandria and Syene are on the same line of longitude?

Posidonius (135 – 51 B.C.) also measured the circumference of the Earth. He assumed that Rhodes, a small Greek island off the Turkish coast, lies on the same line of longitude as Alexandria, and that the two are 5000 stadia apart. Instead of observing the sun, Posidonius looked at the stars. One in particular, Canobus, is not visible in mainland Greece. In Rhodes, when it is seen, it sets immediately after rising. In Alexandria, on the same night, this star is observed to rise to a height of one-quarter of a sign of the zodiac (that is, 1/48th of a circle or 7.5°) before it sets. Posidonius therefore concludes that the circumference of the Earth is 240,000 stadia.

---

[2]Stade – singular of stadia.

Exercise 2.9. Use Eratosthenes' measurements (5000 stadia and 7.2°), Aristarchus' values (moon and sun span .5°, the angle called θ in Figure 2.1 is 3°), and Hipparchus' measurement (the shadow of the Earth is 2.5 times the diameter of the moon), to find D_E, D_M, D_S, R_M, and R_S in stadia.

Exercise 2.10. Of the Greek measurements in Exercise 2.9, only the angle θ is not very accurate. This angle is approximately $10'$ or $\left(\frac{1}{6}\right)^\circ$. Use this value, together with Smart's values (787 km, 7.2°), Aristarchus' values (the moon and sun span .5°), and Hipparchus' value (the shadow of the moon is 2.5 times the diameter of the moon) to find D_E, D_M, D_S, R_M, and R_S. Compare these values with today's accepted values given in Table 2.1. What is the percent error in each?

Polar diameter of the Earth:	$D_E = 12,720$ km
Equatorial diameter of the Earth:	$12,760$ km
Diameter of the moon:	$D_M = 3475$ km
Diameter of the sun:	$D_S = 1,392,000$ km
From the Earth to the moon at *perigee* (closest):	$356,000$ km
From the Earth to the moon at *apogee* (farthest):	$407,000$ km
From the Earth to the sun:	$R_S = 149,000,000$ km

Table 2.1

2.3 The Babylonians to Kepler, A Time Line of Astronomy

The ancient Egyptians set the month at 30 days, and the year at 365 days or 12 months and 5 extra days. They partitioned both day and night into 12 hours, so the length of an hour differed from day to night, and from season to season. They invented sun dials and water clocks.

The Babylonians (2000 – 600 B.C.) named the constellations and the 12 signs of the zodiac (Aquarius, Pisces, Aries, etc.). They further divided each zodiac sign into 30 degrees. They used base 60 and invented the angular measuring system of degrees, minutes, and seconds. They made accurate records of the cycles of the moon, sun, and at least three planets (Venus, Mars, and Jupiter) and were able to predict eclipses.

Thales (624 – 547 B.C.) predicted the solar eclipse of May 28, 585 B.C. This was a rather spectacular prediction, since it occurred during a battle between the Lydians and the Medes, and it was a total eclipse. His prediction was almost certainly a result of access to Babylonian records. Thales was also probably the first known nerd. Plato writes "A case in point is that of Thales, who, when he was star-gazing and looking upward, fell into a well, and was rallied (so it is said) by a clever and pretty maidservant from Thrace because he was eager to know what went on in the heaven, but did not notice what was in front of him, nay, at his very feet."

Anaximander (611 – 546 B.C.) proposed that the Earth is suspended freely, without support, but also proposed that it is a short cylinder. He also speculated on the size and distances of the sun and moon. He claimed $R_M = 19D_E$, $R_S = 28D_E$, and $D_S = D_E$. He also hinted at a theory of evolution.

Pythagoras (ca. 572 – 500 B.C.) proposed that the Earth is spherical. He recognized that the morning and evening stars are the same planet (Venus). He also proposed that space or void exists. Some of his followers, the Pythagoreans, believed that the Earth moves on its axis.

Anaxagoras (ca. 500 – 428 B.C.) claimed that the moon is illuminated by the sun, and that it is of an earthy nature.

Empedocles (ca. 495 – 435 B.C.) proposed that light moves.

Plato (ca. 427 – 347 B.C.) placed the Earth at the center of the universe, but in his later years, is said to have regretted that concept.

Heraclides of Pontus (ca. 388 – 315 B.C.) claimed that the apparent rotation of the universe is caused by the rotation of the Earth. He also believed that Mercury and Venus rotate about the sun.

Aristotle (384 – 322 B.C.) was a very influential philosopher, whose "services to astronomy," according to Heath, "consist largely of thoughtful criticisms, generally destructive, of opinions held by earlier astronomers; ...." [H2].

Aristarchus of Samos (310 – 230 B.C.) proposed the heliocentric theory or Copernican theory of the universe. He found few followers, and Cleanthes thought it the duty of Greeks to indict Aristarchus on the charge of impiety. Aristarchus also calculated the relative diameters of the moon, Earth, and sun, and of the distances to the moon and sun.

Euclid (4th century B.C.) believed the Earth is at the center of the universe.

Archimedes (287 – 212 B.C.) wrote *Psammites* (Sand-reckoner), in which he attempted to calculate the number of grains of sand in the universe. It is in this work that Archimedes writes of Aristarchus.

Eratosthenes of Cyrene (ca. 275 – 195 B.C.) measured the circumference of the Earth and put it at 252,000 stadia.

Hipparchus (ca. 161 – 126 B.C.) was an astronomer of great note in his time. He placed the Earth at the center of the universe. He observed the precession of the equinoxes. If we observe the night time sky on the day of the summer solstice, the visible stars will slowly change from year to year.

This is because the axis of the Earth (which makes an angle of $23.5°$ to the zodiac plane) slowly revolves, like a wobbling top. It makes a complete revolution every $26,000$ years. It is a wonder that this phenomenon was ever noticed by Hipparchus. He figured that the equinoxes precessed by $2°$ in the 160 years between his measurements and those of Timocharis. This puts his figure for a complete revolution at $28,800$ years. Hipparchus also put the year at 365 days, 5 hours, 55 minutes, and 12 seconds (about $6\frac{1}{2}$ minutes too long), and the lunar month at 29.530585 days (it is actually 29.530596 days). He figured the sun is $1245 D_E$ away, $12\frac{1}{3} D_E$ in diameter, and that the moon is $33\frac{2}{3} D_E$ away and $D_E/3$ in diameter, where D_E is the diameter of the Earth. Hipparchus also constructed tables of lengths of chords – that is, he developed a version of trigonometry.

Posidonius (135 – 51 B.C.) measured the circumference of the Earth and put it at $240,000$ stadia.

Geminus (1st century B.C.) noted that the time between the solstices and equinoxes are different – between the spring equinox, summer solstice, fall equinox, winter solstice and spring equinox, there are respectively $94\frac{1}{2}$ days, $92\frac{1}{2}$ days, $88\frac{1}{8}$ days, and $90\frac{1}{8}$ days. He explained this phenomenon by proposing that the Earth is offset from the center of circle about which the sun revolves. (The phenomenon was accurately explained by Kepler.) Geminus also proposed that the stars do not all lie on one sphere, but that some are closer and some are further away.

Ptolemy (ca. 100 – 170 A.D.) is the astronomer for which the Ptolemaic theory of the universe is named. This theory placed the Earth at the center, where it did not move (neither laterally nor on an axis). The motions of the planets and stars were explained by compositions of circular movements – they move in circular paths around a center which also moves in a circular motion. The theory had its roots in Heraclides' beliefs and was fully developed by the time of Hipparchus. The Ptolemaic theory prevailed until and even during the time of Copernicus. The complexity of this model motivated in part the development of spherical geometry.

Nicolaus Copernicus (1473 – 1543) is usually credited with developing the heliocentric theory of the universe. He found the idea in Greek writings and developed it in detail using the geometry of Ptolemy.

Galilei Galileo (1564 – 1642) invented the telescope and discovered four of Jupiter's moons. He supported the Copernican theory, a belief he was forced by the Church to publicly renounce.

Johannes Kepler (1571 – 1630) proposed that the orbit of the planets are ellipses with the sun at one of the foci (Kepler's First Law). He further proposed that a ray from the sun to a planet sweeps out equal areas of the ellipse in equal times (Kepler's Second Law). This means, for example, that when the Earth is furthest from the sun, it moves slower in its orbit than when it is closest to the sun. These two laws explain the phenomena observed by Geminus – that the times between solstices and equinoxes differ depending on the season. Kepler also proposed that $T^2 = kR^3$, where T is the period of a planet, R is the (mean) distance from the planet to the sun, and k is constant for all planets (Kepler's Third Law).

1781: Sir William Herschel discovers Uranus.

1833: Thomas Henderson calculates the distance to Alpha Centauri.

1846: Johanne Galle sights Neptune. The planet is discovered independently by John Couch Adams and Urbain Leverrier, who predict its location based on observed irregularities of the motions of Uranus.

1930: Clyde Tombaugh discovers Pluto.

Chapter 3

Constructions Using a Compass and Straightedge

In this chapter, we investigate constructions using only a straightedge and compass. We will very quickly discover that we can construct an equilateral triangle, square, and regular hexagon. The regular pentagon is also constructible. However, it is not possible to construct either the regular 7-gon or 9-gon. This is the central question we consider: *What regular polygons can be constructed using only a straightedge and compass?* The ancient Greeks asked the same question, as well as the following three specific questions:

1. Can we trisect an arbitrary angle?

2. Is it possible to double the cube? That is, given the side s of a cube, is it possible to construct a side of length $\sqrt[3]{2}s$? A cube with side $\sqrt[3]{2}s$ has double the volume of the cube with side s.

3. Is it possible to square the circle? That is, given a circle of radius r, is it possible to construct s so that the square with side s has the same area as the circle of radius r?

Though the ancient Greeks were convinced that these constructions are not possible, they were unable to prove this. In the early nineteenth century, more than two millennia later, Gauss proved that the first two constructions are not possible. In 1882, Ferdinand Lindemann proved that π is transcendental, from which it follows that the third construction is also not possible.

Though we will not prove any of these results in this chapter, we will investigate a key ingredient – the algebra of constructions. In this chapter, proofs that these constructions are not possible are left as exercises (for those readers familiar with fields). In Chapter 14, we solve these problems after introducing the reader to the required tools from modern algebra.

3.1   The Rules

In our constructions using a straightedge and compass, we follow these rules:

0. We start with two distinct points in the plane.

1. We can draw a line through any two already constructed points.

2. We can draw a circle with center an already constructed point, and through another already constructed point.

3. We can construct the points which are at the intersection of two distinct constructed lines, two distinct constructed circles, or a constructed line and a constructed circle.

Definition 12. *Constructible.* A figure is *constructible* if we can construct it by applying step 0 and a finite number of steps $1-3$ as outlined above. The sequence of steps is called a *construction*. In particular, if the figure is a line, circle, or point, we call it a *constructible line, constructible circle*, or *constructible point*. The two points in step 0 are called the *base* points.

Thus, we are using a compass to draw circles, and a straightedge to draw lines. We avoid the term 'ruler' since our straightedge has no markings on it – that is, we are not allowed to measure lengths with the straightedge.

3.2   Some Examples

Recall that we use the notation $\mathcal{C}_P(r)$ to represent the circle centered at P with radius r.

Theorem 3.2.1. *We can construct an equilateral triangle.*

Proof. Let O and P be the two base points in our construction. Construct $\mathcal{C}_O(|OP|)$ and $\mathcal{C}_P(|OP|)$. These two circles intersect at Q (and another point). Note that $|OQ| = |PQ| = |OP|$. Thus, $\triangle OPQ$ is an equilateral triangle (see Figure 3.1(a).) $\square$

Theorem 3.2.2. *We can construct a square.*

Proof. Let O and P be the two base points. Construct the line OP and $\mathcal{C}_P(|OP|)$, which intersect again at Q. Construct $\mathcal{C}_O(|OQ|)$ and $\mathcal{C}_Q(|OQ|)$. Let these two circles intersect at R, and let PR intersect the original circle $\mathcal{C}_P(|OP|)$ at S and T. Then $OSQT$ is a square (see Figure 3.1(b).) $\square$

Theorem 3.2.3. *We can construct a regular hexagon.*

Proof. Again, we start with the base points O and P. Let $\mathcal{C}_P(|OP|)$ inter-
sect OP again at Q. Construct $\mathcal{C}_O(|OP|)$ which intersects $\mathcal{C}_P(|OP|)$ at R
and R', and construct $\mathcal{C}_Q(|PQ|)$, which intersects $\mathcal{C}_P(|OP|)$ at S and S'.
Then $QSROR'S'$ is a regular hexagon (see Figure 3.1(c).) $\square$

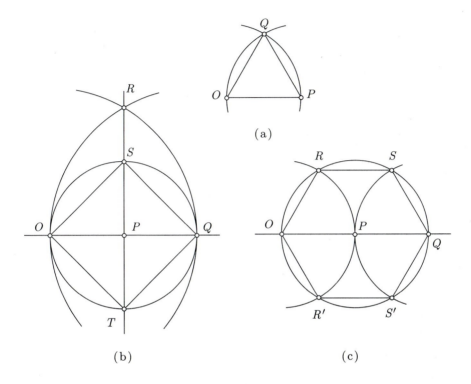

Figure 3.1. Constructions of the equilateral triangle, square, and
regular hexagon.

Exercise 3.1. If the length $|OP|$ is one, then the square constructed in
Figure 3.1(b) has sides of length $\sqrt{2}$. Describe how to construct a square
with sides of length one.

3.3 Basic Results

Some constructions, for example bisecting an angle and bisecting a line
segment, require a few steps and are used so often that it is useful to
establish that these construction can be done so that we will not have to
always repeat the process.

Lemma 3.3.1. *If $\angle BAC$ is a constructed angle, then we can bisect it.*

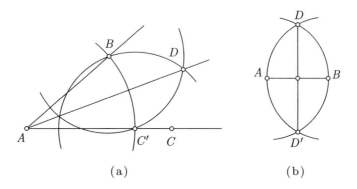

(a)   (b)

Figure 3.2

Proof. Construct $\mathcal{C}_A(|AB|)$, and let it intersect the ray AC at C', as in Figure 3.2(a). Construct $\mathcal{C}_B(|BC'|)$ and $\mathcal{C}_{C'}(|BC'|)$, which intersect at D (and another point). Then AD bisects $\angle CAB$. To see this, consider the triangles $\triangle ADB$ and $\triangle ADC'$. Note that $|AB| = |AC'|$, $|BD| = |C'D|$, and AD is shared, so by SSS, $\triangle ABD \cong \triangle AC'D$. In particular, we get $\angle BAD = \angle C'AD$. $\qquad\square$

Lemma 3.3.2. *We can construct the perpendicular bisector of any arbitrary line segment AB.*

Proof. Construct $\mathcal{C}_A(|AB|)$ and $\mathcal{C}_B(|AB|)$, which intersect at D and D', as in Figure 3.2(b). Then DD' is the perpendicular bisector of AB. To see this, let us first label the intersection of DD' with AB with E. Note that $|AD| = |BD|$, $|AD'| = |BD'|$, and DD' is shared, so by SSS, $\triangle ADD' \cong \triangle BDD'$. In particular, $\angle ADD' = \angle BDD'$. Thus, by SAS, $\triangle ADE \cong \triangle BDE$, so $|AE| = |BE|$, as desired, and $\angle BED = \angle AED$. Since these two angles are adjacent angles, they must both be right angles. $\qquad\square$

In our definition of constructibility, we have only allowed a collapsible compass. That is, we cannot *a priori* lift our compass and move it without losing the radius at which it is set. This next lemma establishes that we may in fact pick up the compass without losing the radius.

Lemma 3.3.3. *Suppose A, B, and C are constructed points. Then we can construct $\mathcal{C}_A(|BC|)$.*

Proof. We will consider only the case when $|AB| > 2|BC|$ and leave the rest as an exercise. Construct AB, as in Figure 3.3. Construct M, the midpoint of AB. Construct $\mathcal{C}_B(|BC|)$, which intersects AB at D. Construct $\mathcal{C}_M(|MD|)$, which intersects AB again at D'. Then, $|AD'| = |BD| = |BC|$, so $\mathcal{C}_A(|AD'|) = \mathcal{C}_A(|BC|)$. $\qquad\square$

The following three results are also often useful.

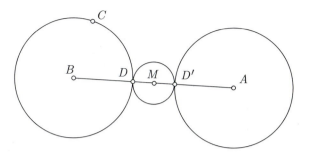

Figure 3.3

Lemma 3.3.4. *Given a line AB and a point C, we can construct the line through C which is parallel to AB.*

Lemma 3.3.5. *Given $\triangle ABC$ and two points A' and D, it is possible to construct a triangle $\triangle A'B'C'$ which is congruent to $\triangle ABC$ and such that B' is on the line $A'D$.*

Corollary 3.3.6. *We can reproduce a constructed angle on any constructed line.*

Exercise 3.2. Given a constructed line segment AB, describe how to find the perpendicular at A.

Exercise 3.3. Finish the proof of Lemma 3.3.3. That is, prove that we can construct $\mathcal{C}_A(|BC|)$ even if $|AB| \leq 2|BC|$.

Exercise 3.4. Prove Lemma 3.3.4.

Exercise 3.5. Prove Lemma 3.3.5 and its corollary.

Exercise 3.6. Show how to construct a regular octagon (8-gon) using a straightedge and compass.

Exercise 3.7** (Mohr Constructions). Show that any constructible point can be constructed using only a compass. This result is due to Mohr (1672).

Exercise 3.8**. Prove that any constructible point can be constructed using only a straightedge and a compass with a fixed radius. Is there any difference between such a compass and the lid of a jar?

3.4 The Algebra of Constructible Lengths

The realization that there exists an algebra associated with constructions, and the development of field theory, are the two main accomplishments that allowed the mathematical community to answer two of the three major questions of antiquity concerning constructions. The third was not answered until it was shown that π is transcendental.

In our rules of construction, we start with two points. We normalize our measurement of length by defining the distance between those two points to be equal to one unit. We then say a length a is a *constructible length* if there exist two constructible points P and Q so that $|PQ| = a$. In this section, we will show that the sum, product, or quotient of constructible lengths is also constructible.

Lemma 3.4.1. *Suppose a and b are constructible lengths. Then there exist constructible points A, C, and C' so that $|AC| = a + b$ and $|AC'| = |a - b|$.*

Proof. Let A and A' be the points which define the length a (so $|AA'| = a$), and let B and B' be the points which define b. Draw the line AA'. We know that we can draw the circle $\mathcal{C}_{A'}(b)$, which intersects AA' at C and C'. Let C be on the opposite side of A' as A. Then $|CA| = a + b$, and $|CA'| = |a - b|$. □

Lemma 3.4.2. *Suppose a and b are constructible lengths. Then ab is a constructible length.*

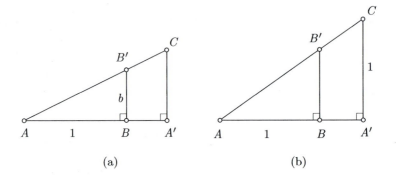

Figure 3.4

Proof. On a constructible line with constructible point A, find the points B and A', respectively, a distance 1 and a away from A in the same direction, as in Figure 3.4(a). Let B' be a distance b away from B on the perpendicular to AA' at B. Let C be the intersection of AB' and the perpendicular to AA' at A'. Then, by similar triangles,

$$\frac{b}{1} = \frac{|A'C|}{a},$$

so $|A'C| = ab$, as desired. □

Lemma 3.4.3. *Suppose a is a constructible length. Then $\dfrac{1}{a}$ is a constructible length.*

Proof. On a constructible line with constructible point A, find the points B and A', respectively, a distance 1 and a away from A in the same direction, as in Figure 3.4(b). Let C be the point a distance 1 away from A' on the line perpendicular to AA'. Let the perpendicular to AA' at B intersect AC at B'. Then

$$\frac{|BB'|}{1} = \frac{1}{a},$$

as desired. $\qquad\square$

Note that to construct ab or $1/a$, we only need similar triangles, and not necessarily right angle triangles.

Corollary 3.4.4. *If x is a positive rational number, then x is a constructible length.*

Proof. Since x is a positive rational number, there exist positive integers r and s so that $x = r/s$. We can construct both r and s by adding 1 to itself enough times. We can then find $1/s$, and multiply this by r to get x. $\qquad\square$

There is one more operation we can do – we can take square roots.

Lemma 3.4.5. *Suppose a is a constructible length. Then $\sqrt{a}$ is a constructible length.*

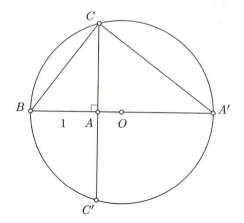

Figure 3.5

Proof. Let A and A' be points a distance a apart. Find the point B a distance 1 away from A on AA' and on the opposite side of A from A', as in Figure 3.5. Let O be the midpoint of BA', and construct the circle centered at O and with radius $|OA'|$. Construct the perpendicular to AA' at A, and let this intersect the circle at C and C'. Since CC' is a chord which is perpendicular to a diameter BA', we know A is the midpoint of CC'. Thus, the power of the point A is

$$|AC||AC'| = |AC|^2 = 1 \cdot a,$$

so $|AC| = \sqrt{a}$, as desired. $\square$

Exercise 3.9. Construct $\sqrt{5}$. (For some numbers, like 5, there are easier ways to construct $\sqrt{5}$ than the above construction.)

Exercise 3.10. Construct $\sqrt{3}$.

Exercise 3.11. The geometric mean of two positive numbers a and b is $\sqrt{ab}$. Find a simple construction to find the geometric mean of two lengths a and b.

Exercise 3.12. Recall the Argand plane model of the complex numbers $\mathbb{C}$. Suppose the two points we begin our constructions with are the points 0 and 1. Then, any point which is constructible represents a complex number. We call such a number a *constructible number*. Hence, we can talk about the sum, difference, product, and quotient of constructible numbers. Prove that if P and Q are constructible numbers, then so are $P+Q$, $P-Q$, $P \cdot Q$, $1/P$, and $\sqrt{P}$. For the last, recall that any complex number P can be expressed as $P = re^{i\theta}$, where (r, θ) is the polar coordinate representation of P. Then, $\sqrt{P} = \sqrt{r}e^{i\theta/2}$.

The following exercises are for those who have studied modern algebra and know some theory of fields. These questions are not easy, but not too hard either – they just require some background knowledge.

Exercise 3.13. Let $\mathcal{C} \subset \mathbb{C}$ be the set of constructible numbers. Prove that $\mathcal{C}$ is a field.

Exercise 3.14. Suppose we have constructed a set of points S, and that $K = \mathbb{Q}[S]$ is the smallest field which contains S. There are only three ways of constructing new points – finding the intersection of two lines defined by points in S; the intersection of a line and circle defined by points in S; and the intersection of two circles defined by points in S. In any of these three cases, show that there exists a $D \in K$ so that S together with the newly constructed points all lie in the field $K[\sqrt{D}]$. Conclude that $[\mathbb{Q}[S] : \mathbb{Q}]$ is a power of two for any set of constructible points S.

Exercise 3.15. Prove that if it is possible to double the cube, then it is possible to construct $\sqrt[3]{2}$. Let S be the set of points required to construct $\sqrt[3]{2}$. Then $\mathbb{Q}[\sqrt[3]{2}] \in \mathbb{Q}[S]$. Explain how this shows that it is impossible to double the cube.

Exercise 3.16. Show that to square the circle, one must be able to construct π. In the latter part of the nineteenth century, it was proved that π is transcendental. Conclude that it is impossible to square the circle.

3.5   The Regular Pentagon

In this section, we will first prove that it is possible to construct the regular pentagon. Then we will give a construction and prove that it is valid.

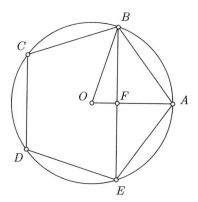

Figure 3.6

Consider the regular pentagon $ABCDE$ inscribed in a circle of radius one centered at O (Figure 3.6). Let the chord BE intersect OA at F. Then

$$|OF| = \cos \angle BOA = \cos 72°.$$

Thus, if we can construct $\cos 72°$, then we can construct F, and hence the regular pentagon.

Recall DeMoivre's theorem, which Euler formulated[1] as

$$e^{i\theta} = \cos \theta + i \sin \theta.$$

We set $\theta = \frac{2\pi}{5} = 72°$, and for convenience, write $\omega = e^{i\frac{2\pi}{5}}$. Note that

$$(e^{i\frac{2\pi}{5}})^5 = e^{i2\pi} = 1$$

so

$$\omega^5 - 1 = 0.$$

---

[1] Euler's formula, stated here, is a generalization of DeMoivre's theorem. At this point, we are really only using DeMoivre's theorem, hence our credit goes to him. See Exercise 3.21 for a precise statement of the weaker version proved by DeMoivre.

If we think of ω as a variable, then $\omega = 1$ is a root of this polynomial. Factoring out $(\omega - 1)$, we get

$$(\omega - 1)(\omega^4 + \omega^3 + \omega^2 + \omega + 1) = 0.$$

Since $\omega \neq 1$, the second factor must be zero. In particular, by dividing by ω^2, we get

$$\omega^2 + \omega + 1 + \omega^{-1} + \omega^{-2} = 0.$$

Let $x = \omega + \omega^{-1} = 2\cos(2\pi/5)$, and note that

$$x^2 = (\omega + \omega^{-1})^2 = \omega^2 + 2 + \omega^{-2}.$$

Thus,

$$x^2 + x = \omega^2 + 2 + \omega^{-2} + \omega + \omega^{-1} = (\omega^2 + \omega + 1 + \omega^{-1} + \omega^{-2}) + 1$$

so

$$x^2 + x - 1 = 0$$

and hence

$$x = \frac{-1 \pm \sqrt{5}}{2}.$$

Since x is positive, we get

$$\cos(2\pi/5) = \frac{-1 + \sqrt{5}}{4}.$$

What do you notice about this number? It involves only square roots and rational operations on integers. Thus, it is constructible! Hence, the regular pentagon is constructible.

There are several ways of constructing the regular pentagon. The following construction is due to H. W. Richmond [C] (see Figure 3.7):

Theorem 3.5.1. *Let $|OA|$ have length 1, and construct the circle with center O and radius 1. Let E be the intersection of this circle with the perpendicular to OA at O. Let B be the midpoint of OE, and let the angle bisector of $\angle ABO$ intersect OA at C. Then*

$$|OC| = \frac{-1 + \sqrt{5}}{4}.$$

Proof. Let $\angle CBO = \alpha$ and let $t = \tan\alpha$. Then

$$t = \tan\alpha = \frac{|OC|}{|OB|} = 2|OC|,$$

and

$$\tan 2\alpha = \frac{1}{\frac{1}{2}} = 2.$$

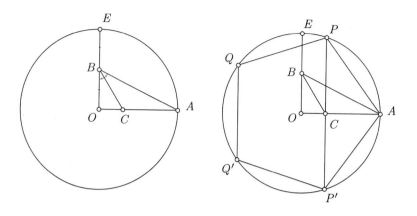

Figure 3.7. Richmond's construction of the regular pentagon.

By the double angle formula

$$\tan 2\theta = \frac{2\tan\theta}{1-\tan^2\theta},$$

we get

$$2 = \frac{2t}{1-t^2}$$
$$1 - t^2 = t$$
$$0 = t^2 + t - 1.$$

So,

$$t = \frac{-1 \pm \sqrt{5}}{2}.$$

Thus, since $|OC| > 0$, we get $|OC| = \frac{-1+\sqrt{5}}{4}$, as desired.

To finish the construction of the regular pentagon, we just find the perpendicular to OA at C which intersects the circle at P and P', and then use the length $|AP|$ to find points Q and Q' to get the regular pentagon $APQQ'P'$. □

Exercise 3.17. As mentioned, this is not the only way of constructing a regular pentagon. Construct a segment of length $\sqrt{5}$, and from this, construct a segment of length $\sqrt{5} - 1$. Rather than try to find a quarter of this, use it as is to construct a regular pentagon inscribed in a circle of radius four.

Exercise 3.18. There is a rather nice proof that $\cos 72° = \frac{-1+\sqrt{5}}{4}$. Consider the isosceles triangle $\triangle ABC$ with $|AB| = |AC| = 1$ and $\angle BAC = 36°$. Find the length of $|BC|$. (Hint: Construct the angle bisector of $\angle ABC$, as in Figure 3.8(a).) The disadvantage of this derivation is that it works only for the regular pentagon. The derivation in the text, though, can be modified to derive the lengths one must construct to construct the regular 7-gon and 9-gon.

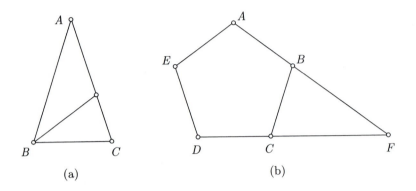

(a)   (b)

Figure 3.8. See Exercises 3.18 and 3.20.

Exercise 3.19. Prove that

$$\cos(\pi/5) = \frac{-1 - \sqrt{5}}{4}.$$

Exercise 3.20. Let $ABCDE$ be a regular pentagon with sides of length 1. The sides AB and CD are extended so that they intersect at F, as in Figure 3.8(b). What is the length $|BF|$?

Exercise 3.21. DeMoivre's theorem actually states

$$\cos n\theta + i \sin n\theta = (\cos \theta + i \sin \theta)^n.$$

Use induction to prove DeMoivre's theorem. Explain why Euler's formula implies DeMoivre's theorem and so can be thought of as a generalization of it.

Exercise 3.22. Show that both $y = Ae^{i\theta}$ and $y = B(\cos \theta + i \sin \theta)$ are solutions to the differential equation

$$\frac{dy}{d\theta} = iy.$$

Solve for A and B if $y(0) = 1$. Conclude Euler's formula

$$e^{i\theta} = \cos \theta + i \sin \theta.$$

Exercise 3.23. Recall the Taylor series for $\sin x$, $\cos x$, and e^x

$$\sin x = x - \frac{x^3}{3!} + \frac{x^5}{5!} - \ldots = \sum_{k=0}^{\infty} \frac{(-1)^k x^{2k+1}}{(2k+1)!}$$

$$\cos x = 1 - \frac{x^2}{2!} + \frac{x^4}{4!} - \ldots = \sum_{k=0}^{\infty} \frac{(-1)^k x^{2k}}{(2k)!}$$

$$e^x = 1 + x + \frac{x^2}{2} + \frac{x^3}{3!} + \ldots = \sum_{k=0}^{\infty} \frac{x^k}{k!}.$$

Use this to prove Euler's formula. You may assume that these formulas are valid for complex values of x.

Exercise 3.24. Use Euler's formula to prove the angle sum formulas:

$$\sin(\alpha + \beta) = \sin \alpha \cos \beta + \cos \alpha \sin \beta$$
$$\cos(\alpha + \beta) = \cos \alpha \cos \beta - \sin \alpha \sin \beta.$$

Hint: $e^{i(\alpha+\beta)} = e^{i\alpha} e^{i\beta}$.

Exercise 3.25. Prove

$$\tan 2\theta = \frac{2 \tan \theta}{1 - \tan^2 \theta}.$$

Exercise 3.26. Let $w = 2\cos(2\pi/7)$. Show that the regular 7-gon is constructible if and only if w is constructible. Find a polynomial $f(x)$ with rational coefficients such that $f(w) = 0$ and $f(x)$ is irreducible over the rationals.
[A]

Exercise 3.27. Let $w = 2\cos(2\pi/9)$. Show that the regular 9-gon is constructible if and only if w is constructible. Find a polynomial $f(x)$ with rational coefficients such that $f(w) = 0$ and $f(x)$ is irreducible over the rationals.

The next exercise is for those who have done Exercises 3.13 and 3.14 in the last section:

Exercise 3.28. Conclude, from the previous two exercises, that the regular heptagon and nonagon are not constructible. Why does this imply that it is impossible to trisect an arbitrary angle?

3.6 Other Constructible Figures

We have seen that it is possible to construct a regular n-gon for $n = 3$, 4, 5, 6, and 8. What other regular n-gons are constructible?

Theorem 3.6.1. *It is possible to construct the regular 2^r-gon for any integer $r \geq 2$.*

Proof? Just keep bisecting angles ....

Theorem 3.6.2. *We can construct the regular 15-gon.*

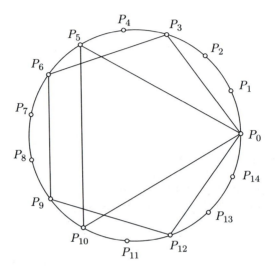

Figure 3.9

Proof. On a unit circle, place the vertices of the regular 15-gon, and label them $P_0, \dots, P_{14}$ (see Figure 3.9). These points exist, but at this point, we do not know if all are constructible. We may assume P_0 is constructible. Note that P_0, P_5, and P_{10} form a regular triangle and so are constructible points. Note also that P_0, P_3, P_6, P_9, and P_{12} form a regular pentagon, so are constructible points. Thus, we can construct the side of the regular 15-gon, since its length is $|P_5P_6|$, and both points are constructible. We use this length to construct all the other points. $\square$

   This same idea can be used in general. What is key is that 3 and 5 are relatively prime.

Theorem 3.6.3. *If m and n are relatively prime, and we can construct the regular m-gon and n-gon, then we can construct the regular mn-gon.*

   The proof of Theorem 3.6.3 makes a reasonable exercise. In contrast, the following two theorems are quite a bit more sophisticated, so we will emphasize that we will not prove either in this chapter. Both results appear in Chapter 14.

Theorem 3.6.4 (Without Proof). *Suppose a length $x > 0$ is the root of an irreducible polynomial of degree n and suppose n has an odd prime factor. Then x is not constructible.*

Theorem 3.6.5 (Without Proof). *Let p be an odd prime. We cannot construct a regular p^r-gon for any $r \geq 2$, and can construct a p-gon if and only if p is of the form*

$$p = 2^{2^k} + 1.$$

In particular, we cannot construct the regular 9-gon and hence cannot trisect $120°$. Thus, we cannot trisect an arbitrary angle.

We call

$$F_k = 2^{2^k} + 1$$

the kth Fermat number. Thus, $F_0 = 3$, $F_1 = 5$, $F_2 = 17$, $F_3 = 257$, and $F_4 = 65537$. All of these are prime, and Fermat incorrectly believed that F_k is prime for all k. In fact, the Fermat numbers F_k for $5 \leq k \leq 21$ are all composite, and it is not known if any other Fermat number is prime [Gu1]. Note that F_{21} has 631,305 digits.

Exercise 3.29. Use Theorem 3.6.4 (but *not* Theorem 3.6.5) to prove that the regular 7-gon is not constructible.

Exercise 3.30. Use Theorem 3.6.4 (but *not* Theorem 3.6.5) to prove that the regular 9-gon is not constructible. [S]

Exercise 3.31. Is $\cos(2\pi/15)$ constructible? Explain.

Exercise 3.32. Suppose x is a root of

$$x^3 - 5x^2 + 7x - 2 = 0.$$

Is the length x constructible? Explain.

Exercise 3.33. Is it possible to construct a triangle $\triangle ABC$ with $\angle BAC = 24°$, $|AB| = \frac{1+\sqrt{5}}{2}$, and $|AC| = \sqrt{1 + \sqrt{2}}$?

Exercise 3.34. Is it possible to construct a triangle $\triangle ABC$ with $\angle BAC = 20°$, $|AB| = 4$, and $|AC| = 2\sqrt{3}$?

Exercise 3.35. Is it possible to construct a triangle $\triangle ABC$ with $|AB| = 1$, $|BC| = 2$, and $|AC| = 1 + \sqrt{5}$? (Caution: You may want to think twice about this question.)

Exercise 3.36. Prove Theorem 3.6.3.

Exercise 3.37. A computer can factor F_5 very quickly, but there exists a nice proof that F_5 is composite. Notice that

$$641 = 5^4 + 2^4 = 5 \cdot 2^7 + 1.$$

Use this to show that 641 divides both $5^4 \cdot 2^{28} + 2^{32}$ and $5^4 \cdot 2^{28} - 1$, and hence divides the difference, which is F_5.

3.7   Trisecting an Arbitrary Angle

We have known for more than a century that it is impossible to trisect
an arbitrary angle using only a straightedge and compass. However, this
impossibility depends very much on the straightedge being only used to
draw straight lines. If we are clumsy enough as to scratch our straightedge,
then these marks can be used to trisect an angle. The following construction
is due to Archimedes.

Theorem 3.7.1. *If we are in possession of a compass and a straightedge
that is notched in two places, then it is possible to trisect an arbitrary angle.*

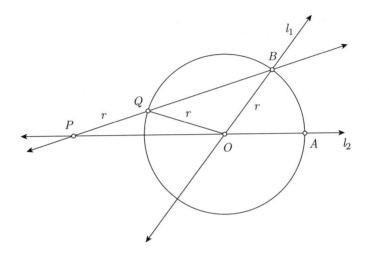

Figure 3.10

Proof. Let the arbitrary angle be described by the lines l_1 and l_2, which
intersect at O. The intersection describes two angles, one of which is acute.
We will trisect the acute angle.

Let r be the distance between the two notches on the straightedge.
Draw $\mathcal{C}_O(r)$, and let this intersect l_1 and l_2 at A and B so that $\angle BOA$ is
the acute angle to be trisected (see Figure 3.10).

Now for the construction which is not in our rules for constructibility:
Place one notch on the line OA at a 'movable' point P and the other at a
'movable' point Q on the circle. Move the straightedge around until it goes
through B, as in the diagram.

Let us now analyze our diagram: Let $\alpha = \angle QPO$. Since $|QP| = r$, we
know $\triangle OPQ$ is isosceles, so $\angle POQ = \alpha$. Thus, $\angle OQB = 2\alpha$, since it is
an external angle to $\triangle OPQ$. Since $\triangle QOB$ is also isosceles, $\angle QBO = 2\alpha$,
and hence $\angle BOQ = 180° - 4\alpha$. But then

$$\angle BOA = 180° - \angle POQ - \angle QOB$$

$$= 180° - \alpha - (180° - 4\alpha)$$
$$= 3\alpha.$$

Thus, $\angle QOP$ is a third of $\angle BOA$, as desired.   □

As Archimedes' trisection algorithm demonstrates, the rules of constructions are not a reflection of the tools that were available to the ancient Greeks. In fact, the ancient Greeks classified problems according to the complexity of their solution. Those which can be solved using only a straightedge and compass are called *plane*. A problem is called *solid* if it can be solved using a compass, straightedge, and one or more conic sections. The third class studied by the ancient Greeks is the class of problems whose solutions require the use of a more complicated curve, such as the *conchoid* (in polar coordinates, $r = a + b \sec \theta$). Such complicated curves were usually considered because there exist some exotic tools which can create them.

Exercise 3.38. Describe how to trisect an arbitrary obtuse angle using a compass and twice notched straightedge.

Exercise 3.39**. Suppose we are given a piece of paper on which there is drawn the parabola $y = x^2$, the points $(0,0)$ and $(1,0)$, and nothing else. Let us call this a piece of *parabola paper* (see Figure 3.11). Come up with a construction of the regular 7-gon using only a straightedge, compass, and the information on the parabola paper.   [H][S]

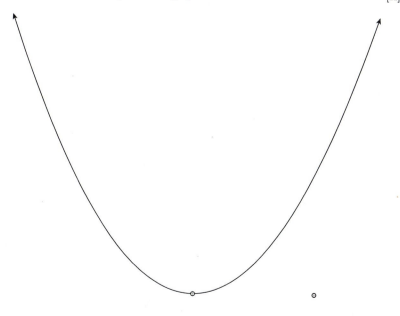

Figure 3.11. Parabola paper (see Exercises 3.39 and 3.40).

Exercise 3.40*. Suppose we are given a piece of parabola paper, as described in the previous exercise. Let $f(x)$ be a polynomial of degree three with rational coefficients and a real root c. Prove that it is possible to construct a length $|c|$ using only a compass, straightedge, and the information on this paper.

Exercise 3.41**. It is clear that it is possible to create a tool which draws ellipses. Using this tool, a straightedge, and compass, come up with a construction of some figure which is not constructible using only a straightedge and compass.

Exercise 3.42. Suppose we place one notch of a twice notched straightedge on the line $x = b$ and align the straightedge so that it passes through the origin $(0, 0)$. Prove that the other notch lies on the conchoid $r = \pm a + b \sec \theta$ where a is the distance between the two notches.

Exercise 3.43**. Prove that it is possible to construct $\sqrt[3]{2}$ using a twice notched straightedge. [H]

Chapter 4

Geometer's Sketchpad

In this chapter, we investigate *Geometer's Sketchpad*, a visual geometry utility created by Nicholas Jackiw and distributed by Key Curriculum Press. This utility is based on the rules of constructions using a straightedge and compass. The fundamental objects it draws are points; circles; and line segments, rays, and lines. These are created using the buttons to the left of the screen (▮ , ▮ , ⊙ , ╱ and ✍).

4.1 The Rules of Constructions

Let us begin with our rules of construction.

Rule 1: We start with two distinct points. Use the second button ▮ to create these. Use the label button ✍ to identify them. Sketchpad labels them A and B.

Rule2: We can draw a line through two constructed points. To do this, click on the segment button ╱ and with the button held down, drag it to the right until the line button ╱ is highlighted. Now, click on point A, and drag the mouse to point B. Notice the commentary in the bottom left corner of the screen. When the cross hair is near the first point A, it reads 'From Point A,' and when it is dragged near the second point B, the commentary reads 'Passing through Point B,' Though Sketchpad will let us draw points and lines anywhere, within our rules of constructions we are only allowed to draw lines which pass through two already constructed points.

Rule 3: We can draw a circle whose center is a constructed point and through another constructed point. To do this, select the circle button ⊙ . Place the cross hair at A and click the mouse. The commentary reads 'Centered at point A.' Holding the button down, drag the cross hair to B. The commentary reads 'Passing through Point B.' Note that it is important to drag the cross hair to B, and to resist the urge to just drag the mouse somewhere (not B) while making sure the circle passes through

B. The problem with this is that even though our intent is that the circle go through B, Sketchpad will not know that that is our intent. For example, in our current sketch, use the select button ![select] and select and drag B someplace else. Note that the circle changes appropriately. Now, construct a circle centered at B, and create a circle by dragging the cross hair to a point which is not B, but so that the circle passes through A. Click on the select button, select A, and move it. Notice that this time the circle we just drew does not appropriately change. It is a circle that goes through a new point that we constructed without obeying our rules.

Use '⟨control⟩ z' to undo the circle we just created. (We may have to use this several times before the circle disappears.)

Rule 4: The points of intersection of constructed lines and circles are constructed points. To construct them, select the select button ![select] and move it to the point of intersection (which does not yet have a big dot). The commentary reads 'Select Point at intersection.' Click the mouse.

Remark: Sketchpad will construct the point of intersection of only two objects. For example, construct a triangle and its three medians. We know they intersect at a common point, but when we move the select button near that point of intersection, Sketchpad sees three objects and hence three different points of intersection. Since it does not know which point you want, it calls it an ambiguous point of intersection. To remedy this, we can select two objects – to do this, click on one median, and while holding the shift key down, select the other median. Now type '⟨control⟩ i'. This constructs the point of intersection of these two medians. Of course, the third median goes through this point too, but Sketchpad will never notice this.

4.2   Lemmas and Theorems

We can 'reproduce' the lemmas and theorems of Section 3.3 in Sketchpad. For example, given three arbitrary points A, B and C, we can construct the angle bisector of $\angle BAC$ in Sketchpad by following the steps of Lemma 3.3.1. The point of stating the lemma, though, is so we will not always have to go through these steps just to find the angle bisector. What we really want to do is program our computer to do this for us. We do this using a *script*. In Sketchpad, pull down the 'File' menu and select 'New Sketch' and 'New Script.' In the script window, press the record button. In the sketch window, select three random points A, B, and C (use the label button to label them); construct the rays AB and AC; draw the circle centered at A and through B; create the point of intersection of the circle with the ray AC and label this point (Sketchpad calls this point D); create the circles centered at B through D and the circle centered at D and through B; select the point of intersection of these two circles (Sketchpad calls it E); create the ray AE – this is the angle bisector. Finally, hide the work we have just done which is important to the proof but not to the final result. We do this

by selecting each object and hitting '⟨control⟩ h' ('h' for hide). All that
should be left are the three original points and the angle bisector. Now,
stop the recording. This script is our lemma. To test it, open a new sketch.
Create three points at random, labeled A, B, and C. Note that the script
says 'Given: Point A, Point B, Point C.' Select the three new points in
that order – to do this, hold down the 'shift' key and select A, B, and C in
that order. Note that all three points are highlighted. Now play the script
by clicking the play button. The computer goes through the construction
step by step. Save this script as 'anglebisector.' As a final exercise, select
B, A, and C in that order, and click 'Fast' on 'anglebisector.' It constructs
the bisector of $\angle ABC$. The point of intersection of these two rays is, of
course, the incenter.

Exercise 4.1. Write a script to construct an equilateral triangle with side
AB. When you are done, your script should say 'Given: Point A, Point B.'
If it asks for more points, try again. Test your script.

Exercise 4.2. Write a script to construct a square with side AB.

Exercise 4.3. Write a script to construct a regular hexagon inscribed in
the circle with center A and through B.

Exercise 4.4. Write a script to construct a perpendicular bisector. Write
two more scripts – one to find the line through A and perpendicular to AB;
and the second to find the midpoint of segment AB.

Exercise 4.5. Write a script to bisect an arbitrary angle. Move around
one of the points that defines an arm of the angle. Does the construction
always work?

Exercise 4.6. Note that the buttons of Sketchpad mimic a 'collapsible
compass.' Write a script which mimics a 'noncollapsing compass' (i.e.,
write a script which models Lemma 3.3.3.) Test it. Does it always work?
Rewrite it so that it always works, or explain why we cannot write a script
which always works.

Finding angle bisectors, segment bisectors, and perpendiculars is very
common in constructions, and after a while, the process becomes tedious.
For all the remaining exercises, we allow the use of the operations under
the pull-down menu 'construct.' Note that all of these (except 'point on
object') can be done using the rules of construction. In particular, the
operation 'Circle by center and radius' is the 'noncollapsing compass.'

Exercise 4.7. Write scripts to add, subtract, multiply, and invert constructible lengths.

Exercise 4.8. Write a script to find square roots.

Exercise 4.9. Use the above script to come up with your own construction
of a regular pentagon.

Exercise 4.10. Write a script to construct a regular pentagon inscribed in $C_A(|AB|)$.

The following is a construction of the regular 17-gon, due to H.W. Richmond [C]. We begin by constructing the diameter through O and P_0, and its perpendicular OA (see Figure 4.1). Find the point B on OA such that $|OB|$ is one-quarter of $|OA|$. Find the point C on OP_0 such that $\angle CBO$ is one-quarter of $\angle P_0BO$. Find the point D on OP_0 such that $\angle DBC = 45°$. Construct the circle with diameter DP_0 and let it intersect OA at E. Construct the circle with center C and through E. Let this circle intersect OP_0 at F and G. The perpendiculars to OP_0 through F and G intersect the original circle at P_3, P_5, P_{12}, and P_{14}. The circle centered at P_3 and going through P_0 intersects the circle again at P_6. The side P_5P_6 is an edge of a regular 17-gon, and can be used to find the rest of the vertices.

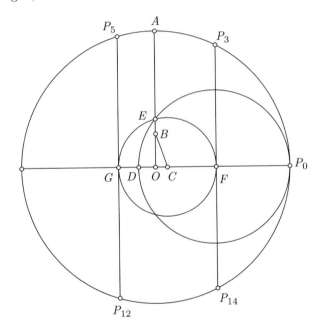

Figure 4.1. Richmond's construction of the regular 17-gon.

Exercise 4.11. Write a script to construct the regular 17-gon.

Exercise 4.12. The point F and the midpoint of DP_0 are almost coincident, but not quite. If O is zero and P_0 is one, what are the values of F and the midpoint of DP_0?

Trivia. A construction of the regular 17-gon was first discovered by Karl Frederic Gauss (1777 − 1855) − at the age of eighteen. Despite the many achievements of Gauss, he is said to have been most proud of this discovery and wished that it appear on his tombstone. His request was not carried out, though a monument with the construction was later erected in his honor.

Figure 4.2. Gauss and his construction of the 17-gon was again honored in the logo of the 30th International Mathematical Olympiad, hosted by West Germany.

4.3 Archimedes' Trisection Algorithm

Sketchpad is fairly faithful to the rules of construction. To demonstrate this, try using Sketchpad to implement Archimedes' trisection algorithm. A script can be written that does everything but the last step – moving the line so that it goes through B (see Theorem 3.7.1). This must be done by hand.

Exercise 4.13. Write a script which does every step of Archimedes' construction, except the one which is not permitted in our rules.

Exercise 4.14. Write a script which satisfies the following: 'Apply the script to points A and B. Move the point C along the circle until the line l intersects D. The resulting figure is a regular 9-gon.'

4.4 Verification of Theorems

Sketchpad can be used to demonstrate many of the results we studied in Chapter 1. Some, such as the Star Trek lemma and power of the point, are a little difficult to demonstrate. Other more complicated results, like the nine point circle, are much easier to demonstrate using Sketchpad.

Exercise 4.15 (The Centroid). Write a script to demonstrate that the medians intersect at a common point. Once you have your sketch, use the select button ![cursor] to move the vertices around.

Exercise 4.16 (The Incircle). Write a script which constructs the incircle of a triangle with vertices A, B, and C. (The 'Given' information should be only three points.)

Exercise 4.17 (The Excircles). Write a script which constructs the excircles.

Exercise 4.18 (The Circumcircle). Write a script which constructs the circumcircle.

Exercise 4.19 (The Orthocenter). Write a script to construct the orthocenter H of $\triangle ABC$. Does it work if the triangle is obtuse? Correct it, if not. Select a vertex in your sketch and move it around. Do you notice a symmetry between the vertices and H? Formulate a conjecture concerning this observation and prove it.

Exercise 4.20 (The Euler Line). Write a script which demonstrates that the circumcenter O, the centroid G, and the orthocenter H are collinear.

Exercise 4.21 (The Nine Point Circle). Write a script which constructs the nine point circle and the nine points which it goes through. Check your script. Does it work if the triangle is obtuse? Are all nine points there? Let H be the orthocenter of $\triangle ABC$. What can you say about the nine point circle for $\triangle HBC$?

Exercise 4.22 (Feuerbach's Theorem). On a triangle $\triangle ABC$, run the script which constructs the nine point circle. On the same triangle, run the script which constructs the incircle and excircles. Formulate a conjecture.

Exercise 4.23. On a triangle $\triangle ABC$, find the center of the nine point circle and the Euler line. Formulate a conjecture.

Exercise 4.24 (The Simson Line). Write a script which constructs the Simson line.

Exercise 4.25 (The Radical Axis). Write a script which finds the radical axis of two circles.

Exercise 4.26 (The Radical Center). Write a script which finds the radical center of three circles. Have this script find the circle which is orthogonal to each of the original three circles.

Exercise 4.27. Write two scripts which demonstrate each direction of the theorem of Menelaus.

Exercise 4.28. Write a script that demonstrates both directions of Ceva's theorem.

4.5 Sophisticated Results

Why do we study geometry? Though some simple geometrical results are practical, most are never used by engineers, scientists, or architects, so one might argue that there is little point in teaching geometry. The end result, though, is not why we teach geometry – it is the path we take to get there. Geometry is a beautiful subject full of elegant and inspired logical arguments. It is a perfect subject to teach a student the skill of creative and logical thinking. How then does Geometer's Sketchpad fit within this philosophy? If the education is in the proofs, then does the use of Sketchpad circumvent the education? I would say not. It can be used as a tool of discovery (see Exercises 4.21 and 4.22 above). It can also be used to make very convincing arguments. Such arguments can be illuminating, can pave the way to understanding the proofs, and can even inspire one to want to understand the proof. This idea can be extended. There are many results in geometry which are either beyond the scope of this text, or which we are not yet ready to prove, but which can be illustrated using Sketchpad. We present a few below. The results which are stated as theorems will be revisited in Chapter 11.

Exercise 4.29. Let $\triangle ABC$ be an arbitrary triangle. Let P be an arbitrary point, and let the perpendiculars from P to the extended sides BC, AC, and AB be labeled X, Y, and Z, respectively. The triangle $\triangle XYZ$ is called the *pedal triangle* for P. Write a script which constructs the pedal triangle for P with respect to a triangle $\triangle ABC$. Move P about. When is $\triangle XYZ$ degenerate?

Exercise 4.30. For $\triangle ABC$ and a point P, write a script which creates the pedal triangle (see Exercise 4.29) for P, finds the area of the pedal triangle, and finds the area of $\triangle ABC$. Make this script find the circumcenter O of $\triangle ABC$ too. Finally, let R be the circumradius, $r = |OP|$ and have this script calculate

$$\frac{R^2 - r^2}{4R^2} |\triangle ABC|.$$

Make a conjecture.

Exercise 4.31. The pedal triangle $\triangle XYZ$ is sometimes called the *derived triangle* or *first derived triangle* of $\triangle ABC$ with respect to the point P. The *second derived triangle* of $\triangle ABC$ with respect to P is the pedal triangle of $\triangle XYZ$ with respect to P. The third derived triangle is the derived triangle of the second derived triangle. Use the script found in Exercise 4.29 to find the third derived triangle of $\triangle ABC$ with respect to a point P. Investigate and formulate a conjecture concerning this triangle.

Exercise 4.32. Write a script which finds the point of tangency of the excircles to the sides of the triangle $\triangle ABC$. Label these points D, E, and F on sides BC, AC, and AB, respectively. Verify that the three segments

AD, BE, and CF are coincident. This point of coincidence is called the Nagel point N. Find the incenter I and centroid G of the triangle $\triangle ABC$ too. Formulate a conjecture about the three points N, I, and G.

Theorem 4.5.1 (Pappus' Theorem). *Let P_1, P_2, and P_3 be three points on the line l_1, and let Q_1, Q_2, and Q_3 be three points on the line l_2. Let R be the intersection of P_2Q_3 and P_3Q_2; let S be the intersection of P_1Q_3 and P_3Q_1; and let T be the intersection of P_1Q_2 and P_2Q_1. Then R, S, and T are collinear. (See Figure 4.3.)*

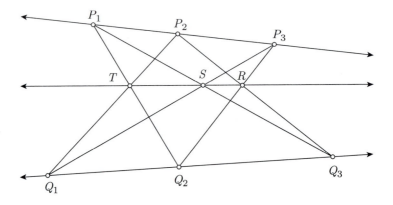

Figure 4.3

Exercise 4.33. Write a script which demonstrates Pappus' theorem.

Theorem 4.5.2 (Pascal's Theorem). *Let P_1, P_2, P_3, Q_1, Q_2, and Q_3 be points on a conic C, as in Figure 4.4. Let R be the intersection of*

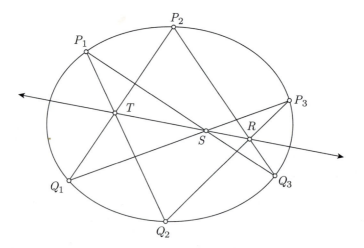

Figure 4.4

P_2Q_3 and P_3Q_2; let S be the intersection of P_1Q_3 and P_3Q_1; and let T be the intersection of P_1Q_2 and P_2Q_1. Then R, S, and T are collinear. (See Figure 4.4.)

Pappus' theorem is really a special case of Pascal's theorem, since two lines may be thought of as a degenerate conic.

Exercise 4.34. Write a script which demonstrates Pascal's theorem for a circle.

Theorem 4.5.3 (Desargues' Theorem). *Let P be a point not on a triangle $\triangle ABC$. Let A', B', and C' be points on the lines PA, PB, and PC, respectively, as in Figure 4.5. Let the (extended) sides BC and $B'C'$ meet at R. Similarly, let AC and $A'C'$ meet at S and let AB and $A'B'$ meet at T. Then R, S, and T are collinear.*

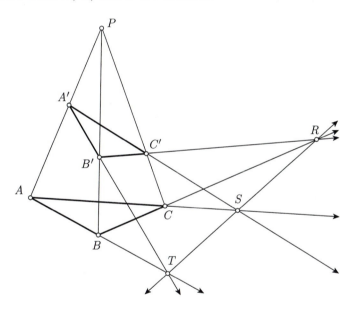

Figure 4.5

Exercise 4.35. Write a script which demonstrates Desargues' theorem.

The proof of Desargues' theorem is actually not too difficult, if thinking in three dimensions comes easily.

Proof of Desargues' Theorem. In Figure 4.5, think of the rays emanating from P as the edges of a pyramid. Then, $\triangle ABC$ can be thought of as the intersection of a plane α with this pyramid. Similarly, $\triangle A'B'C'$ is the intersection of a plane α' with this pyramid.

The intersection of the two planes α and α' is a line l. Since the line BC is on α, and $B'C'$ is on α', the intersection of these two lines must be in the intersection of the two planes. That is, R is on l. Similarly, both S and T are on l. That is, the three points are collinear.   □

Exercise 4.36. Let $ABCDEF$ be a hexagon inscribed in a circle. Let AB and CD intersect at G, and let DE and AF intersect at H. Write a script which shows that BE, CF, and GH are coincident.

Exercise 4.37. State and prove the converse of Desargues' theorem.   [H]

4.6   Parabola Paper

In this section, we construct *parabola paper* (see Exercise 3.39). In Geometer's Sketchpad, select 'Plot Points' under the 'Graph' pull-down menu. Manually insert the points (x, x^2) in .1 increments from .1 to 1.6. That is, enter $(.1, .01)$, $(.2, .04)$, ..., $(1.6, 2.56)$. Notice that the points $(0, 0)$ and $(1, 0)$ are already plotted. Select and move the point $(1, 0)$ to stretch the axis system. Join the points with line segments and hide the points (but not the points $(0, 0)$ or $(1, 0)$). Select the y-axis, mark it as a mirror by double clicking it (or select it and choose 'Mark Mirror' under the 'Transform' menu). Select all (under the 'Edit' menu), and reflect the graph using 'Reflect' under the 'Transform' menu. Hide the axis and any leftover stray marks, leaving only the parabola and the points $(0, 0)$ and $(1, 0)$. Save the Sketch for use in the following exercises.

Exercise 4.38*. Use the parabola paper to construct a regular 7-gon (see Exercise 3.39).

Exercise 4.39*. Use the parabola paper to construct a regular 9-gon.

Exercise 4.40**. Use the parabola paper to construct a regular 13-gon.

Chapter 5

Higher Dimensional Objects

5.1   The Platonic Solids

Recall that we call a polygon a regular polygon if all of its sides and all of its angles are equal. In this section we will investigate the three-dimensional analogues, the *regular polyhedra* or *Platonic solids*.[1]

Definition 13. *Platonic solid.* A polyhedron is called a *regular polyhedron* or *Platonic solid* if it is convex, all of its faces are identical regular polygons, and there are the same number of faces at each vertex.

We are probably most familiar with the *cube*, which has six square faces. There exist also the *tetrahedron*, which has four triangular faces; the *octahedron*, which has eight triangular faces; the *dodecahedron*, which has twelve pentagonal faces; and the *icosahedron*, which has twenty triangular faces (see Figure 5.1).

Unlike the two-dimensional case, there are only a finite number of regular polyhedra.

Theorem 5.1.1. *There are only five Platonic solids – the tetrahedron, cube, octahedron, dodecahedron, and icosahedron.*

Proof. Let us look at the shape of each face and the number of faces at each vertex.

If each face is triangular, then there cannot be more than five faces to a vertex, since if there are six or more, then the sum of the angles at the

---

[1]It is a bit unfortunate that we think of these terms as synonyms. A polyhedron is usually thought of as just the surface of a polyhedral solid, in the same way a sphere is the surface of a ball, and a circle is the boundary of a disc. For now, the distinction is not important, but it will be when we start talking about the topology of the regular polyhedron or solid.

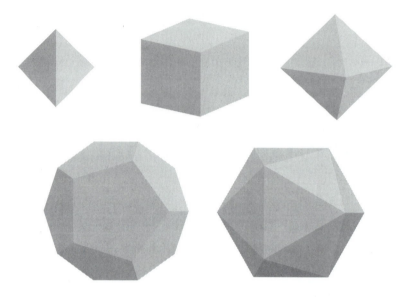

Figure 5.1. The five Platonic solids – the tetrahedron, cube, octahedron, dodecahedron, and icosahedron.

vertex is 60° times the number of faces, which is greater than or equal to 360°. That is, if there are six equilateral triangles coming to a point, then the figure is flat, and if there are more, then the figure has hills and valleys. Of course, there must be at least three faces coming to a vertex, so the possible numbers are three, four, and five. These produce the tetrahedron, octahedron, and icosahedron, respectively.

If each face is a square, then again, four squares produce a flat figure, so there can be only three squares to a vertex, which gives the cube. Furthermore, any figure with regular n-gonal faces for $n \geq 4$ must have exactly three faces to a vertex.

Thus, for pentagons, there can be only three faces to a vertex, and this gives the dodecahedron.

For hexagons, there again can be only three faces, but this gives a flat surface. Thus, there are no Platonic solids with regular n-gonal faces for $n \geq 6$.   □

Exercise 5.1. Prove that the angles in a regular n-gon measure $\left(\frac{n-2}{n}\right) 180°$.

Exercise 5.2. Construct a regular pentagon inscribed in a circle with radius two inches. Make a cutout of this pentagon on card stock (to use as a template). Construct templates of a square, equilateral triangle, and hexagon, each with sides equal in length to the side of the pentagon you constructed.

Figure 5.2. The icosahedron was featured in the logo for the 21st International Mathematical Olympiad hosted by the United States. An icosahedron is also the logo for the Mathematical Association of America.

Exercise 5.3. Construct models of the Platonic solids using the templates constructed in the previous exercise.

Exercise 5.4*. When we talk about regular polygons, we usually mean convex regular polygons. There are, though, polygons whose sides cross, but whose angles and edges are all identical. For example, the five-pointed star in Figure 5.3 is one such *regular star polygon*. Note that this polygon has only five vertices and five sides. In a similar fashion, we can construct the *regular star polyhedra* (see, for example, Figure 5.3). Identify the set of

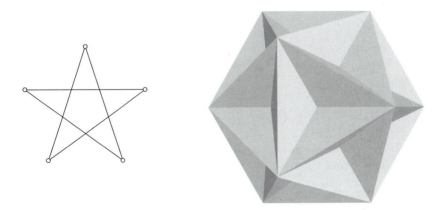

Figure 5.3. A star polygon and a star polyhedron. The vertices of the five pointed star are the vertices of a regular pentagon. The vertices of this star polyhedron are the vertices of a regular icosahedron.

all regular star polyhedra. How many faces does the regular star polyhedron in Figure 5.3 have?

Exercise 5.5. We consider the faces of the star polyhedron in Figure 5.3 to be regular pentagons which intersect. If we instead think of the triangles as the faces, then how many triangular faces are there? Find the lengths of the sides of each of these triangles. These triangular faces are constructible using a straightedge and compass. Construct a template, and use it to construct one of these star polyhedra.

5.2   The Duality of Platonic Solids

There is a natural pairing between Platonic solids. Let us take our favorite Platonic solid. Mark the center of each face, and connect the dots. If done properly, we should have again a Platonic solid. The new solid has as many vertices as the old object has faces, and as many faces as the old object has vertices. Both objects have the same number of edges. If we do this again, we end up with an object which is similar to the one we started with. This gives a duality between Platonic solids.

Count the number of faces, vertices, and edges of each Platonic solid, as in Table 5.1. From this information, it see that the tetrahedron is the dual of itself, the cube is the dual of the octahedron, and the dodecahedron is the dual of the icosahedron.

Platonic Solid	Faces	Vertices	Edges
Tetrahedron	4	4	6
Cube	6	8	12
Octahedron	8	6	12
Dodecahedron	12	20	30
Icosahedron	20	12	30

Table 5.1

5.3   The Euler Characteristic

Let F, V, and E be the number of faces, vertices, and edges of a polyhedron. Euler observed that the quantity

$$\chi = F + V - E$$

depends only on the *topological shape* of the polyhedron. For example, for the five regular polyhedra, this quantity is always two. This quantity χ is called the *Euler characteristic*.

Definition 14. *Topological equivalence.* We say two regions are *topologically equivalent* if we can stretch one shape into the other without tearing or puncturing the object.

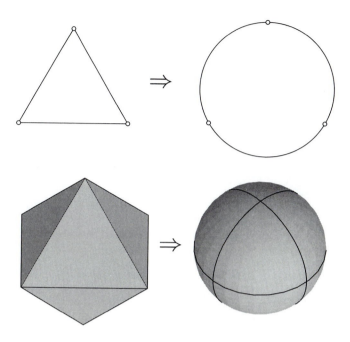

Figure 5.4. Topologically equivalent figures: A triangular region is topologically equivalent to a disc, and an octahedron is topologically equivalent to a sphere. Here, it is important that the octahedron be thought of as a polyhedron, and not a solid.

For examples, see Figures 5.4 and 5.5.

Before we continue, let us formalize the objects we are talking about. A *vertex* is a point. An *edge* is a curve (not necessarily straight) which is bounded by one or two vertices. Two edges cannot intersect except at a vertex. A *face* is a region which is bounded by edges, is topologically equivalent to a disc, and contains no interior vertices or edges.

To prove a result, it is sometimes easier to prove something a little more general. In this case, we will first study the Euler characteristic of *filled finite connected planar graphs*.

A *graph* is a nonempty set of edges and vertices. If this set is finite, then we call it a *finite graph*. A graph is *connected* if for any two vertices, there exists a path of edges which connects them. A graph is *planar* if it can be imbedded in the plane in such a way that no pair of edges cross. A *connected component* of a graph R is a subset S which is connected and is not connected to the remainder $R \setminus S$. See Figure 5.6 for an example.

A finite connected planar graph partitions the plane into a finite number of regions. One of these regions is unbounded, and the rest are all topologically equivalent to a disc. Thus, the bounded regions can be thought of

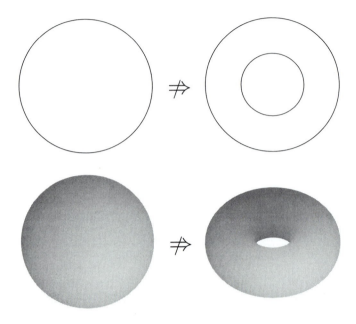

Figure 5.5. Regions which are not topologically equivalent: A disc is not topologically equivalent to an *annulus* or washer since there is no way to deform one into the other without tearing or puncturing either object. Similarly, a sphere is not topologically equivalent to a *torus* or doughnut.

as faces. We will call a finite connected planar graph together with these faces a *filled* finite connected planar graph.

Theorem 5.3.1. *Let R be a filled finite connected planar graph. Then*

$$\chi_R = 1.$$

Proof. We use strong induction on the number of edges. A filled finite connected planar graph R with no edges has only one point (since it is connected and is nonempty). It therefore has no faces, and so

$$\chi_R = 1.$$

Suppose that every filled finite connected planar graph with at most k edges has Euler characteristic one. Let R be a filled finite connected planar graph with $k + 1$ edges. Pick an edge and remove it (but not its bounding vertices). There are two possibilities for the resulting subgraph R'. It is either connected, or has two connected components.

If R' is connected, then by our induction hypothesis, $\chi_{R'} = 1$. By removing the edge, we have either joined the two faces on either side of it,

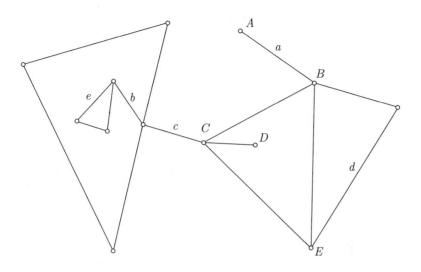

Figure 5.6. A finite connected planar graph. The graph obtained by removing either of the edges d or e is again a finite connected planar graph. The graph obtained by removing a, b, or c gives a finite planar graph which has two connected components. If a is removed, then one of the connected components is the isolated point A. If b is removed, then one of the components lies entirely within a face of the other connected component. Note that the face with vertices B, C, and E is thought of as a pentagon $BCDCE$, so it contains no interior vertices or edges.

or have joined a face on one side of it with the unbounded region on the other side. In either case, the number of faces has decreased by one. The number of edges has remained unchanged, and the number of edges has of course decreased by one. The net effect on the Euler characteristic is nil, so

$$\chi_R = \chi_{R'} = 1.$$

If R' has two connected components, then $\chi_{R'}$ is the sum of the Euler characteristic of each component. Since each component has at most k edges, the Euler characteristic of each is 1, so $\chi_{R'} = 2$. If the regions on either side of the removed edge are different, then there exists another path of edges between the endpoints of the removed edge. That is, removing the edge does not disconnect the graph. Thus, the region on either side of the removed edge is the same. This region may be the unbounded region, or possibly a face. In either case, removing the edge does not change the number of faces. Again, the number of vertices remains unchanged, and the number of edges drops by one, so

$$2 = \chi_{R'} = \chi_R - 1.$$

Thus, $\chi_R = 1$.   □

Corollary 5.3.2. *Suppose a polyhedron R is topologically equivalent to a sphere. Then*

$$\chi_R = 2.$$

Proof. Remove a single face of this polyhedron. What is left, R', can be stretched and flattened so that it lies on the plane. The faces, vertices, and edges of this region are just the faces, vertices, and edges of a filled finite connected planar graph, so $\chi_{R'} = 1$. Since R has one more face then R',

$$\chi_R = \chi_{R'} + 1 = 2.$$ □

Exercise 5.6. What is the Euler characteristic of a polyhedron which is topologically equivalent to a torus? Prove your assertion.

5.4   Semiregular Polyhedra

Definition 15. *Semiregular polyhedron.* We call a polyhedron a *semiregular polyhedron* if it is convex, all of its faces are regular polygons, and each vertex looks identical.

The semiregular polyhedron with which our youth is most familiar is probably the *hypo-truncated icosahedron*, better known as the soccer ball (and also known as the hyper-truncated dodecahedron or more recently, the Buckyball.)

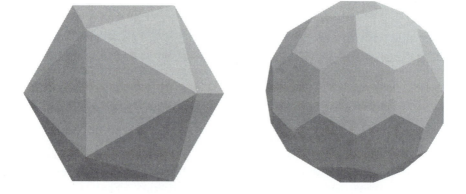

Figure 5.7. A *hypo-truncated icosahedron* or soccer ball (right), created by slicing the vertices off an icosahedron.

We arrive at such a figure by taking an icosahedron and slicing off the corners (see Figure 5.7). If the slices are shallow enough, the resulting face is a pentagon. The formerly triangular faces are triangles with their corners cut off, and if done properly, the resulting faces are regular hexagons. On a soccer ball, the pentagonal faces are black and the hexagonal faces are white. Note that each face is a regular polygon, and each vertex has two hexagonal faces and one pentagonal face.

If we take a deeper slice, the triangular faces become triangles again, and we get the truncated icosahedron (see Figure 5.8. This object can also be obtained by slicing the corners of the dodecahedron, so it is also known as the truncated dodecahedron. If we take an even deeper cut, we get the hyper-truncated icosahedron (hyper means a lot, hypo means a little), which can also be obtained from the dodecahedron by taking shallow cuts. Finally, if we take such a deep cut that the triangular faces are cut away, we get the dodecahedron, as discussed in the section on duality.

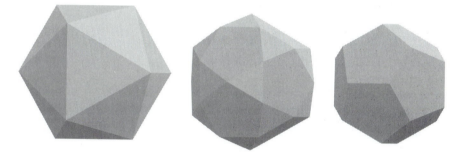

Figure 5.8. The *truncated icosahedron* (middle), created by cutting the corners off an icosahedron. Deeper cuts give us the dodecahedron.

The same can be done with the cube (see Figure 5.10) or octahedron, and with the tetrahedron.

We now have quite a few examples of semiregular polyhedra. They all have, however, a common theme – each polyhedron has the underlying

Figure 5.9. Representatives of the two infinite classes of semiregular polyhedra – a hexagonal right prism (left) and a hexagonal drum (right).

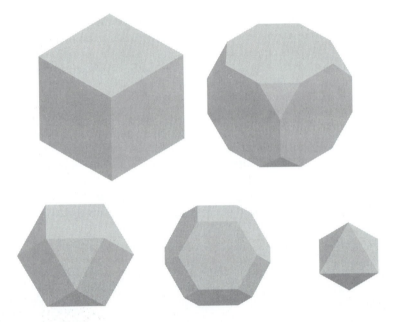

Figure 5.10. Metamorphosis of solids: A cube, hypo-truncated cube, truncated cube, hyper-truncated cube, and the octahedron.

structure of a Platonic solid. One might wonder if this is always the case. It is, with the exception of two infinite classes of semiregular polyhedra (see Figure 5.9). The first class is the class of *right prisms*. At each vertex of such a polyhedron there are two squares and one polygon. The second class is the class of *drums*, which have three triangles and one polygon at each vertex. Furthermore, if we exclude the drums and right prisms, there are only a finite number of semiregular polyhedra. The task of categorizing these polyhedra is a difficult combinatorial problem, but we will begin the task in the following section.

Trivia. Synthetic oil consists of synthetic hydrocarbon molecules in which the carbon atoms are arranged so that they form the vertices of a hypo-truncated icosahedron. The resulting molecule is therefore as spherical as a soccer ball, and the oil is essentially molecule size ball bearings. Since their creation in 1985, these molecules have been called Buckyballs, Fullerene, or Buckminsterfullerene, named after R. Buckminster Fuller. Buckminster Fuller is the American architect who, together with Shoji Sadao, designed the geodesic dome for Expo '67 in Montreal (see Figure 5.11). Geodesic domes are usually made with isosceles triangles, so are not portions of semiregular polyhedra. However, their underlying structure is usually that of a semiregular polyhedron.

GreatBuildings.com Photo ©Lawrence A. Martin.

Figure 5.11. The American pavilion at Expo '67 in Montreal, a geodesic dome designed by Buckminster Fuller and Shoji Sadao.

Exercise 5.7. What is the truncated tetrahedron better known as?

Exercise 5.8*. In Exercise 5.4, we introduced the notion of a regular star polyhedron. There exist, also, semiregular star polyhedra. Find and classify all of them.

Exercise 5.9*. A geodesic dome can be created by making shallow pentagonal and hexagonal based pyramids and gluing these pyramids on the pentagonal and hexagonal faces of a soccer ball. What lengths should we choose for the sides of the faces of the pyramids if we want to make the geodesic dome as spherical as possible? Note that the triangular faces used for the pentagonal pyramids will be different from the triangular faces used for the hexagonal pyramids.

5.5   A Partial Categorization of Semiregular Polyhedra

We can classify (though not uniquely) the semiregular polyhedra by the shape of the vertices. Note first that the angles of six equilateral triangles joined at a vertex sum to 360°, so the greatest number of faces to a vertex is five. Let us consider this case first. Suppose four faces are triangles. If the fifth is a hexagon, then we again have a flat vertex (which gives an interesting tiling).

If the fifth is a pentagon, then we get a figure called the *snub dodecahedron* – something we have not seen yet. If the fifth is a square, we get the *snub cube*, which we also have not seen. And if the fifth is a triangle, we get the icosahedron. Now suppose only three are triangles. Then the fourth and fifth faces are at least squares, but that already gives a flat vertex. Thus, we have exhausted all possibilities.

Let us now assume there are four faces to a vertex. If all four are squares, we have a flat figure. Otherwise, at least one must be a triangle. If there is only one triangle, then two of the other faces must be identical. (Why?) These two other faces cannot be pentagons, since two pentagons, a square and a triangle are already too much (366°). Thus, we have one triangle and two squares. The fourth face cannot be a hexagon (which gives another interesting tiling) so must be a pentagon or square. Both these figures exist. Now let us assume there are two triangles. These triangles cannot be adjacent to each other. (Why?) Note that the other two faces must be identical. (Why?) They cannot both be hexagons, since that gives a flat vertex. If they are pentagons, we get the truncated icosahedron, and if they are squares, we get the truncated cube. Finally, if three of them are triangles, then for each polygon we get a drum.

It is convenient to describe semiregular polyhedra with k-tuples, where k is the number of faces at each vertex. Each entry in the k-tuple is a number representing the number of edges to each face at the vertex, taken in order. For example, the soccer ball has two hexagons and one pentagon at each vertex, so we represent it with a triple $(6, 6, 5)$. Of course, the representations $(6, 5, 6)$ and $(5, 6, 6)$ are equivalent.

Using this representation, we can tabulate the set of semiregular polyhedra which we have already identified, as shown in Table 5.2.

Representation	Name	Remarks
(3,3,3,3,5)	Snub dodecahedron	There are two orientations.
(3,3,3,3,4)	Snub cube	There are two orientations.
(3,4,5,4)		
(3,4,4,4)		There are two.
(3,5,3,5)	Truncated dodecahedron	
(3,4,3,4)	Truncated cube	
(3,3,3,n)	n-gonal drum	An infinite class.

Table 5.2

The number of possibilities to cover is quite a bit higher for the case of three faces to a vertex. It is a doable problem and we include it in the exercises.

Exercise 5.10. How many pentagonal faces are on a snub dodecahedron?

Solution. We can solve this using the Euler characteristic. Let F_3 and F_5 be the number of triangular and pentagonal faces. Thus $F = F_3 + F_5$. Note that each corner of each pentagon is the vertex of exactly one vertex of the polyhedron, so

$$V = 5F_5.$$

Similarly, at each vertex, there are four corners of triangles, so we also have

$$V = \frac{3F_3}{4}.$$

We count edges by noting that each pentagonal face has five, each triangle has three, and we have counted each edge twice, so

$$E = \frac{3F_3 + 5F_5}{2}.$$

Finally, combining this information with the Euler characteristic, we get

$$2 = F + V - E$$
$$= \frac{20F_5}{3} + F_5 + 5F_5 - \frac{20F_5 + 5F_5}{2}$$
$$= \frac{F_5}{6}(40 + 6 + 30 - 60 - 15)$$
$$= \frac{F_5}{6}.$$

Thus, $F_5 = 12$, which is what we would expect, since this figure has pentagonal faces, so should have an underlying dodecahedral structure. $\square$

Exercise 5.11. How many square faces are on the snub cube?

Exercise 5.12. How many square faces are on a semiregular polyhedron represented by $(3, 4, 4, 4)$? The answer might surprise you.

Exercise 5.13. Find formulas for F, V, and E in terms of m and n for a Platonic solid with m n-gons at each vertex.

Exercise 5.14. Using the templates you made earlier, construct a model of your favorite semiregular polyhedron. If you can, involve a young child in the project.

Exercise 5.15. Answer the first 'Why' in this section. That is, if there are four faces to a vertex and only one of them is a triangle, explain why two of the other faces must be identical.

Exercise 5.16. Answer the second and third 'Whys' in this section. That is, if there are four faces to a vertex and two of them are triangles, explain why these triangles cannot be adjacent, and explain why the other two faces must be identical.

Exercise 5.17. Identify the two different semiregular polyhedra represented by $(3, 4, 4, 4)$.

Exercise 5.18*. Classify the semiregular polyhedra with three faces to each vertex.

Exercise 5.19. Note that $(3, 3, 3, 3, 6)$ represents a tiling. Identify the complete set of tilings (both regular and semiregular) with five faces at each vertex. Are there any tilings with six faces to a vertex?

Exercise 5.20*. Consider the set of polyhedra whose faces are all congruent equilateral triangles (but with possibly different numbers of triangles at each face). Show that this set is finite. How many such polyhedra are there?

5.6   Four-Dimensional Objects

How can we make sense of a four-dimensional object? Probably the easiest way is to use Cartesian coordinates and make inferences from what we know about two- and three-dimensional objects.

For example, a square is the object with vertices $(0, 0)$, $(0, 1)$, $(1, 0)$ and $(1, 1)$. A cube is the object with vertices (x, y, z), where x, y, and z run through all combinations of 0 and 1. The four-dimensional analogue, a hyper-cube, is therefore the object with vertices (w, x, y, z), where w, x, y, and z run through all combinations of 0 and 1. A face of the cube is the set of points in the cube with $z = 0$. This is the polygonal region bounded by the vertices $(0, 0, 0)$, $(0, 1, 0)$, $(1, 0, 0)$, and $(1, 1, 0)$. That is, this face

is a square. By analogy, a face of the hyper-cube is the set of points with
$z = 0$, which describes a cube. Thus, just as the cube has square faces, the
hyper-cube has cube faces. What then should the analogue of an edge be?
The set of points with both y and z equal to 0? Then, the analogue of an
edge is a square.

With this analogy in mind, perhaps we should define a regular four-
dimensional hyper-polyhedron as a figure whose faces are Platonic solids.
The natural question one might ask is, are there any more regular four-
dimensional hyper-polyhedra?

The analogue of a circle and sphere in four dimensions is the figure
described by

$$w^2 + x^2 + y^2 + z^2 = r^2,$$

which is the set of points a distance r from the origin. An interesting
question we can answer is this: What is the hyper-volume of the interior of
the hyper-sphere in four dimensions? To answer this question, let us first
recall how one derives the volume of a sphere.

Theorem 5.6.1. *The volume of a sphere with radius r is $4\pi r^3/3$.*

Proof. Let us use the equation for a sphere:

$$x^2 + y^2 + z^2 = r^2.$$

Let us take a slice of this sphere by fixing $z = z_0$. This gives a cross section
with equation

$$x^2 + y^2 + z_0^2 = r^2$$
$$x^2 + y^2 = (r^2 - z_0^2),$$

which is the equation of a circle of radius $\sqrt{r^2 - z_0^2}$, which has area $\pi(r^2 - z_0^2)$. Thus, this cross section has volume

$$\Delta V = \pi(r^2 - z^2)\Delta z$$

(yes, we are using calculus now) and hence the volume of a sphere is

$$V = \int_{-r}^{r} \pi(r^2 - z^2)dz$$
$$= \left[\pi\left(r^2 z - \frac{1}{3}z^3\right)\right]_{-r}^{r}$$
$$= \frac{4}{3}\pi r^3. \qquad \square$$

We can use the same argument with the hyper-sphere.

Theorem 5.6.2. *The hyper volume of the hyper-sphere of radius r in four
dimensions is $\pi^2 r^4/2$.*

Proof. We use the equation

$$w^2 + x^2 + y^2 + z^2 = r^2.$$

We take a cross section $z = z_0$, which has the equation

$$w^2 + x^2 + y^2 = r^2 - z_0^2$$

and hence is a sphere with radius $\sqrt{r^2 - z_0^2}$. Thus, the hyper-volume of a four-dimensional hyper-sphere is

$$V = \int_{-r}^{r} \frac{4}{3}\pi \left(r^2 - z^2\right)^{3/2} dz.$$

We make the substitution $r \sin\theta = z$, and get

$$
\begin{aligned}
V &= 2\int_{0}^{\pi/2} \frac{4}{3}\pi r^4 \cos^4\theta\, d\theta \\
&= \frac{2}{3}\pi r^4 \int_{0}^{\pi/2} (\cos^2 2\theta + 2\cos 2\theta + 1)\, d\theta \\
&= \frac{\pi r^4}{3} \int_{0}^{\pi/2} (\cos 4\theta + 1 + 4\cos 2\theta + 2)\, d\theta \\
&= \frac{\pi^2 r^4}{2}.
\end{aligned}
$$

□

Exercise 5.21. We can unfold the faces of a cube to get the region in Figure 5.12. What do we get if we unfold the hypercube?

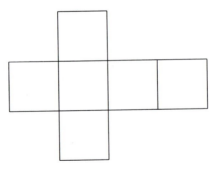

Figure 5.12. An unfolded cube. See Exercise 5.21.

Exercise 5.22. We saw that the dual of the cube is the octahedron, and it can be thought of as having vertices which are the centers of the faces of the cube. Find the centers of the faces of the hyper-cube. Check that this gives a four-dimensional regular hyper-polyhedron.

Exercise 5.23. What does the dual of the hyper-cube look like when it is unfolded?

Exercise 5.24. What is the analogue of a cube (a *hypo-cube*?) in one dimension? What is a *hypo-sphere* in one dimension?

Exercise 5.25. What is the hyper-surface area (a volume) of the hyper-sphere in four dimensions?   [H]

Exercise 5.26. Find the hyper-volume of the interior of a hyper-sphere in five dimensions.   [A]

Figure 5.13. *Crucifixion (Corpus Hypercubus)* by Salvador Dalí, The Metropolitan Museum of Art, Gift of the Chester Dale Collection, 1955. The cross is an unfolded hypercube.

Chapter 6

Hyperbolic Geometry

As has been stressed, the difference between hyperbolic and Euclidean geometry is Euclid's fifth postulate, so let us begin this section by recalling the set of axioms we gave for Euclidean geometry:

1. We can draw a unique line segment between any two points.

2. Any line segment can be continued indefinitely.

3. A circle of any radius and any center can be drawn.

4. Any two right angles are congruent.

The fifth axiom, the parallel postulate, is the axiom we change in this chapter, so let us return to its replacement later.

We also assumed that the usual isometries exist:

6. Given any two points P and Q, there exists an isometry f such that $f(P) = Q$.

7. Given a point P and any two points Q and R which are equidistant from P, there exists an isometry which fixes P and sends Q to R.

8. Given any line l, there exists a map which fixes every point in l and fixes no other points.

And finally, the new fifth postulate:

5. Given any line l and any point P not on l, there exist two distinct lines l_1 and l_2 through P which do not intersect l.

6.1 Models

The biggest problem most students have understanding hyperbolic geometry is getting past the fifth axiom. How can there exist two lines through

P neither of which intersect l? As history demonstrates, mathematicians had the same problem. Since the time of Euclid, mathematicians (Euclid included) have tried to prove that Euclid's fifth postulate follows from Euclid's first four postulates. After all, how can there exist two lines through P neither of which intersect l? The mathematical community very slowly came to the realization that their efforts were in vain. In the nineteenth century, several mathematicians independently arrived at the conclusion that Euclid's fifth postulate does not follow from the first four, and constructed models of geometry in which through any point P not on l, there exist two lines neither of which intersect l.

To make things easier to grasp, let me provide you with a crutch the ancient mathematicians did not have – a model of hyperbolic geometry.

In this model, the entire hyperbolic plane is represented by a disc, not including its boundary. Lines are circular arcs which are perpendicular to the boundary of the disc. Thus, it is indeed possible that for any line l and any point P not on l, there exist two distinct lines through P which do not intersect l – as in Figure 6.1.

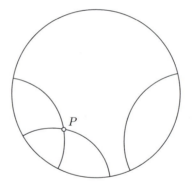

Figure 6.1. A model of hyperbolic geometry, a crutch the ancient mathematicians did not have. In this model, the entire plane is represented by the disc, and lines are arcs of circles which are perpendicular to the boundary.

This model may still not make us feel too comfortable. Perhaps we want lines to be straight. Perhaps a model of the plane should be unbounded. To help us get over this, let me also present a model of the Euclidean plane which looks a little more like the above model of the hyperbolic plane.

Let the Euclidean plane be represented by the plane $z = -1$ in three-space. Consider also the unit sphere

$$x^2 + y^2 + z^2 = 1.$$

For any point $P = (P_0, P_1, -1)$ in the Euclidean plane, draw a line through P and the North Pole of the sphere $N = (0, 0, 1)$, as in Figure 6.2. Let

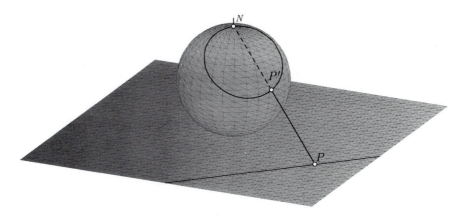

Figure 6.2. An unusual model of Euclidean geometry: The Euclidean plane is represented by the sphere not including the North Pole N. Lines in this model are circles which go through N. The model is derived by projecting points in the plane $z = -1$ to points on the sphere through the point N. Such a projection is called a *stereographic projection*.

this line intersect the sphere at P'. In this way, we can map the entire Euclidean plane to the surface of the sphere, not including the North Pole. Thus, the punctured sphere is a model for the Euclidean plane. What does a line look like in this model? Let l be a line on the plane $z = -1$. The image of l on the sphere is the intersection of the sphere with the plane through $(0, 0, 1)$ and l. In particular, it is the intersection of a plane with a sphere, so is a circle. Furthermore, this circle goes through the North Pole. Thus, we have constructed a model of the Euclidean plane which is bounded, and on which lines are circles through a fixed point. This model is not too dissimilar from the above model of the hyperbolic plane, and I hope it makes you feel more comfortable with that model.

Before moving on, let me emphasize that this model of hyperbolic geometry is meant to be a crutch. It is to help us with our intuition and is not meant to be a statement of fact. The results we desire must be proved from the axioms we have set forth, and not from this model.

It is very important for us to eventually establish the existence of a model for hyperbolic geometry, since such a model is the evidence that hyperbolic geometry exists. However, we may establish many interesting results (as we will do in the following few sections) using just the axioms. For now, we will think of these results as properties of hyperbolic geometry *if* such a geometry exists. Eventually, we will construct a model which satisfies the axioms of hyperbolic geometry, thereby proving that such a geometry exists.

Some of the results we will discover in this chapter are probably quite a bit older than the discovery of hyperbolic geometry (usually attributed to János Bolyai, Karl Friederich Gauss, and/or Nikolai Ivanovich Lobachevsky, all in the early nineteenth century). I say this because of the work of a seventeenth-century Jesuit priest, Girolamo Saccheri. He, like many mathematicians before him, thought Euclid's first four postulates were quite reasonable, but that the fifth looked like a theorem and therefore should follow from the first four. He set out to prove this, using the time honored method of *reductio ad absurdum* or *proof by contradiction*.

Figure 6.3. A *Saccheri quadrilateral*, studied by Girolamo Saccheri in the seventeenth century, and earlier by Omar Khayyám and Nasir Eddin al-Tusi in the eleventh and thirteenth centuries.

Saccheri studied quadrilaterals $ABCD$ with equal sides AB and CD perpendicular to BC, as in Figure 6.3. By symmetry, the angles at A and D must be equal. There are therefore three possibilities. The angles A and D are either obtuse, right, or acute. If these angles are right angles, then the parallel postulate follows (see Exercise 6.5), so Saccheri assumed that they are either obtuse or acute. He assumed first that the angles are obtuse, which in fact leads to a contradiction (see Exercise 6.1). He then assumed that the angles are acute, and though he was able to deduce many strange results [Gre], he was unable to arrive at a contradiction. In this chapter, we too will discover many strange results. Saccheri was, perhaps, referring to these results.

6.2   Results from Neutral Geometry

Some results may be proved without using either version of the fifth axiom. Such a result is called a result in *neutral geometry*. These include several properties we proved for Euclidean geometry. For example, two circles intersect in at most two points. Thus, both SSS and SAS are results in neutral geometry and hence results in hyperbolic geometry. So is ASA, *pons assinorum*, and the following very important result which was also proved in Chapter 1:

Lemma 6.2.1 (Lemma 1.4.2). *Let l be a line and P a point not on l. Then there exists a point Q on l so that PQ is perpendicular to l.*

Let us add one more result to that list:

Theorem 6.2.2 (The Saccheri-Legendre Theorem). *The sum of the angles in a triangle is at most $180°$. That is, in $\triangle ABC$, we have*

$$A + B + C \leq 180°.$$

In the statement of the above theorem, we have again used the abbreviation A for $\angle BAC$, and so on. Though this notation can at times lead to confusion, our use of it here is for the sake of the clarity that brevity can sometimes bring.

To prove Theorem 6.2.2, we require a couple of lemmas:

Lemma 6.2.3. *In any triangle $\triangle ABC$, we have*

$$A + B < 180°.$$

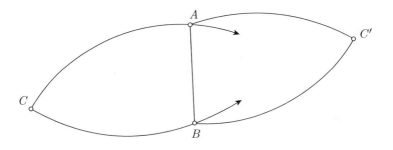

Figure 6.4

Proof. Suppose $A + B \geq 180°$. Construct $\triangle ABC'$ so that $\triangle ABC \equiv \triangle BAC'$, as in Figure 6.4. If $A+B = 180°$, then $\angle CAC' = 180° = \angle CBC'$. That is, the lines CA and CB intersect again at C', a contradiction of the first axiom. Thus, we may assume $A + B > 180°$. But then the ray CA enters $\triangle ABC'$ at A, so must intersect the segment BC'. Similarly, the ray CB must intersect AC', so these two rays must intersect inside $\triangle ABC'$, and again we get that the lines CA and CB intersect twice, contradicting the first axiom. Thus, we must have $A + B < 180°$. $\qquad\square$

Lemma 6.2.4. *Given a triangle $\triangle ABC$, there exists a triangle $\triangle A'B'C'$ such that*

$$C' < \frac{1}{2}C$$

and

$$A' + B' + C' = A + B + C.$$

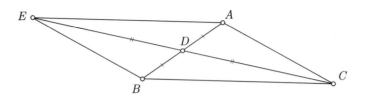

Figure 6.5

Proof. Let D be the midpoint of AB (see Figure 6.5). Let E be the point on CD so that $|CD| = |DE|$ and C and E are on opposite sides of AB. Then, either $\angle BCE$ or $\angle ECA$ is at most $\frac{1}{2}\angle BCA$. Without loss of generality, let us assume

$$\angle ECA \leq \frac{1}{2}\angle BCA.$$

Then choose $A' = A$, $C' = C$, and $B' = E$. By SAS, $\triangle BCD \equiv \triangle AED$, so

$$A' = \angle DAE + \angle DAC = \angle DBC + \angle CAD$$
$$B' = \angle DEA = \angle BCD$$
$$C' = \angle DCA,$$

so

$$A' + B' + C' = A + B + C. \qquad \square$$

Proof of Theorem 6.2.2. By Lemma 6.2.4, we can find a sequence of triangles $\triangle A_k B_k C_k$ such that $\triangle A_0 B_0 C_0 = \triangle ABC$,

$$A_{k+1} + B_{k+1} + C_{k+1} = A_k + B_k + C_k,$$

and

$$C_{k+1} \leq \frac{1}{2}C_k.$$

Thus,

$$A_k + B_k + C_k = A + B + C$$

and

$$C_k \leq \frac{1}{2^k}C.$$

In particular, if

$$A + B + C = 180° + \epsilon$$

for some $\epsilon > 0$, then there exists a k such that $\frac{1}{2^k}C < \epsilon$. But then

$$A_k + B_k = A + B + C - C_k > 180° + \epsilon - \epsilon = 180°,$$

which contradicts Lemma 6.2.3. $\qquad \square$

The reader might be a bit skeptical of the proof of Lemma 6.2.3 – in particular with the statement that a line which enters a triangle must also exit the triangle. To prove such a statement, we need a better set of axioms. Such a set will be presented in Chapter 9 and this particular question will be addressed in Exercises 9.3 and 9.10.

Exercise 6.1. Prove that the sum of angles in a quadrilateral is at most 360°. Conclude that the angles A and D in the Saccheri quadrilateral in Figure 6.3 are at most right angles.   [S]

Exercise 6.2. Suppose $ABCD$ is a rectangle. That is, suppose $ABCD$ is a quadrilateral whose angles are all right angles. Prove that $|AB| = |CD|$ and $|BC| = |DA|$. Prove that the line AB does not intersect the line CD. Let a line l intersect AB perpendicularly at E. Show that l intersects CD at a point F and that $AEFD$ is a rectangle.

Exercise 6.3. Prove that if there exists a triangle whose angles sum to 180°, then there exists a right angle triangle whose angles sum to 180°. Conclude that if there exists a triangle whose angles sum to 180°, then there exists a rectangle.   [H]

Exercise 6.4. Suppose $ABCD$ is a rectangle, and that l is a line through A which is not the line AD. Prove that l intersects the line BC. Conclude that there are no rectangles in hyperbolic geometry.   [S]

Exercise 6.5. Prove that the existence of a rectangle (or a triangle whose angles sum to 180°) implies the Euclidean parallel postulate.   [H]

6.3   The Congruence of Similar Triangles

In Exercise 6.5, we are asked to show that the existence of a triangle whose angles sum to 180° implies the Euclidean parallel postulate. Thus, in hyperbolic geometry, we have the following:

Theorem 6.3.1. *In any triangle $\triangle ABC$, we have*

$$A + B + C < 180°.$$

This has a corollary which we were asked to prove in Exercise 6.1:

Corollary 6.3.2. *The sum of the angles in a quadrilateral is less than 360°.*

These two results imply the rather surprising result:

Theorem 6.3.3. *Suppose $\triangle ABC \sim \triangle A'B'C'$. Then*

$$\triangle ABC \equiv \triangle A'B'C'.$$

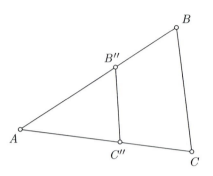

Figure 6.6

As in Euclidean geometry, we say two triangles are *similar* if their angles
are equal.

Proof. Since $\angle BAC = \angle B'A'C'$, there exists an isometry which sends A'
to A, the ray $A'B'$ to the ray AB, and the ray $A'C'$ to the ray AC. Let the
image of B' and C' under this isometry be respectively B'' and C'', as in
Figure 6.6. If these two triangles are not congruent, then we may assume,
without loss of generality, that $B'' \neq B$, and that B'' lies between A and
B. By Exercise 6.6, BC and $B''C''$ cannot intersect. But then, $BCB''C''$
forms a quadrilateral. This quadrilateral has the angles

$$\angle B''BC = \angle ABC,$$
$$\angle C''CB = \angle ACB,$$
$$\angle BB''C'' = 180° - \angle ABC,$$
$$\angle CC''B'' = 180° - \angle ACB,$$

which sum to 360°, which contradicts Lemma 6.3.2. Thus, we must have
$B'' = B$ and $C'' = C$, so the two triangles are congruent. $\qquad\square$

Exercise 6.6. Show that, in the proof of Theorem 6.3.3, the edges BC
and $B''C''$ cannot intersect.   [H]

6.4   Parallel and Ultraparallel Lines

Let l be a line and P a point not on l. Then there exist two lines through
P which do not intersect l. Any line between these two lines also cannot
intersect l, since to do so, such a line must first cross one or the other, and
this would contradict our first axiom. Thus, there exists an infinite set of
lines through P which do not intersect l.

This also demonstrates that there must be two boundary lines – two
lines through P which do not intersect l and such that all lines through P

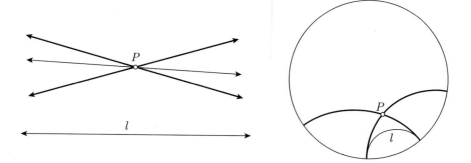

Figure 6.7. Lines through P which are parallel (bold) and ultra-parallel to l. (Abstract picture on the left; our model on the right.)

which do not intersect l lie between these two lines (see Figure 6.7). These boundary lines through P are called the lines through P which are *parallel* to l. All the lines between them are called *ultraparallel* lines.

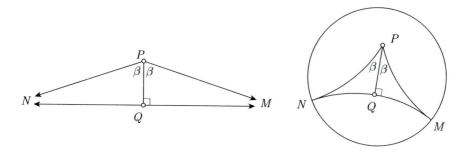

Figure 6.8

Let P be a point not on l, and let l 'intersect infinity' at M and N, as in Figure 6.8. (In our model, M and N are points on the boundary of the disc, which are not points in hyperbolic space. However, this picture is the inspiration for referring to points at infinity.) Then, the rays PM and PN are parallel to l. Let Q be the point on l such that PQ is perpendicular to l. Let $\beta = \angle MPQ$. By symmetry (reflection through PQ), $\angle NPQ = \beta$.
Furthermore:

Theorem 6.4.1. *The angle β depends only on the length $|PQ|$.*

Proof. Let l' be another line, Q' a point on l', and P' a point not on l' such that $P'Q'$ is perpendicular to l' and $|P'Q'| = |PQ|$. Let l' have endpoints M' and N' at infinity. Let $\beta' = \angle M'P'Q'$. We wish to show $\beta' = \beta$. Suppose not. Then, without loss of generality, we may assume $\beta' < \beta$.

Draw the ray r which makes an angle β' with PQ. Since $\beta' < \beta$, this ray must intersect l, say at a point R. By our sixth and seventh axioms, and since $|PQ| = |P'Q'|$, there exists an isometry which sends P to P' and Q to Q'. Let R' be the image of R under this isometry. Since isometries preserve angles, $\angle R'P'Q' = \beta'$, a contradiction, since this says $R'P'$ is parallel to l'. Thus, $\beta' = \beta$. $\qquad\square$

This theorem justifies our writing the angle β as a function of the distance $|PQ|$. We write

$$\Pi(|PQ|) = \beta$$

and call β the *angle of parallelism*. Note that $\Pi(|PQ|) < \pi/2$ for all points P and Q (for otherwise, we would have a contradiction to our fifth axiom).

Theorem 6.4.2. *Two ultraparallel lines have a common perpendicular.*

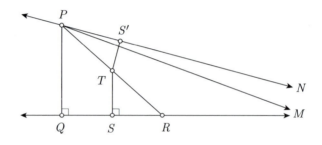

Figure 6.9

Proof. Let l_1 and l_2 be two ultraparallel lines. Let P be a point on l_2 and let Q be the point on l_1 such that PQ is perpendicular to l_1. Let M and N be points at infinity at the ends of l_1 and l_2 on the side where the angle l_2 makes with PQ is smaller, as in Figure 6.9. Since l_2 is ultraparallel to l_1, the $\angle NPQ$ is greater than $\beta = \Pi(|PQ|)$. Let R be a point on l_1 on the M side of Q. Let us vary R. As we move R toward Q, $\angle PRQ$ goes to a right angle. By moving R towards M, we can make $\angle PRQ$ as small as we like (see Exercise 6.7). Thus, we can make $\angle PRM$ as close to $180°$ as we like, which means we can make $\angle RPM$ as small as we like. Hence, if $R = Q$, then $\angle NPR = \angle NPQ$, and as R goes to M, $\angle NPR$ approaches $\angle NPM$. Thus, at some point, we must have $\angle NPR = \angle PRQ$. Fix R so that this is the case. Let T be the midpoint of PR, and let S be the point on l_1 so that TS is perpendicular to l_1. Let S' be the point on l_2 between P and N so that $|PS'| = |RS|$. Then, by SAS, $\triangle PS'T \simeq \triangle RST$. Hence, $\angle S'TP = \angle STR$, and because PTR is a straight line, the angle $\angle S'TS$ must also be $180°$. Further, $\angle PS'T = \angle RST$, which is a right angle. Thus, $S'S$ is a common perpendicular to both l_1 and l_2. $\qquad\square$

Exercise 6.7. Let M be a point at infinity and let $\angle PQM = 90°$. Given $\epsilon > 0$, show that there exists a point R on QM such that $\angle PRQ < \epsilon$. Figure 6.10 is a hint.

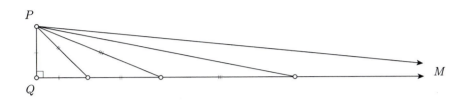

Figure 6.10. See Exercise 6.7.

Exercise 6.8. Prove that if alternate interior angles of a transversal l to two lines l_1 and l_2 are equal, then l_1 and l_2 are ultraparallel.

Exercise 6.9. Two parallel lines do not have a common perpendicular (see Theorem 6.5.1 below). Where does the proof of Theorem 6.4.2 fail in this case? That is, where was it important that the two lines in Theorem 6.4.2 be ultraparallel?

6.5 Singly Asymptotic Triangles

Let M be a point at infinity, and let A and B be points in the hyperbolic plane. We call the region bounded by AB, AM and BM an *asymptotic triangle*. The angle at infinity can either be thought of as not existing, or in light of Exercise 6.7, considered to have the measure zero.

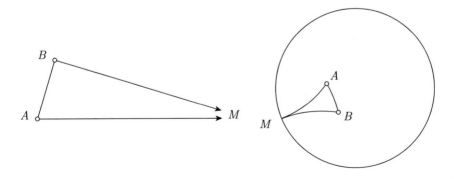

Figure 6.11. Singly asymptotic triangles.

Theorem 6.5.1. *The angles in an asymptotic triangle sum to less than $180°$.*

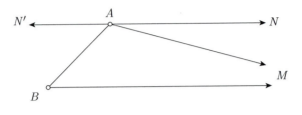

Figure 6.12

Proof. Let $\triangle ABM$ be an asymptotic triangle, where M is a point at infinity, as in Figure 6.12. Construct a line with endpoints N and N' through A such that N is on the same side of AB as M and $\angle N'AB = \angle ABM$. Then, by Exercise 6.8, NN' and BM are ultraparallel lines. Thus, the ray AN is above AM, and hence $\angle NAB > \angle MAB$.   □

Theorem 6.5.2. *Two similar singly asymptotic triangles are congruent. That is, if $\triangle ABM$ and $\triangle A'B'M'$ are two asymptotic triangles such that $\angle MAB = \angle M'A'B'$ and $\angle ABM = \angle A'B'M'$, then $|AB| = |A'B'|$.*

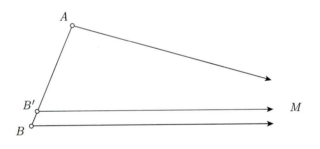

Figure 6.13

Proof. Suppose $|AB| \neq |A'B'|$. Translate and rotate $\triangle A'B'M'$ so that A' is sent to A and $A'B'$ is sent to AB, as in Figure 6.13. Note that M' must go to M. Thus, $\triangle BB'M$ is an asymptotic triangle. But the sum of the angles in this triangle is $180°$, a contradiction.   □

6.6   Doubly and Triply Asymptotic Triangles

Given a line l intersecting infinity at M and N, and a point A not on l, it is possible to define the *doubly asymptotic* triangle $\triangle AMN$.

    However, given two rays AM and AN, is it possible to construct the triangle $\triangle AMN$? Note that the points M and N are not points in hyperbolic geometry, so we cannot use our first axiom to assure the existence of a line MN. In our model, it is painfully clear that the line MN should exist, but do not forget that as yet, our model is only a crutch, and like a poor picture in Euclidean geometry, may lead us astray. In this case, it does not, but we have to work to prove this 'obvious' result.

Theorem 6.6.1. *Two intersecting rays have a unique common parallel.*

Proof. Let OM and OM' be the two rays. Pick A on OM and A' on OM' such that $|OA| = |OA'|$, as in Figure 6.14. Construct AM' and $A'M$. Let a and a' be the angle bisectors of $\angle MAM'$ and $\angle MA'M'$, respectively. Let E be the point of intersection of AM' and $A'M$, and let OE intersect infinity at L. Finally, let $A'M$ intersect a at F.

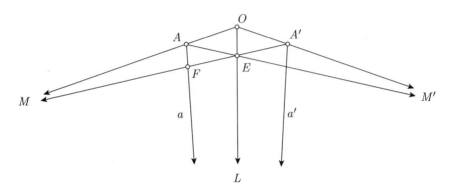

Figure 6.14

We first show a and a' are ultraparallel. Suppose first that a and a' intersect at a point D, which, by symmetry, lies on OL. Consider the two triangles $\triangle ADM$ and $\triangle A'DM$. Since $|AD| = |A'D|$, we know there exists an isometry which fixes D and sends A to A'. Since $\angle MAD = \angle MA'D$, we know AM is sent to $A'M$ under this isometry. Thus, $\triangle MAD \equiv \triangle MA'D$, which is clearly not possible, since $\angle ADM \neq \angle A'DM$. Thus, the point D cannot exist.

Let us now assume a and a' are parallel, so they intersect at infinity at L. This time, let us consider the two triangles $\triangle AFM$ and $\triangle A'FL$. By construction, $\angle MAF = \angle LA'F$, and clearly $\angle A'FL = \angle AFM$, since they are vertical angles. Thus, $\triangle AFM$ is similar to $\triangle A'FL$, and hence are congruent. Thus, $|AF| = |A'F|$, which means $F = E$, which is clearly absurd. Thus, a and a' cannot be parallel.

Since a and a' neither intersect nor are parallel, they must be ultraparallel. Hence, by Theorem 6.4.2, there exists a line which is perpendicular to both a and a'. Let this line intersect a at C and a' at C', as in Figure 6.15.

We now will show that CC' intersects infinity at M and M', so is the mutual parallel we are looking for. Assume CC' intersects infinity at points N and N'. Note that the triangles $\triangle MAC$ and $\triangle MA'C'$ are similar, since

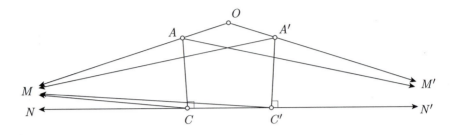

Figure 6.15

$\angle MAC = \angle MA'C'$ and $|AC| = |A'C'|$. Thus, $\angle NCM = \angle NC'M$. If $N \neq M$, then $\triangle MCC'$ is an asymptotic triangle whose angles sum to $180°$, a contradiction. Thus, $N = M$, as desired. Similarly, $N' = M'$. $\qquad\square$

Theorem 6.6.2. *Two similar doubly asymptotic triangles are congruent. That is, if $\triangle AMN$ and $\triangle A'M'N'$ are two doubly asymptotic triangles with $\angle MAN = \angle M'A'N'$, then there exists an isometry which sends $\triangle AMN$ to $\triangle A'M'N'$.*

Proof. Let Q and Q' be the points on MN and $M'N'$ such that AQ and $A'Q'$ are, respectively, perpendicular to MN and $M'N'$. Then $\triangle MAQ$ is similar to $\triangle M'A'Q'$, so they are congruent. That is, there exists an isometry which sends the first triangle to the second. But isometries preserve angles, so under this isometry, we must also have N sent to N'. $\qquad\square$

We can also define triply asymptotic triangles (see Figure 6.16).

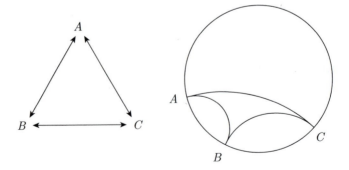

Figure 6.16. Triply asymptotic triangles.

Exercise 6.10. Prove that any two triply asymptotic triangles are congruent.

6.7 The Area of Asymptotic Triangles

In Euclidean geometry, an asymptotic triangle has infinite area. In hyperbolic geometry, the opposite is true. Though we will not define area in hyperbolic geometry in this section, we will establish and use several of the properties which any area function should have. These are

1. Two congruent regions have the same area.

2. A nonempty region has a nonzero area.

3. The area of a disjoint union of regions is the sum of the areas of the regions.

4. The area of a bounded region is finite.

Theorem 6.7.1. *The area of any asymptotic triangle in hyperbolic space is finite.*

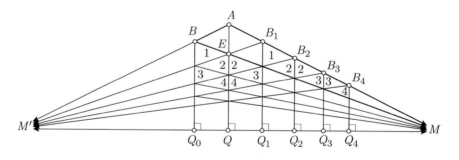

Figure 6.17

Proof. Let $\triangle ABM$ be an asymptotic triangle. Let M' be an endpoint at infinity of AB, as in Figure 6.17. Find the common parallel MM'. Let Q_0 and Q be the points on MM' such that BQ_0 and AQ are perpendicular to MM'. Reflect B through AQ to get B_1, and find Q_1 on MM' such that B_1Q_1 is perpendicular to MM'. Let B_1M' intersect BM at E_1. By symmetry, E_1 is on AQ. Note that $\triangle B_1Q_1M' \equiv \triangle B_1Q_1M$, so the reflection of E_1 through B_1Q_1 is a point B_2 on AM and E_2 at the intersection of B_1A_1 and B_2M'. Repeat, to get B_3, B_4, and so on. Note that the triangles labeled $1, 2, 3, \ldots$ in $\triangle ABM$ are congruent to the corresponding disjoint triangles labeled $1, 2, 3, \ldots$ in the pentagon $BAB_1Q_1Q_0$. That is, the triangle $\triangle ABM$ can be sliced into pieces and reassembled to form a subset of the pentagon $BAB_1Q_1Q_0$. Thus, the area of $\triangle ABM$ is less than or equal to the area of the pentagon, which is bounded and therefore has finite area. $\square$

Exercise 6.11. Prove that the area of any doubly asymptotic triangle is finite.

Exercise 6.12. Prove that the area of all triply asymptotic triangles are finite and equal.

Chapter 7

The Poincaré Models of Hyperbolic Geometry

In this chapter, we investigate two models of hyperbolic geometry developed by Henri Poincaré (1854 – 1912). Both models are in common use today. Both have their advantages and disadvantages.

We first study the upper half plane model and eventually show that it is a model of hyperbolic geometry. The other model is the disc model, which is the crutch introduced in Chapter 6.

7.1 The Poincaré Upper Half Plane Model

Recall that a complex number z can be represented in Cartesian coordinates (x, y) where $z = x + iy$. This planar representation of the complex numbers is called the *Argand plane*. The Poincaré upper half plane model of hyperbolic geometry is the set of points $\mathcal{H}$ above the real axis:

$$\mathcal{H} = \{x + iy : y > 0\},$$

together with the arclength element

$$ds = \frac{\sqrt{dx^2 + dy^2}}{y}.$$

Amazingly, that is the complete description. The remainder of this chapter involves a detailed investigation of this model, which will in particular establish that this is a model for hyperbolic geometry. That is, this model does in fact satisfy the axioms of hyperbolic geometry. In order to show this, we must first discover what lines are, what the distance function is, and what the isometries are.

7.2   Vertical (Euclidean) Lines

Let $\vec{x}(t) = (x(t), y(t))$ be a piecewise smooth parameterized curve between points $\vec{x}(t_0)$ and $\vec{x}(t_1)$. Recall that the arclength element ds in Euclidean geometry is given by $ds = \sqrt{dx^2 + dy^2}$, so the arclength of this curve *in Euclidean space* is given by

$$s = \int_{t_0}^{t_1} \sqrt{\left(\frac{dx}{dt}\right)^2 + \left(\frac{dy}{dt}\right)^2}\, dt.$$

Now let us suppose $\vec{x}(t) = (x(t), y(t))$ represents a curve in the Poincaré upper half plane. Then, because of the different arclength element, the arclength of this curve is given by

$$s = \int_{t_0}^{t_1} \frac{\sqrt{\left(\frac{dx}{dt}\right)^2 + \left(\frac{dy}{dt}\right)^2}}{y}\, dt.$$

Let us consider a special curve, the curve

$$\vec{x}(y) = (x_0, y) \qquad y \in [y_0, y_1].$$

This is the vertical (Euclidean) line segment between (x_0, y_0) and (x_0, y_1). The Poincaré arclength of this curve is

$$s = \int_{y_0}^{y_1} \frac{1}{y}\, dy = \ln y_1 - \ln y_0 = \ln(y_1/y_0).$$

Now, let us consider any piecewise smooth curve $\vec{x}(t) = (x(t), y(t))$ with $\vec{x}(t_0) = (x_0, y_0)$ and $\vec{x}(t_1) = (x_0, y_1)$. So, this curve starts and ends at the endpoints of the previously considered segment. Let us also suppose that $y(t)$ is an increasing function. Then, we get

$$s = \int_{t_0}^{t_1} \frac{\sqrt{\left(\frac{dx}{dt}\right)^2 + \left(\frac{dy}{dt}\right)^2}}{y}\, dt$$

$$\geq \int_{t_0}^{t_1} \frac{\sqrt{\left(\frac{dy}{dt}\right)^2}}{y}\, dt$$

$$\geq \int_{y(t_0)}^{y(t_1)} \frac{dy}{y}$$

$$\geq \ln y(t_1) - \ln y(t_0).$$

That is, this curve is longer (using the Poincaré arclength element) than the vertical line segment which joins the two points. Hence, the shortest path between these two points is a vertical (Euclidean) line segment. Thus, vertical (Euclidean) lines in the upper half plane are lines in the Poincaré model.

Exercise 7.1. What is the distance between the points $3 + i$ and $3 + 5i$ in the Poincaré upper half plane $\mathcal{H}$? Remember, the Poincaré upper half plane $\mathcal{H}$ comes equipped with a non-Euclidean arclength element, so the answer is not 4.

Exercise 7.2. What is the distance between the points $-2 + 2i$ and $-2 + i\frac{1}{7}$ in the Poincaré upper half plane $\mathcal{H}$?

7.3   Isometries

An isometry is a map which preserves lengths. In particular, it preserves arclengths.

For example, the arclength element in Euclidean space is

$$ds = \sqrt{dx^2 + dy^2}$$

and it is preserved under the action of the rotation

$$R(x,y) = (u(x,y), v(x,y)) = \begin{bmatrix} \cos\theta & -\sin\theta \\ \sin\theta & \cos\theta \end{bmatrix} \begin{bmatrix} x \\ y \end{bmatrix}$$
$$= \begin{bmatrix} x\cos\theta - y\sin\theta \\ x\sin\theta + y\cos\theta \end{bmatrix}.$$

To show this, we must show that $du^2 + dv^2 = dx^2 + dy^2$:

$$du^2 + dv^2 = (\cos\theta dx - \sin\theta dy)^2 + (\sin\theta dx + \cos\theta dy)^2$$
$$= \cos^2\theta dx^2 - 2\cos\theta\sin\theta dxdy + \sin^2\theta dy^2$$
$$\quad + \sin^2\theta dx^2 + 2\sin\theta\cos\theta dxdy + \cos^2\theta dy^2$$
$$= dx^2 + dy^2,$$

as desired.

In a similar fashion, to show that a map in the Poincaré upper half plane model is an isometry, we must show that it preserves this slightly different arclength element. That is, a map

$$(u(x,y), v(x,y))$$

is an isometry if

$$\frac{du^2 + dv^2}{v^2} = \frac{dx^2 + dy^2}{y^2}.$$

Some maps are clearly isometries, though the full group[1] of isometries is not obvious. One obvious isometry is the map

$$T_a(x,y) = (u(x,y), v(x,y)) = (x + a, y),$$

---

[1] In Appendix A.3, we remind the reader of the definition of a group. If the reader is not familiar with this concept, then this text reads fine with the substitute definition that the group of isometries is just the collection or set of isometries.

since

$$\frac{du^2 + dv^2}{v^2} = \frac{dx^2 + dy^2}{y^2},$$

as desired. We will refer to this map as *horizontal translation by a*. Note that the word 'horizontal' refers to a Euclidean concept. It is a concept particular to this model and not to hyperbolic geometry.

Another obvious isometry is reflection through the vertical line $x = b$:

$$R_b(x, y) = (u, v) = (2b - x, y).$$

A not so obvious isometry is a map called *inversion in the unit circle*

$$\Phi(x, y) = (u, v) = \left(\frac{x}{x^2 + y^2}, \frac{y}{x^2 + y^2} \right).$$

Let us write $r^2 = x^2 + y^2$. This aides in checking the arclength element:

$$\frac{du^2 + dv^2}{v^2} = \frac{r^4}{y^2} \left(\left(\frac{r^2 dx - 2x^2 dx - 2xy dy}{r^4} \right)^2 \right.$$

$$\left. + \left(\frac{r^2 dy - 2xy dx - 2y^2 dy}{r^4} \right)^2 \right)$$

$$= \frac{1}{y^2} \left(\frac{((y^2 - x^2)dx - 2xy dy)^2 + ((x^2 - y^2)dy - 2xy dx)^2}{r^4} \right)$$

$$= \frac{1}{y^2 r^4} \left((x^4 - 2x^2 y^2 + y^4 + 4x^2 y^2)dx^2 \right.$$

$$\left. -(2xy(y^2 - x^2) + 2xy(x^2 - y^2))dx dy + r^4 dy^2 \right)$$

$$= \frac{dx^2 + dy^2}{y^2}.$$

Exercise 7.3. Prove that the dilation

$$\delta_\lambda(x, y) = (\lambda x, \lambda y)$$

preserves the Poincaré arclength element.

7.4   Inversion in the Circle

This last isometry deserves some investigation. In this section, we will show that the image of a (Euclidean) line under inversion in the unit circle is either a line or circle, and the image of a circle is either a line or a circle. We will also show that angles are preserved under this map.

The image of a point P under inversion in the circle centered at O and with radius r is the point P' on the ray OP and such that

$$|OP'| = \frac{r^2}{|OP|}.$$

In the map $\Phi(x, y)$ considered in the previous section, we took $r = 1$ and $O = 0$.

We will also have occasion to use dilation by λ,

$$\delta_\lambda(x, y) = (\lambda x, \lambda y)$$

which was introduced in Section 1.7. It just shrinks or magnifies the image, so clearly preserves angles and sends lines to lines and circles to circles.

Lemma 7.4.1. *Let l be a line which does not go through the origin O. The image of l under inversion in the unit circle is a circle which goes through the origin O.*

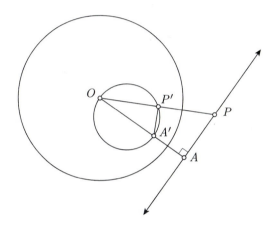

Figure 7.1

Proof. We will prove this for a line l which does not intersect the unit circle, as in Figure 7.1. We leave verification of the other case as an exercise.

Find the point A on l so that OA is perpendicular to l. Let $|OA| = a$. Find the point A' on the ray OA so that $|OA'| = 1/a$. Construct the circle with diameter OA'. We claim that this circle is the image of l under inversion. To see this, let P be an arbitrary point on l and let $|OP| = p$. Let P' be the point of intersection of the ray OP and the circle with diameter OA'. Let $|OP'| = x$. Consider the two triangles $\triangle OAP$ and $\triangle OP'A'$. These two triangles are similar, since they are both right angle triangles and they have a shared angle at O. Hence,

$$\frac{|OP'|}{|OA'|} = \frac{|OA|}{|OP|}$$

$$\frac{x}{(1/a)} = \frac{a}{p}$$

$$x = \frac{1}{p}.$$

Thus, P' is the image of P under inversion in the unit circle. □

Lemma 7.4.2. *Suppose Γ is a circle which does not go through the origin O. Then the image of Γ under inversion in the unit circle is a circle.*

Proof. We will prove this for a circle Γ which does not intersect the unit circle, as in Figure 7.2. We leave verification of the other case as an exercise.

Let the line through O and the center of Γ intersect Γ at points A and B. Let $|OA| = a$ and $|OB| = b$. Let Γ' be the image of Γ under dilation by $1/ab$. We claim that this circle Γ' is the image of Γ under inversion in the unit circle.

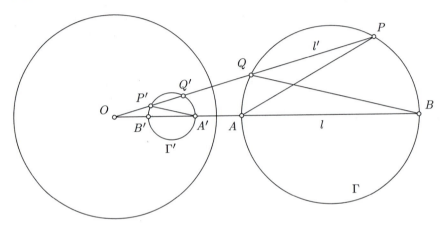

Figure 7.2

Let B' and A' be the images of A and B, respectively (note the reverse order), under this dilation. Then $|OA'| = \dfrac{1}{ab}b = 1/a$, so A' is the image of A under the inversion. Similarly, B' is the image of B under the inversion. Now, let l' be an arbitrary ray through O which intersects Γ at P and Q. Let Q' and P' be the images of P and Q, respectively, under the dilation (again, note the reverse order). We claim that P' is the image of P under the inversion. To see this, note that $\triangle OA'P'$ is similar to $\triangle OBQ$ since one is the dilation of the other. Note that $\angle QBA = \angle QPA$, by the Star Trek lemma, and hence $\triangle OBQ$ is similar to $\triangle OPA$. Hence, $\triangle OA'P' \sim \triangle OPA$ and hence

$$\frac{|OA'|}{|OP|} = \frac{|OP'|}{|OA|}$$

$$\frac{(1/a)}{|OP|} = \frac{|OP'|}{a}$$

$$|OP'| = \frac{1}{|OP|}.$$

Thus, P' is the image of P under the inversion, and hence the circle Γ' is the image of Γ under inversion.   □

Lemma 7.4.3. *Inversions preserve angles.*

Note: We define the angle between two curves at a point of intersection to be the angle between the tangent lines at that point. When we say that angles are preserved under inversion, we mean that the angle between two curves is equal to the angle between the images of these two curves under inversion.

Proof. Let us only consider an angle α created by the intersection of a line l not intersecting the unit circle, and a line l' through O, as in Figure 7.3. Note that any angle can be thought of as the sum or difference of two such angles.

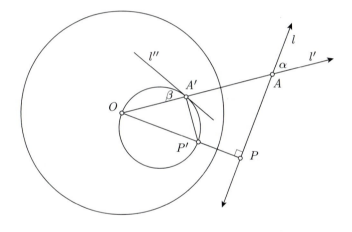

Figure 7.3

Let A be the point of intersection of the lines which create α. Let P be the point on l so that OP is perpendicular to l. Let P' be the image of P under the inversion. Then, the image of l under inversion is the circle Γ whose diameter is OP'. The image of A is the point A' of intersection of Γ with l'. Let l'' be the tangent to Γ at A'. Then the angle β is the image of the angle α under inversion, and so we wish to show $\alpha = \beta$.

To show these two angles are equal, we first note that $\triangle OAP$ and $\triangle OP'A'$ are similar, since they are both right angle triangles and they share the angle at O. Thus, $\angle A'P'O = \angle OAP = \alpha$. But by the tangential case of the Star Trek lemma, $\beta = \angle A'P'O$. Thus, $\alpha = \beta$, as desired.   □

Corollary 7.4.4. *Lines in the Poincaré upper half plane model are (Euclidean) lines and (Euclidean) half circles which are perpendicular to the x-axis (see Figure 7.5).*

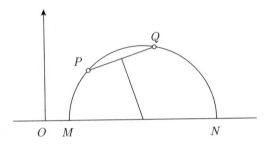

Figure 7.4

Proof. Let P and Q be two arbitrary points in $\mathcal{H}$. Let Γ be the circle through both points P and Q and whose center lies on the x-axis. Let Γ intersect the x-axis at M and N (Figure 7.4 suggests how to find M and N). Consider the map φ which is the composition of the horizontal translation by $-M$ followed by inversion in the unit circle. This map is an isometry since it is the composition of isometries. Note that M is first sent to the origin O and then to ∞ by the inversion. Thus, the image of Γ is a (Euclidean) line. Since the center of the circle is on the real axis, the circle intersects the axis at right angles. Since inversion preserves angles, the image of Γ is a vertical (Euclidean) line. Since vertical lines are lines in the model, and isometries preserve arclength, it follows that Γ is a line through P and Q. $\qquad\square$

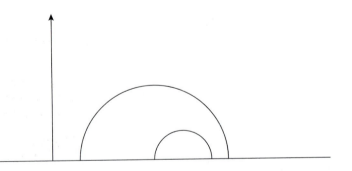

Figure 7.5. Lines in the upper half plane.

Exercise 7.4. Let $P = 4 + 4i$ and $Q = 5 + 3i$. Find M and N, the endpoints of the Poincaré line through P and Q.

Exercise 7.5. Let $P = 12i$ and $Q = 7 + 5i$. Find M and N, the endpoints of the Poincaré line through P and Q.

Exercise 7.6. Finish the proof of Theorem 7.4.1. That is, show that the image of a line l under inversion in the unit circle is a circle through O when l is tangent to the unit circle, or when l intersects the circle in two places, but does not go through O. What is the image of l if l goes through the origin O?

Exercise 7.7. In the proof of Lemma 7.4.2, we assumed that Γ lies outside the unit circle. What are the remaining cases which must be checked? Prove them.

Exercise 7.8. The proof of Lemma 7.4.3 is not quite complete. What other cases must be considered?

Exercise 7.9. Show that the map

$$\Phi_r(x, y) = \left(\frac{rx}{x^2 + y^2}, \frac{ry}{x^2 + y^2} \right)$$

is inversion in the circle with radius r and centered at the origin O.

Exercise 7.10. Prove Lemmas 7.4.1 and 7.4.2 analytically. That is, for Lemma 7.4.1, let l be a line with equation $ax + by = c$ and $c \neq 0$. Show that the image of l under the action of Φ (or Φ_r) gives the equation of a circle which goes through O. Do the same for Lemma 7.4.2.

Exercise 7.11. Let P be a point inside a circle centered at O. Let T be a point where the perpendicular to PO intersects the circle. Let P' be the point where the tangent to the circle at T intersects OP. Show that P' is the image of P under inversion in the circle.

Exercise 7.12. Two circles, as shown in Figure 7.6, intersect at right angles. The circle centered at O has radius one unit. The point R is an arbitrary point on the circle centered at P. Prove that

$$|OS| = \frac{1}{|OR|}.$$

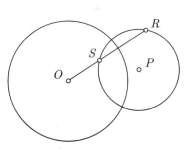

Figure 7.6. See Exercise 7.12.

Exercise 7.13. Write a Geometer's Sketchpad script which draws the Poincaré line through two points in $\mathcal{H}$. The input might be the two points and the line which defines the real axis.

Exercise 7.14. Write a Geometer's Sketchpad script which constructs the inversion of a line in a given circle. Use only the buttons to the left and the 'Construct' pull-down menu. Does your script work for all lines? If not, consider the use of rays instead of lines at some places in your script.

Exercise 7.15. Write a Sketchpad script which demonstrates that inversion in a circle preserves angles.

Exercise 7.16. Write a Sketchpad script which inverts a circle in a given circle. Again, use only the buttons and the 'Construct' pull-down menu.

Exercise 7.17. Write a Sketchpad script which inverts a circle in a given circle. This time, include, in your script, an arbitrary point P on the original circle and its image under inversion. As you drag the point P around the circle, does its inverse move appropriately? If not, rethink your method of constructing the inverse of P.

Exercise 7.18 (The Necklace Theorem or Steiner's Porism). Let the circle Γ' be inside the circle Γ. Insert a circle c_1 externally tangent to Γ' and internally tangent to Γ, as in the diagram. Insert another circle c_2 tangent to c_1, Γ, and Γ'. Continue around Γ'. Suppose that, for some n, we have c_n tangent to c_1 (an example with $n = 6$ is shown in Figure 7.7.)

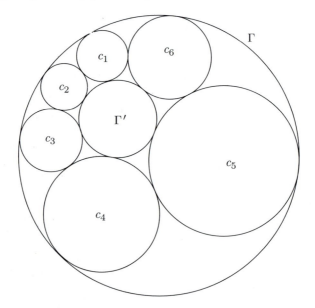

Figure 7.7. A necklace with $n = 6$. See Exercise 7.18.

Prove that if this happens for some n and some pair of circles Γ and Γ', then no matter where we first draw c_1, we will always have c_n tangent to c_1.

Exercise 7.19. Use Sketchpad to draw a picture like the one in Figure 7.7.

Exercise 7.20. Let Γ' and Γ be internally tangent and have centers on the x-axis, as in Figure 7.8. Let c_0 be the circle with center on the x-axis and such that c_0 is tangent to Γ and Γ'. Let c_1 be the circle tangent to Γ, Γ', and c_0, and in general, let c_n be the circle tangent to Γ, Γ', and c_{n-1}. Let h_n be the distance from the center of c_n to the x-axis, and let d_n be the diameter of c_n. Prove that

$$h_n = nd_n.$$

This result is attributed to Pappus.

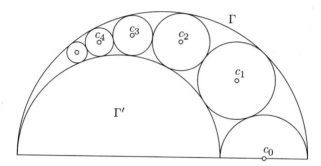

Figure 7.8. See Exercise 7.20.

The following Exercises 7.21 – 7.24 are related:

Exercise 7.21. Suppose three congruent circles are concurrent at H and intersect in pairs at A, B, and C, as in Figure 7.9(a). Write a Sketchpad script which demonstrates that H is the orthocenter of $\triangle ABC$.

Exercise 7.22. Three tangents to a circle centered at I intersect in pairs at A', B', and C', as in the Figure 7.9(b). Write a Sketchpad script which demonstrates that the line $A'I$ goes through the circumcenter of $\triangle B'C'I$.

Exercise 7.23. For A, B, C, and H as in Exercise 7.21 and A', B', C', and I as in Exercise 7.22, show that H is the orthocenter of $\triangle ABC$ if and only if $A'I$ goes through the circumcenter of $\triangle B'C'I$. (Hint: Think inversion in some circle.)

Exercise 7.24. For A', B', C', and I as in Exercise 7.22, prove that $A'I$ goes through the circumcenter of $\triangle B'C'I$. (Hint: Let O be the center of the circumcircle of $\triangle B'C'I$. What is $\angle IC'B'$? What is $\angle IOB'$? What type of triangle is $\triangle IOB'$? What is $\angle A'IO$?) Conclude that for A, B, C, and H as in Exercise 7.21, H is the orthocenter of $\triangle ABC$.

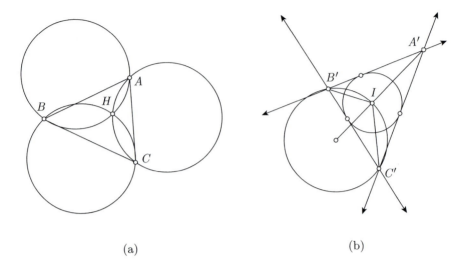

(a)   (b)

Figure 7.9. See Exercises 7.21 – 7.24.

Exercise 7.25*. Two circles Γ and Γ' intersect at A and D. A common tangent intersects Γ and Γ' at E and F, respectively. A line through D parallel to EF intersects Γ and Γ' and C and B, respectively, as in Figure 7.10. Show that the circumcircles of $\triangle BDE$ and $\triangle CDF$ intersect again on the line AD. Hint: Exercise 1.125 might help.

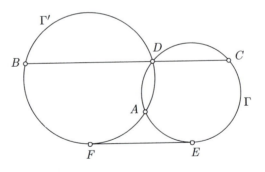

Figure 7.10. See Exercise 7.25.

Exercise 7.26. Check the result in Exercise 7.25 using Sketchpad.

7.5 Inversion in Euclidean Geometry

In this section, we look at inversion as a tool for proving results in Euclidean geometry. The contents of this section are not relevant to the development of the Poincaré upper half plane, so may be safely skipped if understanding $\mathcal{H}$ is the immediate goal of the reader. However, I would like to invite all such readers to return to this section at some time.

There are two more results which are particularly useful when applying inversion to a problem in Euclidean geometry:

Theorem 7.5.1. *Let the images of A and B under inversion in a circle centered at O be A' and B'. Then*

$$\angle OAB = \angle OB'A'.$$

Theorem 7.5.2. *Let the images of A and B under inversion in a circle centered at O with radius r be A' and B'. Then*

$$|A'B'| = \frac{|AB|r^2}{|OA||OB|}.$$

A beautiful example of the power of inversion is the following exercise:

Exercise 7.27 (Ptolemy's Inequality). For an arbitrary quadrilateral $ABCD$, show that

$$|AC||BD| \leq |AB||CD| + |BC||DA|.$$

When does equality hold?

Solution. Try this question before reading this solution. If a hint is needed, read the next line and try again.

Invert in a circle centered at D with radius 1, and apply Theorem 7.5.2. The left side of the inequality becomes:

$$|AC||BD| = \frac{|A'C'|}{|A'D||C'D|}\frac{1}{|B'D|},$$

and the right side becomes

$$|AB||CD| + |BC||DA| = \frac{|A'B'|}{|A'D||B'D|}\frac{1}{|C'D|} + \frac{|B'C'|}{|B'D||C'D|}\frac{1}{|A'D|}.$$

Multiplying through by $|A'D||B'D||C'D|$, the inequality becomes

$$|A'C'| \leq |A'B'| + |B'C'|,$$

which is just the triangle inequality! Furthermore, equality holds if and only if A', B', and C' are collinear with B' between A' and C'. That is, we have equality if and only if A, B, and C lie on a circle through D with B and D on opposite sides of AB. That is, we have equality if and only if $ABCD$ is a convex cyclic quadrilateral. □

Exercise 7.28. Prove Theorems 7.5.1 and 7.5.2.

Exercise 7.29. Let Q be the image of O under reflection through the line l. Prove that the image Q' of Q under inversion in a circle centered at O is the center of the circle Γ which is the image of l under the same inversion.

Exercise 7.30. Four circles are coincident at a point O in such a way that they are externally tangent at O in pairs, as in Figure 7.11(a). These circles intersect again at points A, B, C, and D. Show that

$$|AB||OD||OC| = |CD||OA||OB|.$$

Exercise 7.31. Two circles intersect at A and C. The tangents at A to these circles intersect them again at B and D, as in Figure 7.11(b). Show that

$$|AB||CD| = |AC||AD|.$$

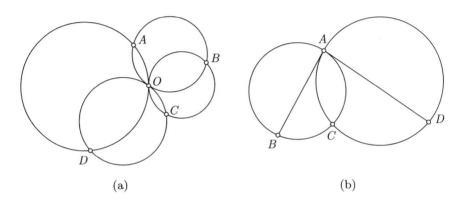

(a)   (b)

Figure 7.11. See Exercises 7.30 and 7.31.

7.6   Fractional Linear Transformations

The group of direct isometries is actually very easy to describe – they are all *fractional linear transformations* of a particular type. Recall that we call an isometry a *direct isometry* if it preserves orientation.

A *fractional linear transformation* is a function of the form

$$T(z) = \frac{az + b}{cz + d}$$

where a, b, c and d are complex numbers, and $ad - bc \neq 0$. The domain of this function is the set of complex numbers $\mathbb{C}$ together with the symbol ∞, which represents a point at infinity. We extend the definition of T to include the following:

$$T(-d/c) = \lim_{z \to \frac{-d}{c}} \frac{az + b}{cz + d} = \infty \qquad \text{if } c \neq 0,$$

$$T(\infty) = \lim_{z \to \infty} \frac{az + b}{cz + d} = a/c \qquad \text{if } c \neq 0,$$

$$T(\infty) = \lim_{z \to \infty} \frac{az + b}{cz + d} = \infty \qquad \text{if } c = 0.$$

We usually represent the fractional linear transformation T with the 2×2 matrix

$$\gamma = \begin{bmatrix} a & b \\ c & d \end{bmatrix},$$

and write $T = T_\gamma$. The matrix representation for T is not unique, since T is also represented by

$$k\gamma = \begin{bmatrix} ka & kb \\ kc & kd \end{bmatrix}$$

for any scalar $k \neq 0$. We say that two matrices γ and γ' are *equivalent* if they represent the same fractional linear transformation. That is, two matrices γ and γ' are equivalent if

$$T_\gamma = T_{\gamma'}.$$

We write $\gamma \equiv \gamma'$.

This matrix representation is more than just a notational convenience. The composition of fractional linear transformations is in fact related to matrix multiplication[2] by the following result (proved in Exercise 7.32):

Theorem 7.6.1.
$$T_{\gamma_1 \gamma_2}(z) = T_{\gamma_1}(T_{\gamma_2}(z)).$$

Note that the fractional linear transformation T_I represented by the identity matrix

$$I = \begin{bmatrix} 1 & 0 \\ 0 & 1 \end{bmatrix}$$

is just the identity function. That is,

$$T_I(z) = \frac{z + 0}{0z + 1} = z.$$

In particular, by Theorem 7.6.1,

$$T_\gamma(T_{\gamma^{-1}}(z)) = T_I(z) = z.$$

Thus,

$$T_\gamma^{-1} = T_{\gamma^{-1}}.$$

Let us verify this directly. To find T_γ^{-1}, let us set $w = T_\gamma(z)$ and isolate z:

$$w = \frac{az + b}{cz + d}$$

---

[2]See Appendix A for a short tutorial on matrices.

$$(cz + d)w = az + b$$
$$czw - az = b - dw$$
$$z = \frac{dw - b}{-cw + a}.$$

Thus, T_γ^{-1} is represented by

$$\begin{bmatrix} d & -b \\ -c & a \end{bmatrix} \equiv \frac{1}{ad - bc} \begin{bmatrix} d & -b \\ -c & a \end{bmatrix} = \gamma^{-1}.$$

Note that we have used the requirement that $ad - bc \neq 0$.

Because of the similarity between matrices and fractional linear transformations, it is common to interchange the role of the two. In particular, it is common to write γz when $T_\gamma(z)$ is meant. We will adopt this convention, and write

$$\gamma z = \begin{bmatrix} a & b \\ c & d \end{bmatrix} z = \frac{az + b}{cz + d}.$$

In light of Theorem 7.6.1, this convention is reasonable, since

$$(\gamma_1 \gamma_2) z = \gamma_1 (\gamma_2 z),$$

so either interpretation is appropriate for $\gamma_1 \gamma_2 z$. This notation, though, is not without its flaws, and we must warn that in general, $\gamma(kz) \neq k\gamma z$, since

$$\gamma(kz) = \frac{akz + b}{ckz + d}$$

but

$$k\gamma z = (k\gamma)z = \gamma z = \frac{az + b}{cz + d}.$$

Let us now recall the notations for various sets of matrices:

$$M_{2 \times 2}(R) = \left\{ \begin{bmatrix} a & b \\ c & d \end{bmatrix} : a, b, c, d \in R \right\}$$
$$GL_2(R) = \{\gamma \in M_{2 \times 2}(R) : \det(\gamma) \neq 0\}$$
$$SL_2(R) = \{\gamma \in GL_2(R) : \det(\gamma) = 1\},$$

where R can be any of the complex numbers $\mathbb{C}$, the reals $\mathbb{R}$, the rationals $\mathbb{Q}$, or in the last case, even the integers $\mathbb{Z}$. The first set is the set of matrices, the second is called the *general linear group*, and the third is the *special linear group*.

There is another group, which is less well known, called the *projective special linear group* and denoted $PSL_2(R)$. This is the group $GL_2(R)$ modulo the equivalence relation mentioned above. The group $PSL_2(\mathbb{C})$ is isomorphic to the group of fractional linear transformations. Though this is more precisely the group we are studying, in the following we will stick with the more familiar groups $GL_2(R)$ and $SL_2(R)$.

As mentioned earlier, we introduced the group of fractional linear transformations so that we may describe the group of direct isometries on the upper half plane. We will show that any 2×2 matrix with real coefficients and determinant equal to one represents a fractional linear transformation which is an isometry of the Poincaré upper half plane. That is, the group $\mathrm{SL}_2(\mathbb{R})$, when viewed as a group of fractional linear transformations, is a group of isometries on the upper half plane.

Lemma 7.6.2. *The horizontal translation by a,*

$$T_a(x, y) = (x + a, y),$$

can be thought of as a fractional linear transformation which is represented by an element of $\mathrm{SL}_2(\mathbb{R})$.

Proof. As a map of complex numbers,

$$T_a(z) = z + a,$$

which is generated by

$$\tau_a = \begin{bmatrix} 1 & a \\ 0 & 1 \end{bmatrix}. \qquad \square$$

Lemma 7.6.3. *The map*

$$\varphi(x, y) = \left(\frac{-x}{x^2 + y^2}, \frac{y}{x^2 + y^2} \right),$$

which is inversion in the unit circle followed by reflection through $x = 0$, *can be thought of as a fractional linear transformation which is represented by an element of* $\mathrm{SL}_2(\mathbb{R})$.

Proof. As a function of complex numbers, the map φ is just

$$\varphi(z) = \varphi(x + iy) = \frac{-x + iy}{x^2 + y^2} = \frac{-(x - iy)}{(x + iy)(x - iy)} = \frac{-1}{z}.$$

This map is generated by

$$\sigma = \begin{bmatrix} 0 & -1 \\ 1 & 0 \end{bmatrix}. \qquad \square$$

Theorem 7.6.4. *The group* $\mathrm{SL}_2(\mathbb{R})$ *is generated by* σ *and the maps* τ_a *as a ranges over* $\mathbb{R}$.

Proof. Our proof is constructive. Note that

$$\sigma \tau_r = \begin{bmatrix} 0 & -1 \\ 1 & 0 \end{bmatrix} \begin{bmatrix} 1 & r \\ 0 & 1 \end{bmatrix}$$

$$= \begin{bmatrix} 0 & -1 \\ 1 & r \end{bmatrix}$$

so

$$\sigma\tau_s\sigma\tau_r = \begin{bmatrix} 0 & -1 \\ 1 & s \end{bmatrix} \begin{bmatrix} 0 & -1 \\ 1 & r \end{bmatrix}$$
$$= \begin{bmatrix} -1 & -r \\ s & rs-1 \end{bmatrix}$$

and

$$\sigma\tau_t\sigma\tau_s\sigma\tau_r = \begin{bmatrix} 0 & -1 \\ 1 & t \end{bmatrix} \begin{bmatrix} -1 & -r \\ s & rs-1 \end{bmatrix}$$
$$= \begin{bmatrix} -s & 1-rs \\ st-1 & rst-r-t \end{bmatrix}.$$

Thus, if

$$\gamma = \begin{bmatrix} a & b \\ c & d \end{bmatrix} \in \mathrm{SL}_2(\mathbb{R})$$

and $a \neq 0$, then set $s = -a$, solve $b = 1-rs = 1+ra$ and $c = st-1 = -at-1$, which gives

$$r = \frac{b-1}{a} \qquad \text{and} \qquad t = \frac{-1-c}{a}.$$

Note that since $\det \gamma = 1$, this forces $d = rst - r - t$. Thus, if $a \neq 0$, then γ can be written as a product involving only σ and translations. If $a = 0$, then $c \neq 0$, since $ad - bc = 1$, and hence

$$\sigma\gamma = \begin{bmatrix} -c & -d \\ a & b \end{bmatrix},$$

which can be written as a suitable product. Thus, $\mathrm{SL}_2(\mathbb{R})$ is generated by the translations and σ. $\qquad\square$

Corollary 7.6.5. *The group $\mathrm{SL}_2(\mathbb{R})$, when thought of as a group of fractional linear transformations, is a subgroup of isometries of the Poincaré upper half plane.*

Corollary 7.6.6. *If $\gamma \in \mathrm{GL}_2(\mathbb{R})$ and $\det \gamma > 0$, then γ is an isometry of the Poincaré upper half plane.*

Proof. Note that

$$\frac{1}{\sqrt{\det \gamma}}\gamma \in \mathrm{SL}_2(\mathbb{R}),$$

and

$$(k\gamma)z = \gamma z. \qquad\square$$

Theorem 7.6.7. *The image of a circle or line in $\mathbb{C}$ under the action of a fractional linear transformation $\gamma \in \mathrm{SL}_2(\mathbb{C})$ is again a circle or line.*

Proof. Note that in the proof of Theorem 7.6.4, the numbers r, s, and t need not be real. Thus, we can solve for a, b, c, and d, even if these numbers are complex. Now note that both σ and τ_a (where a can now be complex) send lines and circles to lines and circles, so every element of $SL_2(\mathbb{C})$ sends lines and circles to lines and circles. $\qquad\square$

Exercise 7.32. Prove Theorem 7.6.1. That is, prove that if γ_1 and γ_2 are two invertible 2×2 matrices, then

$$T_{\gamma_1 \gamma_2}(z) = T_{\gamma_1}(T_{\gamma_2}(z)),$$

or equivalently, that

$$(\gamma_1 \gamma_2)z = \gamma_1(\gamma_2 z).$$

Exercise 7.33. Prove that if $\gamma \equiv \gamma'$, then

$$\gamma' = k\gamma$$

for some scalar k.

Exercise 7.34. In the upper half plane model $\mathcal{H}$, carefully draw the asymptotic triangle with vertices i, $1 + i$, and 1. Is the map

$$\gamma = \begin{bmatrix} 1 & -1 \\ 1 & 0 \end{bmatrix}$$

an isometry of $\mathcal{H}$? In the same diagram, carefully draw the image of the asymptotic triangle under the action of γ.

Exercise 7.35. In the Poincaré upper half plane $\mathcal{H}$, carefully draw the triangle with vertices i, $-1 + i$, and $1 + i$. In the same diagram, carefully draw the image of this triangle under the isometry

$$\gamma = \begin{bmatrix} 2 & 1 \\ 1 & 1 \end{bmatrix}.$$

[S]

Exercise 7.36. Let $P = \dfrac{8 + i}{13}$, $Q = \dfrac{13 + i}{20}$, and $\gamma = \begin{bmatrix} 2 & -1 \\ -3 & 2 \end{bmatrix}$. What are γP and γQ? Sketch P, Q, and their images. Is γ an isometry? Why? Use all this information to find the distance between P and Q in $\mathcal{H}$.

Exercise 7.37. Let $P = 2 + 4i$ and $Q = \dfrac{6 + 4i}{3}$ be two points in the Poincaré upper half plane. Let

$$\gamma = \begin{bmatrix} 1 & 2 \\ -1 & 2 \end{bmatrix}.$$

What are γP and γQ? What is the Poincaré distance from P to Q?   [S]

Exercise 7.38†. Suppose T is a fractional linear transformation such that $T(1) = 1$, $T(0) = 0$, and $T(\infty) = \infty$. Prove that T is the identity map. That is, show that $T(z) = z$ for all z.

Exercise 7.39. Show that the dilation $\delta_\lambda(z) = \lambda z$ is an isometry of $\mathcal{H}$. Find an isometry which sends $a + ib$ to i by composing a dilation δ_λ and a horizontal translation τ_r for some λ and r.

Exercise 7.40. Write the dilation $\delta_\lambda(x, y) = (\lambda x, \lambda y)$ as a product of σ's and τ_a's.

Exercise 7.41. Suppose $\gamma \in \mathrm{SL}_2(\mathbb{R})$. Show that γ is a direct isometry. Hint: Show that τ_a and σ are direct isometries.

7.7 The Cross Ratio

An important tool when working with fractional linear transformations is the cross ratio.

Definition 16. *Cross ratio.* Let a, b, c, and d be four elements of $\mathbb{C} \cup \{\infty\}$, at least three of which are distinct. We define the *cross ratio* of a, b, c, and d to be
$$(a, b; c, d) = \frac{a - c}{a - d} \Big/ \frac{b - c}{b - d}.$$

The algebra for the element ∞ and for division by zero is the same as for fractional linear transformations.

If we fix three distinct elements a, b, and $c \in \mathbb{C} \cup \{\infty\}$, and think of the fourth element as a variable z, then we get a fractional linear transformation:
$$T(z) = (z, a; b, c) = \frac{z - b}{z - c} \Big/ \frac{a - b}{a - c}.$$

This is the unique fractional linear transformation T with the property that $T(a) = 1$, $T(b) = 0$, and $T(c) = \infty$ (see Exercise 7.44). This observation is the motivation for defining the cross ratio. It can also be used to solve some very general problems, as we will see in the following exercises.

Exercise 7.42. Find the fractional linear transformation which sends 1 to 1, $-i$ to 0, and -1 to ∞.

Solution. Set
$$w = (z, 1; -i, -1)$$
$$= \frac{z + i}{z + 1} \Big/ \frac{1 + i}{1 + 1}$$
$$= \frac{2z + 2i}{(1 + i)(z + 1)},$$

or in the matrix notation,

$$w = \begin{bmatrix} 2 & 2i \\ 1+i & 1+i \end{bmatrix} z. \qquad \square$$

Exercise 7.43. Find the fractional linear transformation which fixes i, sends ∞ to 3, and 0 to $-1/3$.

Solution. Let

$$\gamma_1 z = (z, i; \infty, 0)$$

and

$$\gamma_2 w = (w, i; 3, -1/3).$$

Then, $\gamma_1(i) = 1$, $\gamma_1(\infty) = 0$, and $\gamma_1(0) = \infty$. Also, $\gamma_2(i) = 1$, $\gamma_2(3) = 0$, and $\gamma_2(-1/3) = \infty$, so in particular, $\gamma_2^{-1}(1) = i$, $\gamma_2^{-1}(0) = 3$, and $\gamma_2^{-1}(\infty) = -1/3$. Hence, if we compose these two functions, we get

$$\gamma = \gamma_2^{-1}\gamma_1$$

with the property that $\gamma(i) = i$, $\gamma(\infty) = 3$, and $\gamma(0) = -1/3$, as desired.

Let us set $w = \gamma(z)$. Then

$$w = \gamma_2^{-1}\gamma_1(z)$$
$$\gamma_2(w) = \gamma_1(z)$$
$$(w, i; 3, -1/3) = (z, i; \infty, 0).$$

Solving, we get

$$\frac{w-3}{w+1/3} \Big/ \frac{i-3}{i+1/3} = \frac{z-\infty}{z-0} \Big/ \frac{i-\infty}{i-0}$$

$$\frac{(3i+1)w - 9i - 3}{3(i-3)w + i - 3} = \frac{i}{z}$$

$$z = \frac{(-9i-3)w - 1 - 3i}{(3i+1)w - 9i - 3}$$

$$= \frac{-3(3i+1)w - (3i+1)}{(3i+1)w - 3(3i+1)}$$

$$= \frac{3w+1}{-w+3}$$

$$= \begin{bmatrix} 3 & 1 \\ -1 & 3 \end{bmatrix} w.$$

Thus,

$$w = \begin{bmatrix} 3 & -1 \\ 1 & 3 \end{bmatrix} z.$$

In the last step, we used that

$$\begin{bmatrix} 3 & 1 \\ -1 & 3 \end{bmatrix}^{-1} = \frac{1}{10}\begin{bmatrix} 3 & -1 \\ 1 & 3 \end{bmatrix} \equiv \begin{bmatrix} 3 & -1 \\ 1 & 3 \end{bmatrix}.$$

Note that the entries of γ are all real, so this fractional linear transformation is in fact an isometry of the upper half plane. $\qquad\square$

Exercise 7.44. Suppose a, b, and c are distinct elements of $\mathbb{C} \cup \{\infty\}$, and that T is a fractional linear transformation such that $T(a) = 1$, $T(b) = 0$, and $T(c) = \infty$. Prove that T is unique. [H]

Exercise 7.45†. There is an important result hidden in the solution to Exercise 7.43. Suppose γ is a fractional linear transformation and that $\gamma a = a'$, $\gamma b = b'$, and $\gamma c = c'$. Prove that

$$(z, a; b, c) = (\gamma z, a'; b', c').$$

Exercise 7.46†. Find the fractional linear transformation which sends 0 to $-i$, 1 to 1, and ∞ to i. What is the image of $\mathcal{H}$ under this map? If l is a vertical line or half circle perpendicular to the real axis, what is its image under this map? (The image of $\mathcal{H}$ under this map is the Poincaré disc model $\mathcal{D}$ of hyperbolic geometry introduced as our 'crutch' in Chapter 6.) [S]

Exercise 7.47. The map $r(z) = e^{i\theta}z$ rotates the complex plane an angle θ about the origin. Compose this map with the inverse of the map found in Exercise 7.46 to get a map of $\mathcal{H}$. Is this new map an isometry?

Exercise 7.48. A circle centered at P and with radius r is the set of points a distance r away from P. Prove that a circle in the Poincaré model is a (Euclidean) circle which lies entirely within the model. [H]

Exercise 7.49. Suppose $w = \gamma z$ where $\gamma \in \mathrm{SL}_2(\mathbb{R})$. Suppose also that

$$(z, A; B, C) = (w, B; C, A)$$

for all $z \in \mathcal{H}$ and some distinct points A, B, and $C \in \mathcal{H}$. What can be said about $\triangle ABC$? What can be said about γ? [A]

Exercise 7.50. Let $(a, b; c, d) = \lambda$. Show that $(a, b; d, c) = 1/\lambda$ and $(a, c; b, d) = 1 - \lambda$. Use this to find $(a, x; y, z)$ for the other three permutations (x, y, z) of $\{b, c, d\}$.

Exercise 7.51. Suppose $\lambda = (a, b; c, d) = (a, c; d, b)$. What is λ?

7.8 Translations

If this model is to model hyperbolic space, then our sixth axiom must hold:

6. Given any two points P and Q, there exists an isometry f such that $f(P) = Q$.

So let $P = a + ib$ and $Q = c + id$. There are many choices for such an isometry. Let us choose the isometry f which also fixes ∞. Since $f(\infty) = \infty$ and $f(P) = Q$, the line through P and ∞ must be sent to the line through Q and ∞. Thus, the vertical line at a is sent to the vertical line at c. In particular, $f(a) = c$. Thus, we must have

$$(w, c + id; c, \infty) = (z, a + ib; a, \infty)$$
$$\frac{w - c}{id} = \frac{z - a}{ib}$$
$$w = \frac{d(z - a)}{b} + c$$
$$= \begin{bmatrix} d & bc - ad \\ 0 & b \end{bmatrix} z.$$

Since this matrix has real coefficients and its determinant is positive (since $b > 0$ and $d > 0$), it is an isometry of the Poincaré upper half plane.

This map is in fact a translation. Recall that translations are direct isometries which have no fixed points (and rotations are direct isometries which have exactly one fixed point). In Exercise 7.41, we showed that every element of $\mathrm{SL}_2(\mathbb{R})$ is a direct isometry. Thus, to show that this map is a translation, we must show that it has no fixed points. A map γ has a fixed point if there exists a point z_0 such that $\gamma z_0 = z_0$. In this case, if we solve for z, we get

$$\frac{dz_0 + bc - ad}{b} = z_0$$
$$z_0 = \frac{ad - bc}{d - b}.$$

But note that a, b, c, and d are all real so if $b \neq d$, then z_0 is real and hence is not in the upper half plane. Thus, this map has no fixed points in $\mathcal{H}$ and hence is a translation. If $b = d$, then $z_0 = \infty$, and again, there are no solutions in the upper half plane, so this map is a translation.

In this geometry, we classify translations depending on how many fixed points there are on the line at infinity (that is, in $\mathbb{R} \cup \infty$.) Let

$$\gamma = \begin{bmatrix} a & b \\ c & d \end{bmatrix}.$$

Then $\gamma(z) = z$ if

$$cz^2 + (d - a)z - b = 0.$$

If $c \neq 0$, then this is a quadratic with discriminant

$$\Delta = (d - a)^2 + 4bc.$$

Hence, γ has a fixed point in $\mathcal{H}$ if $\Delta < 0$, and no fixed points if $\Delta \geq 0$. If $\Delta = 0$, then γ has exactly one fixed point on the line at infinity. We call such a translation a *parabolic translation*. If $\Delta > 0$, then we call γ a *hyperbolic translation*.

Exercise 7.52. Suppose $\gamma = \begin{bmatrix} a & b \\ c & d \end{bmatrix} \in \mathrm{SL}_2(\mathbb{R})$ and $c = 0$. Under what condition(s) is γ a parabolic translation?

Exercise 7.53. Suppose $\gamma \in \mathrm{SL}_2(\mathbb{R})$ and $\gamma \infty = \infty$. Prove that γ is a translation.

Exercise 7.54. Is horizontal translation by a (the map τ_a) a parabolic or hyperbolic translation?

Exercise 7.55. Prove that a map $\gamma \in \mathrm{SL}_2(\mathbb{R})$ is either the identity on $\mathcal{H}$ or has at most one fixed point in $\mathcal{H}$. Thus, every element in $\mathrm{SL}_2(\mathbb{R})$ is either a rotation or translation.

Exercise 7.56. Suppose γ is a hyperbolic translation. Show that there exists exactly one line l in $\mathcal{H}$ such that for any $P \in l$, we have $\gamma P \in l$. Show that no such line exists if γ is a parabolic translation. What is the corresponding statement for translations in Euclidean geometry?

Exercise 7.57. Prove that for any two distinct points P and Q in $\mathcal{H}$, there exists an infinite number of translations f such that $f(P) = Q$. Contrast this with Exercise 1.20.

Exercise 7.58. Find a map $\gamma \in \mathrm{SL}_2(\mathbb{R})$ which sends $1 + i$ to i and ∞ to 1. Does this map have a fixed point?

Solution. Let γ be the isometry which we are looking for. We do not seem to have enough information, since we only know the image of two points. However, since γ is an isometry, it also has some special properties. In particular, it sends lines to lines (that is, lines in $\mathcal{H}$).

Let us look at the line through $i + 1$ and ∞ (see Figure 7.12). This line goes through 1, too. The image of this line includes the point i and 1, so it is the line through i and 1. This is the (Euclidean) half circle perpendicular to the real axis that goes through 1 and i. This circle also goes through -1. Thus, the last piece of information we need is that 1 is sent to -1. Thus,

$$(z, i + 1; 1, \infty) = (w, i; -1, 1)$$

$$\frac{z - 1}{z - \infty} \Big/ \frac{i + 1 - 1}{i + 1 - \infty} = \frac{w + 1}{w - 1} \Big/ \frac{i + 1}{i - 1}$$

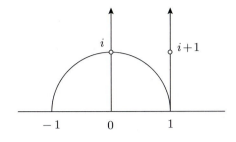

Figure 7.12

$$\left(\frac{z-1}{i}\right)\left(\frac{i+1}{i-1}\right) = \frac{w+1}{w-1}$$

$$\begin{bmatrix} -1 & 1 \\ 0 & 1 \end{bmatrix} z = \begin{bmatrix} 1 & 1 \\ 1 & -1 \end{bmatrix} w$$

$$\begin{bmatrix} -1 & -1 \\ -1 & 1 \end{bmatrix}\begin{bmatrix} -1 & 1 \\ 0 & 1 \end{bmatrix} z = w$$

$$w = \begin{bmatrix} 1 & -2 \\ 1 & 0 \end{bmatrix} z.$$

Thus, we should choose $\gamma = \begin{bmatrix} 1 & -2 \\ 1 & 0 \end{bmatrix}$, except that this matrix does not have determinant equal to 1, so is not in $\mathrm{SL}_2(\mathbb{R})$. Since two matrices which are multiples of each other give the same fractional linear transformation, we merely divide through by the square root of the determinant:

$$\gamma = \frac{1}{\sqrt{2}}\begin{bmatrix} 1 & -2 \\ 1 & 0 \end{bmatrix} = \begin{bmatrix} 1/\sqrt{2} & -\sqrt{2} \\ 1/\sqrt{2} & 0 \end{bmatrix}.$$

To see whether γ has a fixed point, we solve

$$\gamma z = z$$
$$\frac{z-2}{z} = z$$
$$0 = z^2 - z + 2$$
$$z = \frac{1 \pm i\sqrt{7}}{2}.$$

Since the point $z = \frac{1+i\sqrt{7}}{2}$ is in $\mathcal{H}$, this map has a fixed point. Thus, γ is a rotation. $\square$

Exercise 7.59. Find the isometry of γ of $\mathcal{H}$ in $\mathrm{GL}_2(\mathbb{R})$ which sends $2i$ to $3i + 4$ and 2 to -5. Does this isometry have any fixed points in $\mathcal{H}$?

Exercise 7.60. Find an isometry $\gamma \in \mathrm{GL}_2(\mathbb{R})$ of $\mathcal{H}$ which sends $2i$ to $3i$ and ∞ to -1.   [S]

7.9 Rotations

We can similarly find the fractional linear transformation which fixes $P = a+ib$ and rotates counter clockwise through an angle θ. To do this, first find the (Euclidean) line through P which makes an angle θ with the vertical line. Find the perpendicular to this line, and find where it intersects the x-axis. The circle centered at this intersection and through P is the image of the vertical line under the rotation. Let this circle intersect the x-axis at M and N. Then the rotation is given by

$$(w, P; M, N) = (z, P; a, \infty).$$

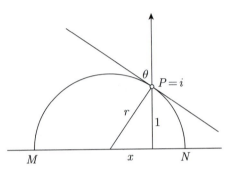

Figure 7.13

Let us do this in the case when $P = i$, as in Figure 7.13. Let the center of the half circle be $-x$, and let the radius of the circle be r. Note that these are Euclidean measures, and not lengths in the Poincaré model. Then $r \cos\theta = x$, $r \sin\theta = 1$, $M = -r - x$, and $N = r - x$.

After much algebraic manipulation, we find $w = \rho_\theta z$ where

$$\rho_\theta = \begin{bmatrix} \cos\frac{\theta}{2} & \sin\frac{\theta}{2} \\ -\sin\frac{\theta}{2} & \cos\frac{\theta}{2} \end{bmatrix}.$$

Exercise 7.61. Fill in the algebraic manipulation that establishes the above result.

Exercise 7.62. Use the above result to find the map which rotates an angle θ about a point P. (Hint: Translate P to i, then do the rotation, and translate back.)

Exercise 7.63. Find an isometry which fixes $1 + i$ and sends 1 to 2.   [S]

Exercise 7.64. Find an isometry in $GL_2(\mathbb{R})$ which sends $i + 1$ to i and ∞ to 1. Is this isometry a rotation or translation?   [A]

Exercise 7.65. The map

$$\gamma = \begin{bmatrix} 1 & -5 \\ 1 & 3 \end{bmatrix}$$

is a rotation of $\mathcal{H}$. What is the center of this rotation?

Exercise 7.66. Find an isometry $\gamma \in \text{GL}_2(\mathbb{R})$ which fixes i and sends ∞ to 1. The isometry γ fixes i and hence is a rotation about i through some angle. What is that angle?

Exercise 7.67. Recall that all triply asymptotic triangles are congruent. Find three distinct isometries in $\text{GL}_2(\mathbb{R})$ which send the triply asymptotic triangle with vertices 1, 2, and 3 to the triply asymptotic triangle with vertices 0, -1, and ∞. Remember, for an element of $\text{GL}_2(\mathbb{R})$ to be an isometry, it must have a positive determinant. Identify which of these isometries are rotations and which are translations.

7.10 Reflections

The set of isometries which do not preserve orientation in the Poincaré upper half plane model $\mathcal{H}$ also have a very nice description. Recall that the reflection through the imaginary axis is given by

$$R_0(x, y) = (-x, y),$$

which can be expressed using complex coordinates as

$$R_0(z) = -\overline{z}.$$

Here, the bar indicates the complex conjugate of z.

Note that we can write

$$R_0(z) = \mu \overline{z},$$

where

$$\mu = \begin{bmatrix} -1 & 0 \\ 0 & 1 \end{bmatrix}.$$

The reflection through the line l in $\mathcal{H}$ can be found by first moving the line l to the imaginary axis using an appropriate isometry γ_1, then reflecting through the imaginary axis, and moving the imaginary axis back to l. Thus, this reflection becomes

$$\gamma_1^{-1} \mu \overline{\gamma_1 z} = \gamma_1^{-1} \mu \gamma_1 \overline{z}.$$

Note that μ^2 is the identity, and that $\mu \gamma \mu \in \text{SL}_2(\mathbb{R})$ for every $\gamma \in \text{SL}_2(\mathbb{R})$, since $\det \mu = -1$. Thus,

$$\gamma_1^{-1} \mu \gamma_1 \overline{z} = \gamma_1^{-1} (\mu \gamma_1 \mu) \mu \overline{z} = \gamma_2 \mu \overline{z} = \gamma_2(-\overline{z}),$$

where $\gamma_2 \in \text{SL}_2(\mathbb{R})$. Thus, every reflection can be written in the form $\gamma(-\overline{z})$ for some $\gamma \in \text{SL}_2(\mathbb{R})$. In fact, we can prove the following:

Theorem 7.10.1. *Every isometry f of $\mathcal{H}$ which is not direct can be written in the form*

$$f(z) = \gamma(-\bar{z})$$

for some $\gamma \in SL_2(\mathbb{R})$. Furthermore, if $\gamma = \begin{bmatrix} a & b \\ c & d \end{bmatrix}$, then $f(z)$ is a reflection if and only if $a = d$.

Exercise 7.68. Find the image of i under reflection through the line l whose endpoints are 1 and 3.

Solution. Let the reflection be

$$f(z) = \gamma(-\bar{z}) = \begin{bmatrix} a & b \\ c & a \end{bmatrix}(-\bar{z}).$$

We will solve for $\gamma \in GL_2(\mathbb{R})$ with positive determinant (instead of in $SL_2(\mathbb{R})$). Since the endpoints of l are 1 and 3, we get the following two equations: From $f(1) = 1$, we get

$$\begin{bmatrix} a & b \\ c & a \end{bmatrix}(-1) = 1$$

$$\frac{-a+b}{-c+a} = 1$$

$$-a+b = -c+a$$

$$0 = 2a - b - c;$$

and from $f(3) = 3$, we similarly get

$$0 = 6a - b - 9c.$$

Subtracting the first equation from the second, we get

$$4a - 8c = 0,$$

so $a = 2c$. Plugging this into the first equation, we get

$$0 = 4c - b - c,$$

so $b = 3c$. Let us set $c = 1$, so $a = 2$ and $b = 3$. Thus,

$$\gamma = \begin{bmatrix} 2 & 3 \\ 1 & 2 \end{bmatrix}.$$

It is a fluke that $\det \gamma = 1$. We now solve for $f(i)$:

$$f(i) = \begin{bmatrix} 2 & 3 \\ 1 & 2 \end{bmatrix}(-\bar{i}) = \frac{2i+3}{i+2} = \left(\frac{2i+3}{2+i}\right)\left(\frac{2-i}{2-i}\right) = \frac{8+i}{5}. \qquad \square$$

Exercise 7.69. Find the reflection of $1+i$ through the line l with endpoints 2 and 5.

Exercise 7.70. Find a formula for the reflection through the line l with endpoints -1 and 1.

Exercise 7.71. Find a formula for the reflection through the line l which goes through $3i$ and $1+4i$.

Exercise 7.72. Prove Theorem 7.10.1.

7.11   Lengths

We can also use the cross ratio to establish a formula for the distance between two points or the length of a line segment. First, let us suppose two points $P = a + ib$ and $Q = a + ic$ lie on the same vertical line. Then, the distance between P and Q is given by

$$|PQ| = \left| \int_b^c \frac{dy}{y} \right|$$
$$= |\ln(c/b)| .$$

Now, suppose P and Q are two points in $\mathcal{H}$ which do not lie on a vertical line. Then there exists a half circle with center on the x-axis which goes through both P and Q. Let this half circle have endpoints M and N. Since isometries preserve distances, let us consider the image of P and Q under the isometry σ which sends P to i and the line PQ to a vertical line. That is, the map which sends P to i, M to 0, and N to ∞. Note that the image of Q under this map lies on this line, so $\sigma(Q) = 0 + ic$ for some c, and that

$$|PQ| = |\ln(c/1)| = |\ln c|.$$

Finally, note that
$$(\sigma z, i; 0, \infty) = (z, P; M, N)$$
and in particular, since $\sigma Q = ic$ and $(\sigma z, i; 0, \infty) = \dfrac{\sigma z}{i}$, we get

$$c = (Q, P; M, N),$$

so
$$|PQ| = |\ln(Q, P; M, N)|.$$

Exercise 7.73. Show that $|PQ|$ is independent of which endpoint we call M and which we call N. That is, show that

$$|\ln(Q, P; M, N)| = |\ln(Q, P; N, M)|.$$

Show also that $|PQ| = |QP|$.

Exercise 7.74. Find the Poincaré distance $|PQ|$ between $P = 4 + 4i$ and $Q = 5 + 3i$.

Solution. In order to find the Poincaré length $|PQ|$, we must first find M and N. We do this by finding the center x of the (Euclidean) circle which describes the line PQ, as in Figure 7.14. Note that x lies on the real axis

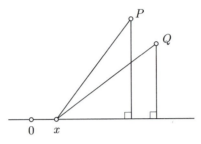

Figure 7.14

(since the circle is perpendicular to the real axis), so x is real. Also, the Euclidean distance between x and P is the same as that between x and Q. Thus,

$$(4 - x)^2 + 4^2 = (5 - x)^2 + 3^2$$
$$x^2 - 8x + 16 + 16 = x^2 - 10x + 25 + 9$$
$$2x = 2.$$

Hence, the center is at $x = 1$, and the radius of the circle is

$$r = \sqrt{(4 - 1)^2 + 4^2} = 5,$$

so M and N are 1 ± 5. Thus,

$$|PQ| = |\ln(4 + 4i, 5 + 3i, -4, 6)|$$
$$= \left| \ln \left(\frac{4 + 4i + 4}{4 + 4i - 6} \Big/ \frac{5 + 3i + 4}{5 + 3i - 6} \right) \right|$$
$$= \left| \ln \left(\frac{(8 + 4i)(-1 + 3i)}{(-2 + 4i)(9 + 3i)} \right) \right|$$
$$= \left| \ln \left(\frac{4(2 + i)(-1 + 3i)}{2i(i + 2)(-3i)(3i - 1)} \right) \right|$$
$$= |\ln(4/6)| = \ln(3/2).$$

Note that the cross ratio is real and positive, as it must be. If this had not been the case, then we would know we had made an error in our calculations. $\square$

Exercise 7.75. Find the Poincaré distance between $P = 1 + 3i$ and $Q = 8 + 4i$. Write your answer in the form $\ln(a/b)$ where a and b are positive integers.

Exercise 7.76. Find the distance $|PQ|$ between $P = i$ and $Q = \dfrac{1 + i\sqrt{3}}{2}$ in the Poincaré upper half plane $\mathcal{H}$.

Exercise 7.77. Find the Poincaré distance between $12 + 5i$ and $5 + 12i$. Write your answer in the form $\ln(a/b)$ where a and b are integers.   [S]

7.12   The Axioms of Hyperbolic Geometry

Up to this point, we have been careful not to refer to the Poincaré upper half plane $\mathcal{H}$ as the hyperbolic plane. In this section, we will establish that $\mathcal{H}$ satisfies the axioms of hyperbolic geometry and hence is a model of hyperbolic geometry.

Recall, from Chapter 6, the axioms of hyperbolic geometry:

1. We can draw a unique line segment between any two points.

2. Any line segment can be continued indefinitely.

The Poincaré upper half plane $\mathcal{H}$ satisfies these two axioms, since there exists a half circle (or vertical line) through any two points in the plane. This line has infinite length in both directions.

3. A circle of any radius and any center can be drawn.

This is essentially by definition.

4. Any two right angles are congruent.

We saw that the isometries preserve the Euclidean angular measure in the Poincaré upper half plane, so let us define angular measure in $\mathcal{H}$ to be the same as the Euclidean angular measure. Then, any two right angles are congruent.

6. Given any two points P and Q, there exists an isometry f such that $f(P) = Q$.

7. Given a point P and any two points Q and R such that $|PQ| = |PR|$, there exists an isometry which fixes P and sends Q to R.

8. Given any line l, there exists a map which fixes every point in l and leaves no other point fixed.

We established these in Sections 7.8, 7.9 and 7.10.
Finally, the fifth postulate:

5. Given any line l and any point P not on l, there exist two distinct lines l_1 and l_2 through P which do not intersect l.

This is clear, since given any half circle perpendicular to the x-axis (the line l), and a point P not on l, there exist two parallel lines l_1 and l_2, as shown in Figure 7.15. Then any line through P and between l_1 and l_2 does not intersect l.

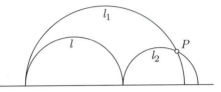

Figure 7.15. Two lines l_1 and l_2 which go through P and are parallel to l.

7.13 The Area of Triangles

In Chapter 6, we saw that the area of an asymptotic triangle is finite. We also saw that all triply asymptotic triangles are congruent. Hence, the area of a triply asymptotic triangle is a constant. What is that area in the Poincaré upper half plane model of hyperbolic geometry?

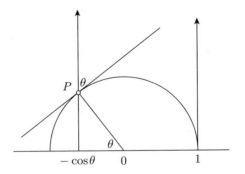

Figure 7.16

We evaluate this by first looking at the doubly asymptotic triangle with vertices at $P = e^{i(\pi-\theta)}$ in $\mathcal{H}$, and vertices at infinity of 1 and ∞, as in Figure 7.16. The angle at P for such a triangle has a measure of θ.

The area element for the Poincaré upper half plane model of hyperbolic space is derived by taking a small (Euclidean) rectangle with sides oriented horizontally and vertically. The sides approximate hyperbolic line segments, since the rectangle is very small. The area is therefore the product of the height and width (measured with the hyperbolic arclength element). The vertical sides of the rectangle have Euclidean length Δy, and since y is essentially unchanged, the hyperbolic length is $\dfrac{\Delta y}{y}$. The horizontal sides have Euclidean length Δx and hence hyperbolic length $\dfrac{\Delta x}{y}$. Thus, the area element is given by $\dfrac{dx\,dy}{y^2}$. We now use this to find the area of the triangle in Figure 7.16:

Theorem 7.13.1. *The area of a doubly asymptotic triangle $\triangle PMN$ with points M and N at infinity and with angle $\angle MPN = P$ has area*

$$|\triangle PMN| = \pi - P,$$

where the angle P is measured in radians.

Proof. Let the angle at P have measure θ. Then $\triangle PMN$ is similar to the doubly asymptotic triangle in Figure 7.16 and hence is congruent to it. But the area of that triangle is given by

$$A(\theta) = \int_{-\cos\theta}^{1} \int_{\sqrt{1-x^2}}^{\infty} \frac{dy\,dx}{y^2}$$

$$= \int_{-\cos\theta}^{1} \frac{dx}{\sqrt{1-x^2}}$$

$$= \pi - \theta. \qquad \square$$

As an immediate consequence, we get the following result:

Corollary 7.13.2. *The area of a triply asymptotic triangle is π.*

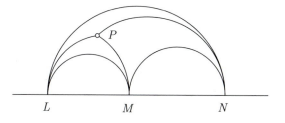

Figure 7.17

Proof. Let $\triangle LMN$ be a triply asymptotic triangle, and let P be a point in the interior, as in Figure 7.17. Then

$$|\triangle LMN| = |\triangle PLM| + |\triangle PMN| + |\triangle PNL|$$

$$= (\pi - \angle MPL) + (\pi - \angle MPN) + (\pi - \angle NPL)$$
$$= 3\pi - 2\pi. \qquad \square$$

Corollary 7.13.3. *Let $\triangle ABC$ be a triangle in $\mathcal{H}$, with angles A, B, and C. Then, the area of $\triangle ABC$ is*

$$|\triangle ABC| = \pi - A - B - C,$$

where the angles A, B, and C are measured in radians.

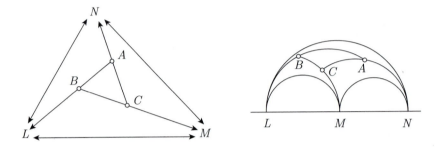

Figure 7.18. An arbitrary triangle in an abstract model (left) and in the Poincaré upper half plane (right).

Proof. Figure 7.18 depicts two arbitrary triangles $\triangle ABC$. One is an abstract picture like those we were drawing in Chapter 6, while the other is in the Poincaré model $\mathcal{H}$. Continue the edges of the triangle indefinitely as rays AB, BC, and CA. Let the points at infinity on these rays be, respectively, L, M, and N. Find the common parallels LM, MN, and NL. These lines form a triply asymptotic triangle, whose area is π. Thus,

$$|\triangle ABC| = \pi - |\triangle ALN| - |\triangle BLM| - |\triangle MCN|$$
$$= \pi - (\pi - (\pi - A)) - (\pi - (\pi - B)) - (\pi - (\pi - C))$$
$$= \pi - A - B - C. \qquad \square$$

Exercise 7.78. Consider the doubly asymptotic triangle $\triangle AMN$ in $\mathcal{H}$ where $A = \dfrac{8+i}{5}$, $M = \dfrac{5}{3}$, and $N = 2$. What is the image of $\triangle AMN$ under the isometry

$$\gamma = \begin{bmatrix} 2 & -3 \\ -1 & 2 \end{bmatrix}?$$

Use this to find the hyperbolic area of $\triangle AMN$. [S]

Exercise 7.79. Find the area in $\mathcal{H}$ of the doubly asymptotic triangle with vertices i, 1, and $1 + \sqrt{2}$.

Exercise 7.80. Draw the asymptotic triangle $\triangle ABM$ in $\mathcal{H}$ with $A = i$, $B = i\sqrt{3}$, and $M = 1$. What is the Poincaré length $|AB|$? What is the area of $\triangle ABM$?

7.14 The Poincaré Disc Model

In Poincaré disc model $\mathcal{D}$ of hyperbolic geometry is the crutch we introduced in Chapter 6. In Exercise 7.46, we found the map

$$\phi = \begin{bmatrix} 1 & -i \\ -i & 1 \end{bmatrix},$$

which sends the upper half plane $\mathcal{H}$ onto the unit disc. Under this map, lines and circles perpendicular to the real line are sent to circles which are perpendicular to the boundary of $\mathcal{D}$. Thus, hyperbolic lines in this model are the portions of Euclidean circles in $\mathcal{D}$ which are perpendicular to the boundary of $\mathcal{D}$.

When dealing with this model, we usually express points in polar coordinates. That is,

$$\mathcal{D} = \{re^{i\theta} : 0 \leq r < 1\}.$$

The arclength element is (see Exercise 7.81)

$$ds = \frac{2\sqrt{dr^2 + r^2 d\theta^2}}{1 - r^2}.$$

The group of proper isometries in $\mathcal{D}$ has a description similar to the description on $\mathcal{H}$. It is the group

$$\Gamma = \left\{ \gamma \in \mathrm{SL}_2(\mathbb{C}) : \gamma = \begin{bmatrix} a & b \\ \bar{b} & \bar{a} \end{bmatrix} \right\}.$$

All improper isometries of $\mathcal{D}$ can be written in the form

$$\gamma(-\bar{z}),$$

where $\gamma \in \Gamma$.

Exercise 7.81. Prove that the arclength element in $\mathcal{D}$ is

$$ds = \frac{2\sqrt{dr^2 + r^2 d\theta^2}}{1 - r^2}.$$

[H][S]

Exercise 7.82. Prove that Γ is the group of proper isometries of $\mathcal{D}$. [S]

Exercise 7.83. Characterize the set of reflections in $\mathcal{D}$. [H][A]

Exercise 7.84. What is the area element in $\mathcal{D}$? [A]

Exercise 7.85. Let P and Q be two points in $\mathcal{D}$ and let M and N be the endpoints of the line PQ. What is the formula for the length $|PQ|$ in terms of P, Q, M, and N. [A]

7.15 Circles and Horocycles

Lines, in the Poincaré upper half plane model, are lines and circles which are perpendicular to the real axis. It took a little bit of investigation before we came to that conclusion. We might also wonder, then, what do circles look like in this model? They are, in fact, circles which lie entirely in $\mathcal{H}$. To see this, we first note that a Euclidean circle centered at 0 in the Poincaré disc model is a hyperbolic circle.

So suppose that C is a hyperbolic circle in $\mathcal{H}$ centered at P. There exists an isometry $\gamma \in \mathrm{GL}_2(\mathbb{R})$ which sends P to i. After composing with ϕ, we have a circle centered at 0 in the disc model, which is a Euclidean circle. The composition of $\phi\gamma$ is in $\mathrm{GL}_2(\mathbb{C})$, and hence so is its inverse. But such maps send circles and lines to circles and lines. Thus, C must be either a Euclidean circle or line. Since this image cannot intersect ∞, it must be a circle. Thus, hyperbolic circles in $\mathcal{H}$ are Euclidean circles in $\mathcal{H}$.

In Euclidean geometry, as the radius of a circle goes to infinity, the circle itself looks more and more like a line. In the Poincaré upper half plane, the limiting case is a (Euclidean) circle tangent to the real axis, which we know is not a line (see Figure 7.19) . Such an object is called a *horocycle*.

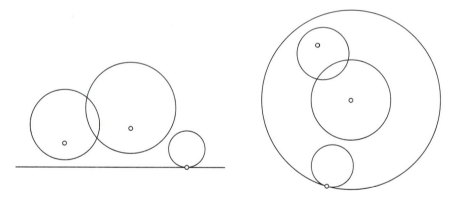

Figure 7.19. Circles and horocycles, together with their centers, in the Poincaré upper half plane and disc models of hyperbolic geometry.

One way of describing a circle centered at a point P is to call it the locus of a point under all rotations centered at P. Similarly, a horocycle can be thought of as the locus of a point under the action of all parabolic translations $\gamma \in \mathrm{SL}_2(\mathbb{Z})$ which fix a point P on the boundary of $\mathcal{H}$. The point P is the point of tangency and is thought of as the center of the horocycle. Poincaré lines through P are thought of as radii. Like the radii of a circle, they intersect the horocycle at right angles. With this interpretation, a parabolic translation exhibits properties similar to a rotation. We sometimes think of parabolic translations as being a boundary case between hyperbolic translations and rotations.

Exercise 7.86. In the upper half plane model, describe the horocycles centered at ∞. Justify your answer.

Exercise 7.87. Use Geometer's Sketchpad to draw a circle and several of its radii in $\mathcal{H}$.

Exercise 7.88*. Show that the area A of a circle of radius r in $\mathcal{H}$ is

$$A = 4\pi \sinh^2(r/2),$$

where $\sinh(x) = \frac{e^x - e^{-x}}{2}$ is the hyperbolic sine function. Note that, unlike triangles, the area of a hyperbolic circle can be arbitrarily large.

Exercise 7.89*. Use the result found in Exercise 7.88 to find an equation for the circumference of a circle in hyperbolic geometry. [H]

Exercise 7.90. The area of a triangle is at most π. The area of a circle is unbounded. Hence, there must be a largest circle that can be inscribed in a triangle. What is the diameter of this circle? [S]

Exercise 7.91. What is the diameter of the largest circle that can be inscribed in a quadrilateral?

Exercise 7.92. What is the diameter of the largest circle that can be inscribed in a pentagon?

Exercise 7.93. In Euclidean geometry, all triangles have a circumcircle. In hyperbolic geometry, this is not the case. Draw a (nonasymptotic) triangle $\triangle ABC$ in $\mathcal{H}$ which does not have a circumcircle.

7.16   Hyperbolic Trigonometry

We are probably all familiar with the hyperbolic trigonometric functions. These are

$$\cosh x = \frac{e^x + e^{-x}}{2},$$

$$\sinh x = \frac{e^x - e^{-x}}{2},$$

$$\tanh x = \frac{\sinh x}{\cosh x} = \frac{e^x - e^{-x}}{e^x + e^{-x}},$$

and so on. We may also be familiar with some identities, such as

$$\cosh^2 x - \sinh^2 x = 1$$

$$\cosh(x + y) = \cosh x \cosh y + \sinh x \sinh y.$$

Hyperbolic trigonometric functions are very similar to the usual trigonometric functions, and are in fact related by Euler's formula:

$$\cos\theta = \frac{e^{i\theta} + e^{-i\theta}}{2} = \cosh i\theta,$$

$$\sin\theta = \frac{e^{i\theta} - e^{-i\theta}}{2i} = -i\sinh i\theta.$$

The hyperbolic trig functions have useful applications in mathematics, engineering, and physics, but these functions alone are not what hyperbolic trigonometry is about. Hyperbolic trigonometry is the geometry of triangles in hyperbolic geometry.

Let us begin our investigation with the geometry of right triangles. By side-angle-side, all triangles $\triangle ABC$ with right angle at C and sides of length $|AC| = b$ and $|BC| = a$ are congruent. Thus, we should be able to find the angles A and B, and the length c of the side AB. In Euclidean geometry, we find these values using the Pythagorean theorem and the trigonometric functions. In hyperbolic geometry, we have similar results.

Theorem 7.16.1 (The Hyperbolic Pythagorean Theorem). *Let $\triangle ABC$ be a right angle triangle in hyperbolic geometry with right angle at C and sides of length a, b, and c opposite points A, B, and C. Then*

$$\cosh c = \cosh a \cosh b.$$

Before we prove this theorem, let us establish a model right triangle which will facilitate the proofs of the following two theorems.

We first recall that a point $P = ie^p$ in $\mathcal{H}$ is a distance p away from i. Recall too that

$$\phi = \begin{bmatrix} 1 & -i \\ -i & 1 \end{bmatrix}$$

sends $\mathcal{H}$ to $\mathcal{D}$. Note that $\phi(i) = 0$, and that

$$\phi(ie^p) = \frac{ie^p - i}{e^p + 1} = i\tanh(p/2).$$

Thus, a point which is a hyperbolic distance p away from zero in $\mathcal{D}$ is a Euclidean distance $\tanh(p/2)$ away from zero.

Hence, without loss of generality, we may choose our right triangle $\triangle ABC$ with right angle at C and sides of length $|AC| = b$ and $|BC| = a$ to be the triangle in the Poincaré disc model $\mathcal{D}$ with vertices $C = 0$, $A = \tanh(b/2)$, and $B = i\tanh(a/2)$. A better choice though is one with A at zero (see Figure 7.20). So let us find the proper isometry γ of $\mathcal{D}$ which sends A to 0 and the line AC to itself. That is, let $\gamma(A) = 0$, $\gamma(1) = 1$ and $\gamma(-1) = -1$. Then,

$$(z, A; 1, -1) = (\gamma z, 0; 1, -1)$$

from which we get

$$\gamma = \begin{bmatrix} -1 & A \\ A & -1 \end{bmatrix}.$$

Applying this to B, we get

$$\gamma(B) = \frac{-B+A}{AB-1} = \frac{-i\tanh(a/2)+\tanh(b/2)}{i\tanh(a/2)\tanh(b/2)-1}.$$

So, in the following, let us use the triangle $\triangle ABC$ in $\mathcal{D}$ with

$$A = 0$$
$$B = \frac{-i\tanh(a/2)+\tanh(b/2)}{i\tanh(a/2)\tanh(b/2)-1}$$
$$C = -\tanh(b/2).$$

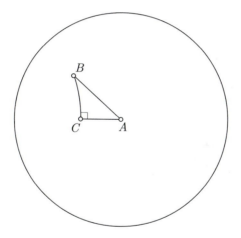

Figure 7.20. A well chosen right angle triangle in the Poincaré disc.

Proof of the hyperbolic Pythagorean theorem. Let $\triangle ABC$ be the right triangle described above. Since B is a distance $\tanh(c/2)$ away from zero, we get

$$\tanh(c/2) = \left| \frac{-B+A}{AB-1} \right|$$
$$\tanh^2(c/2) = \frac{\tanh^2(a/2)+\tanh^2(b/2)}{\tanh^2(a/2)\tanh^2(b/2)+1}.$$

Note that $\operatorname{sech}^2 x = 1 - \tanh^2 x$ (see Exercise 7.94), so

$$\operatorname{sech}^2(c/2) = 1 - \tanh^2(c/2)$$

$$= \frac{\tanh^2(a/2)\tanh^2(b/2) - \tanh^2(a/2) - \tanh^2(b/2) + 1}{\tanh^2(a/2)\tanh^2(b/2) + 1}$$

$$= \frac{(\tanh^2(a/2) - 1)(\tanh^2(b/2) - 1)}{\tanh^2(a/2)\tanh^2(b/2) + 1}$$

$$\cosh^2(c/2) = (\tanh^2(a/2)\tanh^2(b/2) + 1)\cosh^2(a/2)\cosh^2(b/2)$$

$$= \sinh^2(a/2)\sinh^2(b/2) + \cosh^2(a/2)\cosh^2(b/2)$$

$$= 2\cosh^2(a/2)\cosh^2(b/2) - \cosh^2(a/2) - \cosh^2(b/2) + 1.$$

Note also that $\cosh 2x = 2\cosh^2 x - 1$ (see Exercise 7.95), so

$$\cosh c = 2\cosh^2(c/2) - 1$$

$$= 4\cosh^2(a/2)\cosh^2(b/2) - 2\cosh^2(a/2) - 2\cosh^2(b/2) + 1$$

$$= (2\cosh^2(a/2) - 1)(2\cosh^2(b/2) - 1)$$

$$= \cosh a \cosh b. \qquad \square$$

To find the angles A and B, we use the following analogue of the definitions of the sine and cosine functions:

Theorem 7.16.2. *Let $\triangle ABC$ be a right angle triangle in hyperbolic geometry with right angle at C and sides of length a, b, and c opposite angles A, B, and C. Then*

$$\sin A = \frac{\sinh a}{\sinh c} \qquad and \qquad \cos A = \frac{\cosh a \sinh b}{\sinh c}.$$

Unfortunately, our choice of notation has now caught up with us. We have used and will use the notation A, B, and C to represent both the points A, B, and C and the angles A, B, and C. I hope that this does not cause too much confusion, and that the intended use will be obvious from the context.

Proof. We again use the triangle described above. Since A is at zero, the hyperbolic lines AB and AC are Euclidean lines. Thus, we can use regular trigonometry to find the angle at A.

Rationalizing the denominator for B, we get

$$B = \frac{(\tanh^2(a/2) + 1)\tanh(b/2) + i\tanh(a/2)(\tanh^2(b/2) - 1)}{\tanh^2(a/2)\tanh^2(b/2) + 1}.$$

Note that

$$\tanh^2(b/2) - 1 = -\operatorname{sech}^2(b/2)$$

$$\tanh^2(a/2) + 1 = \frac{\sinh^2(a/2) + \cosh^2(a/2)}{\cosh^2(a/2)} = \frac{\cosh a}{\cosh^2(a/2)}.$$

Thus,

$$B = \frac{\cosh a \tanh(b/2) \cosh^2(b/2) - i \tanh(a/2) \cosh^2(a/2)}{\sinh^2(a/2) \sinh^2(b/2) + \cosh^2(a/2) \cosh^2(b/2)}$$
$$= \frac{\cosh a \sinh b - i \sinh a}{2(\sinh^2(a/2) \sinh^2(b/2) + \cosh^2(a/2) \cosh^2(b/2))}.$$

Hence

$$\cos A = \frac{\cosh a \sinh b}{\sqrt{\cosh^2 a \sinh^2 b + \sinh^2 a}}$$
$$\sin A = \frac{\sinh a}{\sqrt{\cosh^2 a \sinh^2 b + \sinh^2 a}}.$$

We note that

$$\cosh^2 a \sinh^2 b + \sinh^2 a = \cosh^2 a \cosh^2 b - \cosh^2 a + \sinh^2 a$$
$$= \cosh^2 a \cosh^2 b - 1$$
$$= \cosh^2 c - 1$$
$$= \sinh^2 c,$$

where $\cosh a \cosh b = \cosh c$ by the hyperbolic Pythagorean theorem. Thus,

$$\cos A = \frac{\cosh a \sinh b}{\sinh c} \qquad \text{and} \qquad \sin A = \frac{\sinh a}{\sinh c}. \qquad \square$$

With these two theorems, we can solve SSS, SAS, and ASA problems if one of the angles is a right angle. In general, given the three pieces of information required of either SSS, SAS, or ASA, we should be able to determine all the sides and angles of the triangle. In Euclidean geometry, this is done using the Law of Sines, the Law of Cosines, and in the case of ASA, we use the fact that the angles sum to π. In hyperbolic geometry, we have some similar results.

Theorem 7.16.3 (The Hyperbolic Law of Sines). *Let $\triangle ABC$ be a triangle in hyperbolic geometry with sides a, b, and c opposite the angles A, B, and C. Then*

$$\frac{\sinh a}{\sin A} = \frac{\sinh b}{\sin B} = \frac{\sinh c}{\sin C}.$$

Theorem 7.16.4 (The Hyperbolic Law of Cosines for Sides). *Let $\triangle ABC$ be a triangle in hyperbolic geometry with sides a, b, and c opposite the angles A, B, and C. Then*

$$\cosh c = \cosh a \cosh b - \sinh a \sinh b \cos C.$$

The proofs of both these theorems are almost identical to the proofs in Euclidean geometry, so we leave the proofs as exercises.

Note that, just like the Law of Cosines in Euclidean geometry, the hyperbolic Law of Cosines includes the hyperbolic Pythagorean theorem as its first term. This is not too surprising, since hyperbolic and Euclidean geometry look alike locally. We can see this in the Law of Cosines. Recall that the Taylor series for the exponential function is

$$e^x = 1 + x + \frac{x^2}{2!} + \frac{x^3}{3} + \dots.$$

For very small values of x, the higher order terms contribute very little to the value of e^x. Thus, we can write the Law of Cosines as

$$1 + \frac{c^2}{2} + (\text{higher order terms}) = 1 + \frac{a^2}{2} + \frac{b^2}{2} - ab \cos C + (\text{higher order terms}).$$

Ignoring the higher order terms, we get

$$c^2 = a^2 + b^2 - 2ab \cos C,$$

which is the Law of Cosines in Euclidean geometry.

In hyperbolic geometry the sum of the angles is not constant so we need a different result to solve ASA problems.

Theorem 7.16.5 (The Hyperbolic Law of Cosines for Angles). *Let $\triangle ABC$ be a triangle in hyperbolic geometry with sides a, b, and c opposite the angles A, B, and C. Then*

$$\cos C = -\cos A \cos B + \sin A \sin B \cosh c.$$

Recall that, by Theorem 6.3.3, similar triangles are congruent in hyperbolic geometry. Thus, we should be able to solve AAA problems. This is in fact done using the hyperbolic Law of Cosines for angles.

Exercise 7.94. Prove $\text{sech}^2 x = 1 - \tanh^2 x$. Compare this with the identity $\sec^2 x = 1 + \tan^2 x$.

Exercise 7.95. Prove the multiple angle formulas

$$\cosh(x + y) = \cosh x \cosh y + \sinh x \sinh y$$
$$\sinh(x + y) = \sinh x \cosh y + \sinh y \cosh x.$$

Use the first to find formulas for $\cosh^2 x$ and $\sinh^2 x$.

Exercise 7.96. Let $\triangle ABC$ be a right angle triangle with right angle at C. Prove that

$$\cos A = \cosh a \sin B.$$

Exercise 7.97. Let $\triangle ABC$ be a right angle triangle with right angle at C. Prove that

$$\cot A \cot B = \cosh c.$$

Exercise 7.98. Prove the hyperbolic Law of Sines.

Exercise 7.99. Prove the hyperbolic Law of Cosines for sides.

Exercise 7.100. Prove the hyperbolic Law of Cosines for angles.

Exercise 7.101. Verify that the Pythagorean theorem, sine and cosine rules, and the Law of Sines are consistent with the corresponding rules in Euclidean geometry if a, b, and c are very small lengths.

Exercise 7.102. What does the Law of Cosines for angles say if c is very small?

Exercise 7.103* (The Extended Law of Sines in Hyperbolic Geometry). Suppose a triangle $\triangle ABC$ has a circumcircle with radius R. Prove that

$$\tanh R = \frac{\tanh(a/2)}{\cos\left(\frac{B+C-A}{2}\right)} = \frac{\tanh(b/2)}{\cos\left(\frac{A+C-B}{2}\right)} = \frac{\tanh(c/2)}{\cos\left(\frac{A+B-C}{2}\right)}.$$

Exercise 7.104*. Suppose $ACBD$ is a quadrialteral with right angles at A, B, and C. Let $a = |BC|$ and $b = |AC|$. Prove

$$\cos D = \sinh a \sinh b.$$

Exercise 7.105* (Heron's Formula in Hyperbolic Geometry). Since the area of a triangle is determined by its angles, and since the sides of a triangle determine the angles, there must exist a formula for the area of a triangle in terms of its sides. That is, there must be a version of Heron's formula in hyperbolic geometry. State and prove this formula. [H][A]

7.17 The Angle of Parallelism

In Section 7.16, our investigation of the geometry of triangles omitted the class of asymptotic triangles. In a singly asymptotic triangle $\triangle ABM$ with M at infinity, we know $M = 0$. We also know AM and BM are infinite, but knowing that is not helpful, since not all singly asymptotic triangles are congruent. However, given two of A, B, and $m = |AB|$, we can solve for the third. The relation between them is, in fact, the Law of Cosines for angles.

Theorem 7.17.1 (Law of Cosines for Asymptotic Triangles). *Let $\triangle ABM$ be a singly asymptotic triangle with M at infinity and $m = |AB|$. Then*

$$1 = -\cos A \cos B + \sin A \sin B \cosh m.$$

Though this looks like a special case of the Law of Cosines, it is not covered by the proof sought for in the previous section. We will establish this relation in the special case when B is a right angle, and leave the general result as an exercise. In the special case when B is a right angle, the angle A is the angle of parallelism $\Pi(m)$ defined in Chapter 6.

Theorem 7.17.2. *The angle of parallelism* $\Pi(m)$ *is given by*

$$\sin \Pi(m) = \operatorname{sech} m.$$

Proof. The angle of parallelism $\Pi(m)$ is the angle at A of a singly asymptotic triangle $\triangle ABM$ with a right angle at B. Since all such triangles are congruent, we can calculate $\Pi(m)$ by investigating a conveniently chosen singly asymptotic triangle.

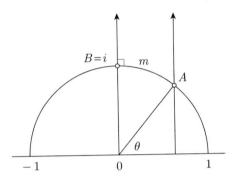

Figure 7.21

Let us choose $B = i$ and $M = \infty$ in $\mathcal{H}$, as in Figure 7.21. Since the angle at B is a right angle, A must lie on the Euclidean half circle centered at 0 and with radius 1. Thus, $A = e^{i\theta} = \cos\theta + i\sin\theta$ for some θ. Note that the angle $\angle MAB$ is the angle between the tangent to the circle and the vertical line, and that this is the same as the angle made by the real axis and the Euclidean line joining A and 0. Hence, $\theta = A$. (Note again the dual role of the symbol A. Sometimes A is meant to be the point A, and at other times, it is the angle $\angle MAB$.)

Let us now find the length of m:

$$m = |\ln(B, A; 1, -1)|.$$

For notational simplicity, let us write $c = \cos A$ and $s = \sin A$. Then,

$$
\begin{aligned}
m &= |\ln(i, c + is; 1, -1)| \\
&= \left| \ln\left(\frac{i-1}{i+1} \middle/ \frac{c+is-1}{c+is+1} \right) \right| \\
&= \left| \ln\left(\frac{i(1+i)(c+is+1)(c-1-is)}{(i+1)((c-1)^2 + s^2)} \right) \right| \\
&= \left| \ln\left(\frac{i(c^2 - (1+is)^2)}{c^2 - 2c + 1 + s^2} \right) \right| \\
&= \left| \ln\left(\frac{i(c^2 - 1 - 2is + s^2)}{2 - 2c} \right) \right|
\end{aligned}
$$

$$= \left| \ln \left(\frac{s}{1-c} \right) \right|.$$

Hence,

$$
\begin{aligned}
\cosh m &= \frac{e^m + e^{-m}}{2} \\
&= \frac{1}{2} \left(\frac{s}{1-c} + \frac{1-c}{s} \right) \\
&= \frac{s^2 + (1-c)^2}{2s(1-c)} \\
&= \frac{2 - 2c}{2s(1-c)} \\
&= \frac{1}{s} = \frac{1}{\sin A},
\end{aligned}
$$

or equivalently,

$$\sin \Pi(m) = \operatorname{sech} m. \qquad \square$$

Exercise 7.106. Prove the Law of Cosines for asymptotic triangles.

Exercise 7.107. Prove that

$$\cos \Pi(m) = \tanh m.$$

Exercise 7.108. Prove that $\triangle ABC$ has a circumcircle if and only if

$$C < \Pi(a/2) + \Pi(b/2).$$

Exercise 7.109. There is a different proof of Theorem 7.17. Consider the triangle $\triangle ABC$ with right angle at C. Solve for $\sin A$ in terms of a and b, and take the limit as b goes to infinity.

7.18 Curvature

Notice that the arclength element for the Poincaré upper half plane model

$$ds = \frac{\sqrt{dx^2 + dy^2}}{y}$$

contains no units. Thus, we get exactly the same model if we use instead the arclength element

$$ds = k \frac{\sqrt{dx^2 + dy^2}}{y}.$$

With this new arclength element, the area of a triply asymptotic triangle becomes $k^2\pi$. The quantity k is just a scaling factor and may even contain units, like miles or kilometers. The *curvature* of a particular model of hyperbolic space is the quantity $-1/k^2$. Thus, if k is very large, then the curvature is very small. By observing distant stars (as in Exercise 1.1, Chapter 1), Kulczycki decided that if we live in a hyperbolic space, then $k > 9.6 \times 10^{14}$ km [Gre]. In Exercise 1.1 of Chapter 1, we assumed that our universe is Euclidean so that we would know that the triangle formed by the Earth at the spring and fall solstices and the distant star has angles which sum to 180°. Using Kulczycki's result, for Alpha Centauri (the nearest star, 4.3 light years away), the sum of the angles differs from 180° by less than one-trillionth of a degree. Though the third angle in the triangulation for this star is tiny – about one one-thousandth of a degree – it is very large compared to the difference from 180° and the sum of the angles in the triangle. Thus, for this star (and other near stars), we may as well assume our universe is Euclidean.

Exercise 7.110. How does scaling the arclength element by k change the Pythagorean theorem?

Exercise 7.111. How does scaling the arclength element by k change the Law of Cosines for sides?

Exercise 7.112. How does scaling the arclength element by k change the formula for the area of a circle?

Chapter 8

Tilings and Lattices

In our study of regular and semiregular polyhedra, we noted that some choices led to flat vertices. These choices give tilings of the plane. In this chapter, we investigate the regular and semiregular tilings of the Euclidean plane. Like the regular polyhedra, there are only a finite number of such tilings. All regular tilings induce a lattice in the plane, where the individual tiles are fundamental domains. However, the fundamental domains or individual tiles need not be regular, and in fact, lend themselves quite well to artistic expression.

Finally, we will investigate regular tilings in the hyperbolic plane. Unlike the Euclidean plane, there exist an infinite number of regular tilings of hyperbolic space. There are also other tilings which do not have analogues in the Euclidean plane.

8.1 Regular Tilings

We say a tiling is regular if each tile is a regular n-gon for a fixed n. There exist tilings with equilateral triangles, squares, and regular hexagons, and no others (see Figure 8.1).

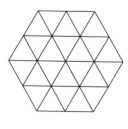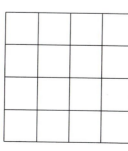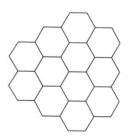

Figure 8.1. The three regular tilings.

Exercise 8.1. Prove that these three tilings are the only regular tilings of Euclidean space.

Exercise 8.2. The idea of duality introduced for Platonic solids extends to regular tilings. What are the duals of each of the regular tilings?

8.2 Semiregular Tilings

A tiling is semiregular if each tile is a regular n-gon and each vertex is identical. The number of tiles to a vertex is largest if the corners of each tile have the smallest possible angular measure – namely $60°$. Thus, there are at most six tiles to a vertex, and this can happen only if each tile is triangular, so we have the regular tiling pictured above. So let us first study the possibility that there are exactly five tiles to a vertex. Not all five can be triangles, so consider first that four are triangles. The last must then be a hexagon, since the remaining corner must have an angular measure of $120°$.

The representation of semiregular polyhedra introduced in Chapter 5 can be used for tilings too. Hence, the tiling with four triangles and a hexagon at each vertex is represented by $(3, 3, 3, 3, 6)$ (see Figure 8.2).

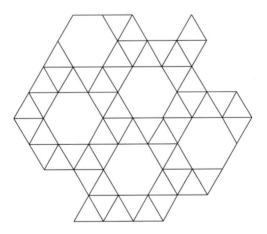

Figure 8.2. The tiling $(3, 3, 3, 3, 6)$.

If there are three triangles to a vertex, then the other two must at least be squares and cannot be larger than squares. This gives the tilings in Figure 8.3

If there are two triangles to a vertex, then the other three are at least squares, so each vertex has more than $360°$. For the same reason, there cannot be fewer than two triangles at each vertex.

Let us now consider the possibility that there are four tiles to a vertex. At most two can be triangles. Note that the angular measure of the corner

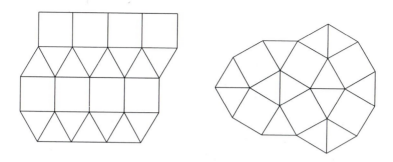

Figure 8.3. The two tilings with three triangles and two squares to a vertex.

of a regular n-gon is $\frac{(n-2)180°}{n}$. Thus, if two tiles are triangles and the other two tiles are respectively an n-gon and an m-gon, then we get the equation

$$\frac{2(180°)}{3} + \frac{(n-2)180°}{n} + \frac{(m-2)180°}{m} = 360°$$

$$\frac{2}{3} + 1 - \frac{2}{n} + 1 - \frac{2}{m} = 2$$

$$\frac{1}{3} = \frac{1}{n} + \frac{1}{m}.$$

The only solution to this last equation in the integers is $n = m = 6$. This gives the tiling $(3, 6, 3, 6)$ in Figure 8.4, and also the tiling $(3, 3, 6, 6)$.

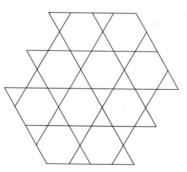

Figure 8.4. The tiling $(3, 6, 3, 6)$.

Now, suppose only one tile is a triangle. If two others are squares, then the last must be a hexagon (see Figure 8.5).

If only one other is a square, then the two remaining tiles are at least pentagons, and the angles sum to more than 360°. If no tiles are triangles, then the rest are at least squares, and four squares already have angles which sum to 360°.

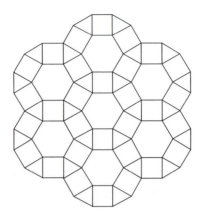

Figure 8.5. The tiling $(3, 4, 6, 4)$.

Finally, suppose there are three tiles to a vertex – an n-gon, m-gon, and l-gon. Then, we get the equation

$$1 - \frac{2}{n} + 1 - \frac{2}{m} + 1 - \frac{2}{l} = 2$$

$$1 = \frac{2}{n} + \frac{2}{m} + \frac{2}{l}.$$

This makes for an interesting number theoretic question. If we further assume $l \leq m \leq n$, then we get the solutions $(l, m, n) = (3, 7, 42)$, $(3, 8, 24)$, $(3, 9, 18)$, $(3, 10, 15)$, $(3, 12, 12)$, $(4, 5, 20)$, $(4, 6, 12)$, $(4, 8, 8)$, $(5, 5, 10)$, and $(6, 6, 6)$. Of these, only $(4, 6, 12)$, $(4, 8, 8)$, and $(6, 6, 6)$ are possible.

Exercise 8.3. Show that if a semiregular tiling has three tiles to a vertex and includes an n-gon for n odd, then the other two tiles must be identical. Explain why this eliminates all possibilities for (l, m, n) above, except those which we claimed are possible.

Exercise 8.4. Construct the tiling with a square, hexagon, and dodecagon, at each vertex.

Exercise 8.5. Construct the tiling with a square and two octagons at each vertex.

Exercise 8.6. Construct the tilings $(3, 3, 6, 6)$ and $(3, 6, 4, 4)$.

8.3   Lattices and Fundamental Domains

The notion of a lattice comes from linear algebra. A lattice Λ in a vector space V is the set

$$\Lambda = \{a_1 \vec{v}_1 + \ldots + a_n \vec{v}_n : a_i \in \mathbb{Z}\},$$

where $\{\vec{v}_1, ..., \vec{v}_n\}$ is a basis for V. For example, in the Euclidean plane, the set of integer pairs (m, n) forms a lattice. In geometry, we have a more general notion of a lattice.

Let G be a subgroup of the group of isometries in Euclidean geometry (or hyperbolic geometry). We say G is a *discrete* subgroup if for any point P, the image of P under G is isolated. That is, if

$$\inf\{d(P, Q) : Q \in G(P), Q \neq P\} > 0.$$

A set F is a *fundamental domain* if for any point Q, there exists an isometry $\sigma \in G$ and a unique point $P \in F$ so that $Q = \sigma P$. We call Λ a *lattice* if there exists a point P and a discrete subgroup G of isometries such that $\Lambda = G(P)$ and the fundamental domain for G has finite area.

Let Λ be a lattice in the Euclidean plane viewed as a vector space. Then Λ is generated by two independent vectors $\vec{v}_1$ and $\vec{v}_2$. This lattice, under our new definition, is given by the group G generated by the two translations by $\vec{v}_1$ and $\vec{v}_2$, and is the image of the origin. For example, the set of integer pairs is a lattice generated by the group G which is generated by the two translations by $\vec{v}_1 = (1, 0)$ and $\vec{v}_2 = (0, 1)$. It can also be described by the group G' generated by translation by $(1, 0)$ and rotation by $90°$ about the origin. Four fundamental domains for G are illustrated in Figure 8.6. Note that F_1 and F_3 are also fundamental domains for G', but F_2 and F_4 are not. (Actually, only part of the boundary of these figures belong to the fundamental domain. For example, F_1 should include, perhaps, the left and top edges and the top left lattice point, but neither the bottom nor

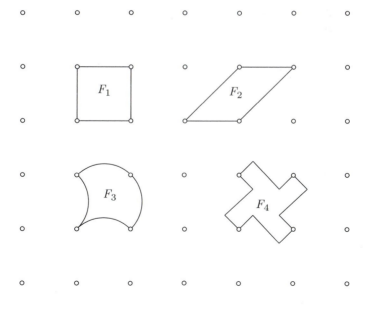

Figure 8.6

right edges, nor the other three lattice points on its boundary. We must make these exclusions so that every point is the image of a *unique* point in the fundamental domain. We will not mention these distinctions again.)

The fundamental domains are the generalizations of tiles.

8.4   Tilings in Hyperbolic Space

We call a polygon in hyperbolic geometry a *regular polygon* if all edges have the same length and every angle has the same measure. In Euclidean geometry, the angles of a regular n-gon depend only on n. In hyperbolic geometry, the angles depend on both n and the length of each side. The longer the edge, the smaller the angle. The angle is largest when the edges are small, which is when the polygon resembles a polygon in Euclidean space. That is, the upper bound on the angles is the angle in a regular Euclidean polygon. Thus, there exists a regular n-gon whose angles are equal to α for any α such that $0 < \alpha < \dfrac{(n-2)180°}{n}$.

So, for example, we cannot tile the hyperbolic plane with 'squares' such that four squares are at each vertex, since it is impossible to construct a regular 4-gon with every angle equal to 90°. However, it is possible to tile

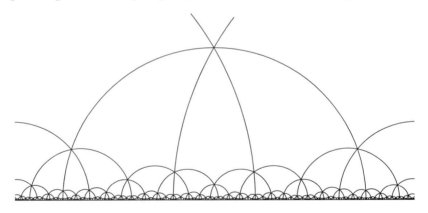

Figure 8.7. A tiling of the Poincaré upper half plane model with six squares to a vertex.

the hyperbolic plane with regular 4-gons such that five or more squares are at each vertex.

A regular tiling of the upper half plane with six squares to a vertex is shown in Figure 8.7. The same tiling in the disc model is shown in Figure 8.8.

Recall that the dual tiling is derived by joining the centers of the tiles. The dual tiling to Figure 8.8 is shown in Figure 8.9, which has four hexagons at each vertex.

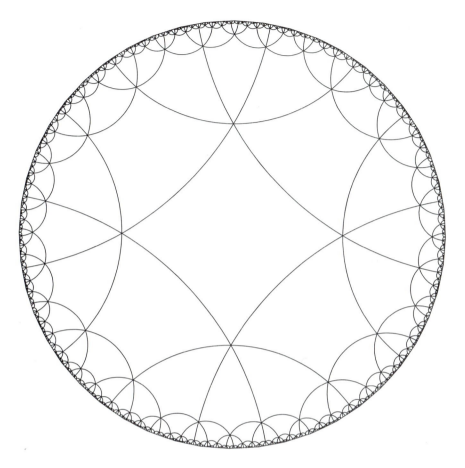

Figure 8.8. The same tiling as in Figure 8.7, but in the Poincaré disc model.

In Figure 8.10, we depict a semiregular tiling with two squares and two hexagons at each vertex. This tiling was obtained from Figure 8.9 by joining the midpoints of the edges of the tiles in that tiling.

Unlike Euclidean geometry, in hyperbolic geometry we have a notion of asymptotic triangles. We can in fact tile the hyperbolic plane with such triangles, as shown in Figure 8.11.

One of the most common tilings in modern mathematics is of interest because of its related group of isometries. This is the tiling whose vertices are the image of the sixth root of unity $\rho = e^{2\pi i/6} = \frac{1+\sqrt{3}}{2}$ under the action of $\mathrm{SL}_2(\mathbb{Z})$. The tiling is shown in Figure 8.12.

Exercise 8.7. What is the area of each square tile in Figure 8.8?

Exercise 8.8. What is the area of each hexagonal tile in Figure 8.9?

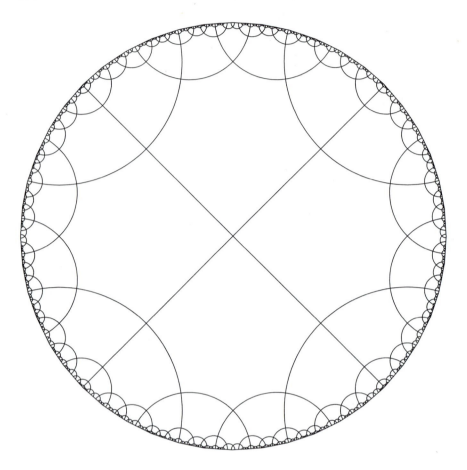

Figure 8.9. The dual to the tiling in Figure 8.8 – four hexagons at each vertex. The vertices in this tiling are the centers of the squares in the tiling in Figure 8.8.

Exercise 8.9. What is the area of each regular triangular tile if $\mathcal{H}$ can be tiled with seven of these triangles at each vertex?   [S]

Exercise 8.10. In this exercise, we find the vertices of a square $ABCD$ centered at $P = i$ which gives a tiling of $\mathcal{H}$ with six squares to a vertex. Let one of the vertices be at $A = ia$. Where is the diagonal vertex C? Use rotation through $\pi/2$ (see Section 7.9) to find the other two vertices. Find a formula in terms of a to find the angle $\angle PAB$. Use this to solve for a.
[A]

Exercise 8.11. Write a Geometer's Sketchpad script which inverts a point in a circle. The input can vary, but one possibility is the circle of inversion and the point.

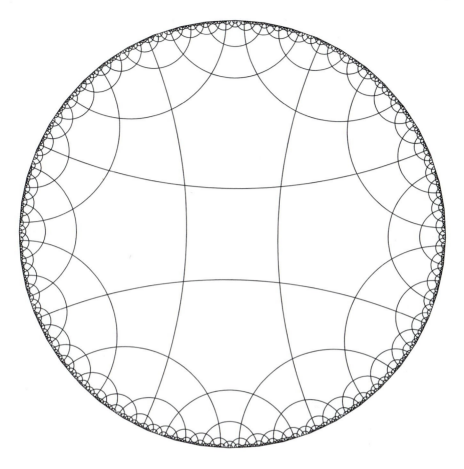

Figure 8.10. A semiregular tiling, with two squares and two hexagons at each vertex. This tiling was obtained from the tiling in Figure 8.9 by joining the midpoints of the sides of the hexagon.

Exercise 8.12. Write a script that draws the line through two points in $\mathcal{H}$. Suggested input is the two points and the real line.

Exercise 8.13. The solution for r found in Exercise 8.10 is a constructible length. Construct the point $A = ir$ in a Geometer's Sketchpad sketch. Construct the other three vertices. (Hint: Construct the diagonal point C and the circumcircle for the square.) Use the scripts found in Exercises 8.11 and 8.12 to tile around one of the vertices.

Exercise 8.14. Use Exercise 8.10 and the map $\phi = \begin{bmatrix} 1 & -i \\ -i & 1 \end{bmatrix}$ which sends $\mathcal{H}$ to the disc model to find the vertices of the square centered at 0 which tiles the plane with six squares to a vertex. Repeat Exercises 8.11, 8.12,

Figure 8.11. A tiling of the Poincaré disc model with triply asymptotic triangles.

and 8.13 to tile a portion of the disc. Compare your sketch with Figure 8.8.
[A]

Exercise 8.15*. Find a if $A = ia$ is the vertex of a regular n-gon in $\mathcal{H}$ centered at i and of the appropriate shape to tile $\mathcal{H}$ with m tiles at each vertex. [S]

Exercise 8.16. Tile the disc with four hexagons to a vertex, but with one hexagon centered at 0.

Exercise 8.17*. Tile the disc with three squares and three triangles at each vertex.

Exercise 8.18*. Let the regular pentagon $ABCDE$ in $\mathcal{H}$ be centered at i and let $A = ir$. If the length r is chosen so that this pentagon can tile $\mathcal{H}$

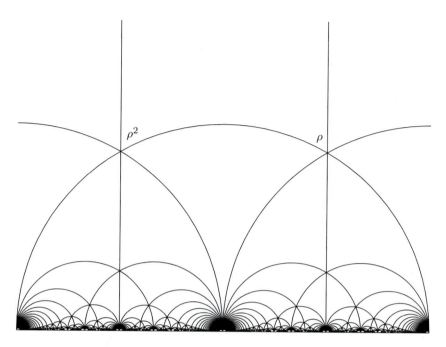

Figure 8.12. The tiling of the Poincaré upper half plane induced by the action of $\mathrm{SL}_2(\mathbb{Z})$. The tiles are singly asymptotic triangles. The usual choice for the fundamental domain is the triangle with vertices at ∞, the sixth root of unity $\rho = e^{2\pi i/6} = \frac{1+i\sqrt{3}}{2}$, and the cube root of unity $\rho^2 = e^{2\pi i/3} = \frac{-1+i\sqrt{3}}{2}$.

with four pentagons to a vertex, then r is constructible. Find it, and use it to tile around the vertex A.

Exercise 8.19. Let the Poincaré disc model be drawn in the complex plane centered at 0. Let $A_1 A_2 \ldots A_n$ be an n-gon with center 0, A_1 real, and such that the hyperbolic plane can be tiled with these n-gons with m such tiles at a vertex. Show that A_1 is constructible (in the Euclidean plane) if and only if both the n-gon and m-gon can be constructed.

Exercise 8.20*. The notion of constructibility is equally valid in hyperbolic geometry. The hyperbolic straightedge is used to make the hyperbolic line through two constructed points, and the hyperbolic compass is used to make the circle centered at a constructed point and going through a constructed point. Suppose that we begin with the points 0 and 1/2 in the Poincaré disc $\mathcal{D}$. Show that every point which is constructible in $\mathcal{D}$ using the hyperbolic tools is also constructible using Euclidean tools. Are all points which are in $\mathcal{D}$ and are constructible using Euclidean tools also constructible using hyperbolic tools?

8.5   Tilings in Art

The master of creative tilings is M.C. Escher (1898 –1972). Creative tiling, in Euclidean geometry, can be fun for anyone and makes an appropriate art or math project for middle school students.

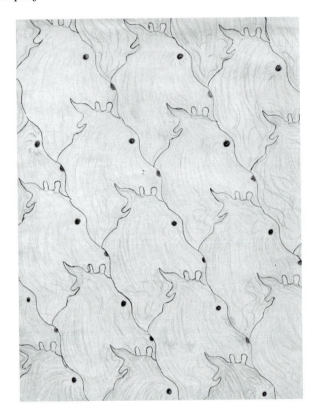

Figure 8.13. *Goats*, 1999, by Sasha Neugebauer, Grade 6, Silvestri Middle School, Las Vegas, Nevada. A tiling with translational symmetry.

Hyperbolic tilings are a little more complicated. Inspired by the work of Coxeter, Escher produced a couple of hyperbolic tilings. These appear in Figures 8.15 and 8.16. *Circle Limit IV* is based on a regular tiling with four hexagons to a vertex. *Circle Limit III* is based on a semiregular tiling with three squares and three triangles to each vertex.

Exercise 8.21. What is the area of each fish in Escher's *Circle Limit III* (Figure 8.16)?   [S]

Exercise 8.22. Create an artistic tiling that has translational symmetry

Figure 8.14. A triangular tiling with rotational symmetry, M.C. Escher's *Symmetry Drawing E21 (Human figures)*.

in two directions. The example in Figure 8.13 was created by Sasha Neugebauer.

Exercise 8.23. Create an artistic tiling that has translational and rotational symmetry.

Exercise 8.24*. Create an artistic tiling in the Poincaré disc.

Figure 8.15. M.C. Escher's *Circle Limit IV*.

Figure 8.16. M.C. Escher's *Circle Limit III*, a semiregular tiling of hyperbolic space.

Figure 8.17. M.C. Escher's *Periodic Space-Filling.* An interesting drawing. Is it a tiling of the Poincaré upper half plane?

Chapter 9

Foundations

Historically, an expansive theory like geometry (or calculus or set theory) must first prove itself useful before it is ever developed in depth. The development of a field does not begin with definitions, but with intuition. A definition is a carefully crafted creation of a mind which already has an intuitive grasp of the subject and has discovered the need of a precisely defined object. In this chapter, our aim is to use our intuitive grasp of geometry to come up with a set of definitions and axioms from which all else can be derived.

9.1 Theories

An *axiom* is a statement which we accept as true without proof. A set of axioms is *consistent* or *coherent* if there are no statements which can be proven to be both true and false. A *theory* is the set of results which follow from a set of definitions and consistent axioms. Two sets can generate the same theory. For example, in Chapter 1 we accepted a set of axioms to define Euclidean geometry, and we also accepted on faith that the intersection of two circles contains at most two points (Lemma 1.3.2). This statement is really a theorem in Euclidean geometry, but since we accepted it without proof, we were in essence thinking of it as an axiom. Thus, there is no unique or canonical choice of axioms one picks to define a geometry, and the particular set of assumptions one chooses is essentially a matter of personal taste. The advantage of accepting Lemma 1.3.2 without proof in Chapter 1 is that we avoided a complicated proof of a result which certainly was not very difficult to accept as being true. There are also disadvantages, the most fundamental of which is that we are asked to accept more on faith.

In Chapter 6, we began a development of the theory of hyperbolic geometry. At that point, we did not know whether the theory even existed, or if our set of axioms led to a contradiction. In Chapter 7, we developed a model and showed that it satisfied the axioms presented in Chapter 6.

The existence of this model is the proof that the set of axioms is consistent, and that the theory of hyperbolic geometry is valid. It also showed that Euclid's fifth postulate could not follow from the first four, since if it did, then the parallel postulate of hyperbolic geometry would contradict the first four postulates, since it clearly contradicts Euclid's fifth postulate. This also points out the disadvantage of accepting an extraneous result like Lemma 1.3.2. To show that the set of axioms which includes Lemma 1.3.2 is consistent, we must find a model which satisfies these axioms. In particular, we must show that Lemma 1.3.2 is satisfied, and this is usually no easier than deriving it from the other axioms.

Another advantage of a simple set of axioms is that such a set is easier to accept on faith – it is easier to believe that they do not lead to a contradiction. Furthermore, if the set of axioms is simple enough, it is more likely that changing only one axiom (like the parallel postulate) might lead to a consistent set of axioms, and consequently a different but equally valid theory. If, in our set of axioms which defines Euclidean geometry, we included as an axiom the theorem that the angles in a triangle sum to 180°, then to 'discover' hyperbolic geometry, we must discard two axioms.

9.2   The Real Line

A fundamental component of the foundations we develop for planar geometry is the existence and properties of the real line, which may be thought of as one-dimensional Euclidean geometry. A formal development of the real line is probably best left to a course in real analysis, so in this section, we will only briefly review the steps, with the understanding that this is a review for the reader. If the reader has never seen such a development, the reader may safely skip this section after accepting that the reals exist and have the following two properties:

1. The reals are *totally ordered* or just *ordered*. That is, for any two real numbers a and b, we can say either $a = b$, $a < b$ or $b < a$. This property gives us a notion of betweeness: c is between a and b if either $a \leq c \leq b$ or $b \leq c \leq a$.

2. The reals are *complete*. Naively, this means that any decimal representation represents a real number. We will use this property to define the completeness of planar geometries.

We develop the reals by starting with the natural numbers, zero, and the operations of addition and multiplication. We note that 0 is the additive identity, and that 1 is the multiplicative identity. We postulate the existence of additive and multiplicative inverses, and the other field axioms, to arrive at the rationals $\mathbb{Q}$. We note that the rationals are ordered.

The rationals have gaps in them. For example, it is possible to find a rational number p/q such that $p^2/q^2 < 2$ and is as close to 2 as we desire, yet it is not possible to find a rational p/q such that $p^2/q^2 = 2$. We fill in these gaps by defining Cauchy sequences.

Definition 17. *Cauchy's criterion.* A sequence $\{a_n\}_{n=1}^{\infty} = \{a_1, a_2, a_3, ...\}$ is *Cauchy* if for any $\epsilon > 0$ there exists an M (which may depend on ϵ) such that for any $m, n > M$, we have

$$|a_m - a_n| < \epsilon.$$

We will be interested only in rational Cauchy sequences. That is, sequences $\{a_n\}_{n=1}^{\infty}$ with $a_n \in \mathbb{Q}$ for all n. Note that a rational Cauchy sequence may not converge in the rationals.

We say two rational Cauchy sequences $\{a_n\}_{n=1}^{\infty}$ and $\{b_n\}_{n=1}^{\infty}$ are equivalent if

$$\lim_{n \to \infty} (a_n - b_n) = 0,$$

and define the equivalence class $[\{a_n\}_{n=1}^{\infty}]$ of the sequence $\{a_n\}_{n=1}^{\infty}$ to be the set of all Cauchy sequences which are equivalent to $\{a_n\}_{n=1}^{\infty}$. We define the reals $\mathbb{R}$ to be the set of all equivalence classes of rational Cauchy sequences.

An example of a rational Cauchy sequence is the sequence

$$\{a_n\}_{n=1}^{\infty} = \{1.4, 1.41, 1.414, 1.4142, ...\}.$$

The general definition of a_n is given by

$$a_n = \frac{a}{10^n},$$

where a is the largest integer such that

$$\frac{a^2}{10^{2n}} < 2.$$

The sequence $\{a_n\}_{n=1}^{\infty}$ is Cauchy since given any $\epsilon > 0$, there exists an M such that

$$\epsilon > 10^{-M} > 0,$$

and for any $n, m > M$,

$$|a_n - a_m| \le 10^{-M} < \epsilon,$$

as desired.

We usually denote the equivalence class which contains this rational Cauchy sequence with $\sqrt{2}$.

9.3   The Plane

As we mentioned before, the set of axioms one accepts on faith can be a matter of personal taste. Euclid's set of postulates for Euclidean geometry are by far the most famous but are not quite adequate. Let us reintroduce them here so that we may use them for guidance and discover both their virtues and failings.

1. We can draw a unique line segment between any two points.

2. Any line segment can be continued indefinitely.

3. A circle of any radius and any center can be drawn.

4. Any two right angles are congruent.

5. If a line meets two other lines so that the sum of the angles on one side is less than two right angles, then the two other lines meet at a point on that side.

These axioms already presuppose several primitive concepts. They presuppose a notion of length; definitions of points, line segments, lines, and circles; and even that the Euclidean plane is two dimensional.

We will start our development with the following definition: Let **E** be a set of elements which we call *points*, and let **E** have several properties which we will define in the following (through Section 9.8). We will call **E** the *Euclidean plane*. The first property we require of **E** is that it contain at least two points.

9.4   Line Segments and Lines

The first notion we want to define on this set is the notion of a line segment. In Chapter 1, we used the definition 'A line segment between two points is the shortest path between the two points.' Not a bad definition, but what do we mean by a *path*, and how do we define *shortest*? So let us begin with a notion of distance.

Definition 18. *Distance.* We call a function d a *distance function* on **E** if

$$d : \mathbf{E} \times \mathbf{E} \to \mathbb{R}$$

and

$$d(P, Q) = d(Q, P)$$
$$d(P, Q) \geq 0 \qquad \text{with equality if and only if } P = Q,$$
$$d(P, R) + d(R, Q) \geq d(P, Q) \qquad \text{(the triangle inequality).}$$

We say the *distance* between P and Q is $d(P, Q)$ and usually denote this distance with $|PQ|$.

A line segment PQ should have the property that for any R on the segment, the triangle $\triangle PRQ$ is degenerate. That is

$$d(P, R) + d(R, Q) = d(P, Q).$$

One might even define a line segment to be exactly the set of all such R. With such a definition, Euclid's first axiom is vacuous. By definition, between any two points P and Q there exists a unique line segment PQ. However, what we have defined to be a line segment does not *a priori* look anything like what we have in mind for a line segment. For example, suppose there exists a pair of points P and Q which we would normally say has two line segments joining it (for example, choose the North and South Pole on a sphere, between which every line of longitude would be considered a line segment). Then our definition of PQ would be (at least) the union of these two line segments (and in the example of the sphere, the line segment would be the whole sphere, which is not even one dimensional!)

We can distinguish between two paths by further demanding that if R_1 and R_2 are on the line segment PQ, then there exists an ordering $\{i, j\}$ of $\{1, 2\}$ so that

$$d(P, R_i) + d(R_i, R_j) + d(R_j, Q) = d(P, Q). \tag{9.1}$$

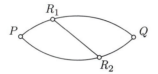

Figure 9.1. How Equation 9.1 distinguishes between two paths.

This gives us a better definition for a line segment. With this refined definition, the line segment PQ always exists, so the only content of Euclid's first axiom is that the line segment is unique. However, the line segments may still not be what we expect. For example, there may not be any point R other than P and Q which produce a degenerate triangle. The line segment PQ would in this case be just the set of two points $\{P, Q\}$, which is not what we have in mind for line segments. To ensure our line segments have content, we will further refine our definition using the completeness of the reals:

Definition 19. *Line segment.* A *line segment* PQ between P and Q is a set of points with the following two properties:

1. For any points R_1 and R_2 in PQ, there is a reordering $\{i, j\}$ of $\{1, 2\}$ such that
$$|PR_i| + |R_iR_j| + |R_jQ| = |PQ|,$$

2. and for any real number r with $0 \le r \le |PQ|$, there exists an R such that
$$|PR| = r.$$

Now, line segments are what we expect, but there is no reason why any should exist, nor that they should be unique. We will therefore adopt Euclid's first axiom as our first axiom:

1. For any two points P and Q in $\mathbf{E}$, the line segment PQ exists and is unique.

Note that we have incorporated the completeness of the reals into our definition of a line segment, and consequently, Axiom 1 indirectly postulates a type of completeness of the plane. In many texts, this property is treated separately in its own set of axioms.

Note also that our notation for line segments implies that they are unique. This is fine in Euclidean geometry, since we have Axiom 1. However, this notation can be misleading in a geometry like spherical geometry.

We now direct our attention to defining a line. Using Euclid's second axiom as inspiration, we should think of a line as being an indefinite extension of a line segment. Thus, a line should have the property that if any two points lie on the line, then the line segment joining those two points should also be on the line. The entire plane $\mathbf{E}$ has this property, so we need a bit more. Using our definition of a line segment as inspiration, let us define a line as follows:

Definition 20. *Line.* A *line* is a set of points $l \subset \mathbf{E}$ so that if P and Q are in l, then $PQ \subset l$. Furthermore, for any point P and any real number $r > 0$, there exist exactly two points R and R' in l so that

$$d(P, R) = d(P, R') = s.$$

The second statement guarantees that our lines are one dimensional, and the first guarantees that a line is an indefinite extension of a line segment. However, we do not know that any lines exist, so we still need an axiom like Euclid's second axiom. Let us adopt it as our second axiom:

2. Every line segment can be realized as a subset of a line.

Exercise 9.1. Show that if $|PR| + |RQ| = |PQ|$, then R lies on the line segment PQ. Hint: Since this is not true in spherical geometry, any solution will require Axiom 1 or 2.

9.5 Separation Axioms

Notice that, so far, we have not addressed the issue of making sure our space $\mathbf{E}$ is two dimensional. Review what we have set up, and check that everything is valid for a three-dimensional space. We will restrict our space $\mathbf{E}$ to two dimensions via two *separation axioms*.

Definition 21. *Same side.* We say two points P and Q not on l are on the *same side* of the line l if the line segment PQ and the line l do not intersect. If there is a point of intersection, then we say P and Q are on *opposite sides* of l.

We further ask that the notion of 'same side' be an equivalence relation on $\mathbf{E} \backslash l$, and that there be only two equivalence classes. That is, we postulate

3. If P and Q are on the same side of l, and Q and R are on the same side of l, then P and R are on the same side of l.

4. For any line l, there exists a point P not on l. If P and Q are on opposite sides of l and Q and R are also on opposite sides of l, then P and R are on the same side of l.

We have specified that there exists a point P not on l, for otherwise $\mathbf{E}$ could be just a line. The line l separates $\mathbf{E}$ into two sides, a property that lines in three dimensions do not have. Note that Euclid's fifth axiom (as he stated it) now makes sense, since we can now talk about a side of a line.

Exercise 9.2. We postulate that there exists a point P not on l. Show that there exists a point Q on the other side of l.

Exercise 9.3 (Pasch's Theorem). Suppose a line l intersects $\triangle ABC$ and that none of A, B, or C lie on l. Show that l intersects exactly two of the three sides of $\triangle ABC$. [S]

Exercise 9.4. Define the *interior* of an angle $\angle BAC$. [A]

Exercise 9.5. Define the interior of a triangle.

Exercise 9.6. Suppose P is inside $\angle BAC$, and that Q is on the ray AP. Prove that Q is inside $\angle BAC$.

Exercise 9.7. Suppose D is on the line segment BC. Prove that D is inside $\angle BAC$.

Exercise 9.8. Suppose the segments BB' and CC' intersect at A. Prove that the interiors of the four angles $\angle BAC$, $\angle CAB'$, $\angle B'AC''$, and $\angle C'AB$ are disjoint.

Exercise 9.9. Suppose the segments BB' and CC' intersect at A. Suppose P is inside $\angle BAC$. Let P' be a point on the line AP such that A is between P and P'. Show that P' is inside $\angle B'AC'$.

Exercise 9.10**. Suppose A lies on l and that there exists a point P on l such that P is inside $\triangle ABC$. Prove that l intersects the side BC. [S]

Exercise 9.11**. Come up with a definition of when two nondegenerate triangles have the same *orientation*. Prove that, by your definition, orientation is an equivalence relation with two equivalence classes on the set of nondegenerate triangles. [H][A]

Exercise 9.12*. Suppose f is an isometry, and f preserves the orientation of a nondegenerate triangle $\triangle ABC$. Prove that f is a direct isometry. That is, prove that f preserves the orientation of all nondegenerate triangles.

Exercise 9.13. We are in the process of defining Euclidean geometry in the plane. How should the separation axioms read in three dimensions? How should they read in one dimension?

9.6 Circles

Our definition of a circle is rather natural:

Definition 22. *Circle.* The *circle* $\mathcal{C}_P(r)$ centered at P and with radius $r > 0$ is the set of points

$$\mathcal{C}_P(r) = \{Q : d(P, Q) = r\}.$$

Note that any circle with any radius exists by definition, so we do not need Euclid's third axiom.

Exercise 9.14. Using the above definition, what is a 'circle' in three dimensions? What is it in one dimension?

Exercise 9.15*. Let $\mathcal{C}_P(r)$ be a circle and suppose R and S are points such that $|PR| < r$ and $|PS| > r$. Show that the line segment RS intersects the circle $\mathcal{C}_P(r)$.

Solution. We expect this result to be true, but what is at the heart of the question is the completeness of the plane. It therefore should not surprise us that a solution involves Cauchy sequences. Our method of proof is common in real analysis. We successively bisect the line segment to hone in on the required point. More precisely, we define several sequences of points R_n, S_n, and T_n recursively as follows: We set $R_0 = R$ and $S_0 = S$. We let T_n be the midpoint of $R_{n-1}S_{n-1}$. If for some n we have $|PT_n| = r$, then we are finished. Otherwise, either $|PT_n| < r$ or $|PT_n| > r$. If $|PT_n| < r$, then we set $R_n = T_n$ and $S_n = S_{n-1}$. Otherwise, we set $R_n = R_{n-1}$ and $S_n = T_n$.

If $|PT_n| \neq r$ for all n, then we can define the infinite sequence of reals $\{|RT_n|\}_{n=1}^{\infty}$. For any n and m with $n > m$, we have

$$||RT_n| - |RT_m|| = |T_nT_m| < 2^{-n}|RS|, \tag{9.2}$$

where we have the first equality since R, T_n, and T_m are collinear, and R is not between T_n and T_m. But from Equation 9.2, it follows that $\{|RT_n|\}_{n=1}^{\infty}$ is Cauchy, so converges to some t. We let T be the point on RS such that $|RT| = t$. Note that the existence of T is guaranteed by the definition of line segments, and because Axiom 1 guarantees that RS is a line segment.

We claim that $|PT| = r$. To see this, we first suppose $|PT| < r$. Then there exists an n such that

$$2|RS|2^{-n} < r - |PT|.$$

We know

$$|TT_n| < 2^{-n}|RS|$$
$$|T_nS_n| < 2^{-n}|RS|.$$

Thus, using the triangle inequality, we get

$$\begin{aligned}
r < |PS_n| &\leq |PT_n| + |T_nS_n| \\
&< |PT| + |TT_n| + 2^{-n}|RS| \\
&< (r - 2|RS|2^{-n}) + 2^{-n}|RS| + 2^{-n}|RS| \\
&< r,
\end{aligned}$$

which is a contradiction. Thus, $|PT| \geq r$. In a similar fashion, we can show $|PT| \leq r$. Thus $|PT| = r$. That is, T is on both the circle $\mathcal{C}_P(r)$ and the line segment RS, as desired. $\qquad\square$

Exercise 9.16*. We say three values a, b, and c are *triangular* if there exists a triangle with sides of length a, b, and c. That is, a, b, and c are triangular if $a \leq b + c$, $b \leq a + c$, and $c \leq a + b$. Suppose $\mathcal{C}_P(r)$ and $\mathcal{C}_Q(s)$ are two circles and that r, s, and $|PQ|$ are triangular. Show that these two circles intersect. [H]

Though the previous exercise guarantees that there is a point of intersection, we also need to know that there are no more than two points of intersection. This was Lemma 1.3.2, whose proof we deferred until now.

Lemma 9.6.1 (Lemma 1.3.2). *Two distinct circles intersect in zero, one, or two points. If there is exactly one point of intersection, then that point lies on the line joining the two centers.*

Proof. Let the two circles have centers A and B. Note that the circles cannot be concentric, for if they are, they cannot have a point of intersection unless they have the same radius, in which case they are not distinct. Thus, we may assume $A \neq B$, and draw the line through these two points.

Suppose there exists exactly one point of intersection C. If C does not lie on AB, then C and its image C' under reflection through AB (the isometry guaranteed by Axiom 8) are distinct. But $|AC| = |AC'|$, so C' is on the circle with center A, and similarly, C' is on the circle with center B. Thus, the circles have two points of intersection, which is a contradiction. Thus, C must lie on the line AB.

To finish the proof, we need only show that there cannot be three points of intersection. Let us assume there exists a point of intersection C on the line AB that lies between A and B. If there exists another point of intersection C' not on AB, then by the triangle inequality,

$$|AC'| + |C'B| > |AB| = |AC| + |CB| = |AC'| + |C'B|,$$

and we arrive at a contradiction. We leave as an exercise the case when C lies on the line AB but does not lie between A and B.

So let us now assume that there are no points of intersection on AB, and that there are at least three points of intersection. Then two of these points of intersection, say C and C', lie on the same side of the line AB. There are two possible cases to consider. Either A and B are on opposite sides of CC', as in Figure 9.2(a), or they are on the same side of CC', as in Figure 9.2(b). We will investigate the first case and leave the other case as an exercise.

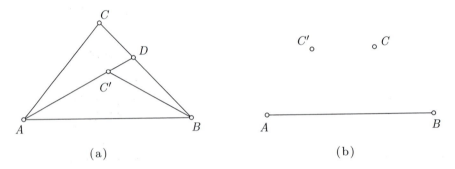

(a)   (b)

Figure 9.2

In the first case, since C and C' are on the same side of AB, and A and B are on opposite sides of CC', one of C or C' is inside the triangle formed by the other three points (see Exercise 9.18). Without loss of generality, we may assume C' is inside $\triangle ABC$. Let AC' intersect BC at D (we are using Exercise 9.10 here). From the triangle inequality and Exercise 9.1 applied to $\triangle BC'D$,

$$|C'D| + |DB| > |C'B|, \tag{9.3}$$

and applied to $\triangle ACD$,

$$|AC| + |CD| > |AD| = |AC'| + |C'D|.$$

But $|AC| = |AC'|$, since both C and C' lie on the circle centered at A. Hence,

$$|CD| > |C'D|$$
$$|BD| + |CD| > |BD| + |C'D|$$
$$|BC| > |BD| + |C'D|. \tag{9.4}$$

Finally, we note that $|BC'| = |BC|$, since they too are both radii of the circle centered at B, and so Equations (9.3) and (9.4) give us a contradiction. □

This result implies SSS, SAS, and ASA, as seen in Section 1.3. Thus, these theorems are results in both Euclidean and hyperbolic geometry.

Exercise 9.17. Fill in the details of the second case in the proof of Theorem 9.6.1.

Exercise 9.18. Suppose C and C' are on the same side of AB, and that A and B are on opposite sides of CC'. Prove that either C or C' is inside the triangle formed by the other three points.

9.7 Isometries and Congruence

In the Euclidean plane, we want every point to look like any other point. We say this by defining a notion of congruence, which, as in Chapter 1, we define via isometries.

Definition 23. *Isometry.* An *isometry* of the plane is a map

$$f : \mathbf{E} \to \mathbf{E}$$

from the plane to itself which preserves distances. That is, f is an isometry if for any P and Q in the plane, we have

$$d(f(P), f(Q)) = d(P, Q).$$

Definition 24. *Congruence.* Two sets of points (defining a triangle, angle, or some other figure) are *congruent* if there exists an isometry which maps one set onto the other.

In particular, we say two angles are congruent (or equal) if there exists an isometry which sends one angle to the other.

Exercise 9.19. We have not yet formally defined angles. Come up with a definition. Does your definition work in the hyperbolic plane? Does it work on the sphere? Come up with a refinement which works in these cases.

Our notion of congruence is completed by the following axiom which guarantees the existence of the isometries we desire:

6. Given any line l, there exists an isometry which fixes every point in l but fixes no other points in the plane.

In Chapter 1, we included two more axioms. They, in fact, follow from this axiom.

Lemma 9.7.1. *Given any points P and Q, there exists an isometry which sends P to Q.*

Proof. If $P = Q$, then the identity works. Otherwise, let us assume $P \neq Q$. By Exercise 9.16, the circles $\mathcal{C}_P(|PQ|)$ and $\mathcal{C}_Q(|PQ|)$ intersect, say at A. Since A cannot be between P and Q, there exists another point of intersection B, by Lemma 1.3.2. By Axiom 6, there exists an isometry f which fixes every point on the line AB and fixes no other points. Since f preserves the distance $|AP|$, the point $P' = f(P)$ must lie on the circle $\mathcal{C}_A(|AP|)$. Similarly, P' lies on $\mathcal{C}_B(|BP|)$. These two circles intersect at two points, P and P'. But Q is also on both these circles, since $|QA| = |QP| = |QB|$. Finally, since P does not lie on AB, $P' \neq P$, so $P' = Q$, as desired.   □

Lemma 9.7.2. *Given a point P and any two points Q and R which are equidistant from P, there exists an isometry which fixes P and sends Q to R.*

We leave the proof as an exercise. We can now define right angles:

Definition 25. *Right angle.* Two lines l_1 and l_2 intersect at *right angles* if any two adjacent angles at the point of intersection are congruent. That is, they intersect at right angles if there exists an isometry which sends an angle to one of its adjacent angles.

Exercise 9.20. Prove that any two right angles are congruent. Thus, we do not need to state Euclid's fourth postulate as an axiom.

The last axiom we need to define Euclidean geometry is the parallel postulate. Since this is the axiom we change when we develop hyperbolic geometry, let us first contemplate how much we have established without that axiom. Note that these results are equally valid in hyperbolic geometry. Such results are sometimes referred to as results in *neutral geometry.*

Let us add two more results to our list of results in neutral geometry. The first was proved in Section 1.4. Though this is the section on parallel lines, the proof does not involve the parallel postulate, so this is a result in neutral geometry.

Lemma 9.7.3 (Lemma 1.4.2). *Let l be a line and P a point not on l. Then there exists a Q on l so that PQ is perpendicular to l.*

Lemma 9.7.4. *A line and circle intersect in either zero, one, or two points. There is exactly one point of intersection if and only if the radius to the point of intersection is perpendicular to the line.*

This is a nice complement to Lemma 9.16. We leave the proof as an exercise.

Exercise 9.21. We have not formally defined angular measure. Come up with a definition.

Exercise 9.22. How should Axiom 6 read in one dimension?

Exercise 9.23. How should Axiom 6 read in three dimensions? Is the full group of isometries in three dimensions guaranteed to exist using this revised version of Axiom 6?

Exercise 9.24. Prove Lemma 9.7.4.

9.8 The Parallel Postulate

Finally, we introduce the parallel postulate, Euclid's fifth axiom:

  5. If a line meets two other lines so that the sum of the angles on one
     side is less than two right angles, then the two other lines meet at a
     point on that side.

Exercise 9.25. Let l_1, l_2, and l_3 be three distinct lines. Prove that if l_1 is parallel to l_2, and l_2 is parallel to l_3, then l_1 is parallel to l_3.   [H]

9.9 Similar Triangles

In our treatment of Euclidean geometry, we saw that the following result and its converse Theorem 1.7.3 are indispensable:

Theorem 9.9.1. *Let $\triangle ABC$ be a triangle. Let B' be on AB, and let the line through B' and parallel to BC intersect AC at C'. Then*

$$\frac{|AB'|}{|AB|} = \frac{|AC'|}{|AC|}.$$

In Chapter 1, we deferred the proof of this result until now. Our proof will be via a sequence of lemmas, some of whose proofs we leave as exercises. We remind the reader of two important results, both of which were proved before we began discussing similar triangles. These are (1) opposite interior angles of a transversal of two parallel lines are equal and (2) angle-side-angle (ASA).

Lemma 9.9.2. *Let $ABCD$ be a parallelogram. Then*

$$|AB| = |CD|.$$

Lemma 9.9.3. *Let B' be the midpoint of $\triangle ABC$, and suppose the line through B' which is parallel to BC intersects AC at C'. Then C' is the midpoint of AC.*

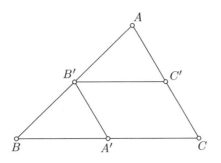

Figure 9.3

Proof. We first note that the line through B' intersects AC, by Pasch's theorem (Exercise 9.3). Similarly, the line through B' which is parallel to AC intersects BC, say at A', as in Figure 9.3. Note that $\angle AB'C' = \angle ABC$, and $\angle BB'A' = \angle BAC$, so $\triangle AB'C' \equiv \triangle B'BA'$, by ASA. In particular,

$$|AC'| = |B'A'|.$$

By Lemma 9.9.2, $|B'A'| = |C'C|$, so

$$|AC'| = |C'C|.$$

That is, C' is the midpoint of AC.   □

Lemma 9.9.4. *Let $ABCD$ be a trapezoid with parallel sides AD and BC. Let B' be the midpoint of AB, and let the line through B' and parallel to BC intersect DC at C'. Then C' is the midpoint of DC.*

Lemma 9.9.5. *Let B' be a point on AB in $\triangle ABC$ such that*

$$\frac{|AB'|}{|AB|} = \frac{p}{2^n}$$

for some positive integers p and n. Let the line through B' and parallel to BC intersect AC at C'. Then

$$\frac{|AB'|}{|AB|} = \frac{|AC'|}{|AC|}.$$

We are now ready to prove Theorem 9.9.1.

Proof of Theorem 9.9.1. Again, we point out that, by Pasch's theorem, the line through B' and parallel to BC intersects AC. Suppose

$$\frac{|AB'|}{|AB|} \neq \frac{|AC'|}{|AC|}.$$

Let $\frac{|AB'|}{|AB|} = \lambda$ and $\frac{|AC'|}{|AC|} = \mu$. Without loss of generality, we may assume $\lambda < \mu$. But then, there exists a p and n such that

$$\lambda < \frac{p}{2^n} < \mu.$$

Let B'' be the point on AB such that

$$\frac{|AB''|}{|AB|} = \frac{p}{2^n},$$

and let the line through B'' and parallel to BC intersect AC at C''. Then, by Lemma 9.9.5,

$$\frac{|AC''|}{|AC|} = \frac{p}{2^n}.$$

Since $\lambda < \frac{p}{2^n}$, we know that A and B' are on the same side of $B''C''$. Since $\frac{p}{2^n} < \mu$, we have that C' and C are on the same side of $B''C''$. But A and C are on opposite sides of $B''C''$, so B' and C' must also be on opposite sides of $B''C''$. Thus, $B'C'$ intersects $B''C''$, a contradiction, since the two lines are parallel. Hence, our assumption must be false, and therefore $\lambda = \mu$. That is,

$$\frac{|AB'|}{|AB|} = \frac{|AC'|}{|AC|}. \qquad \qquad \square$$

Exercise 9.26. Prove Lemma 9.9.2.   [H]

Exercise 9.27. Prove Lemma 9.9.4.

Exercise 9.28. Prove Lemma 9.9.5.

Chapter 10

Spherical Geometry

In Chapters 6 and 7, we developed hyperbolic geometry by first setting forth the axioms of the geometry and then constructing a model which we eventually showed satisfies the axioms. We are already familiar with a model of spherical geometry, so in this chapter our investigation will be in the reverse order. We first study a model and describe a few results and then describe a set of axioms which are satisfied by this model.

10.1   The Area of Triangles

Our model is a sphere $\mathbf{S}$ in 3-space, centered at O and with radius ρ. The distance between two points on the sphere is given by

$$|PQ| = \rho \angle POQ$$

where $\angle POQ \leq \pi$ and is measured in radians. This is just the length of the arc of the great circle between P and Q.

On a sphere, we will have occasion to talk about *antipodal* points. Given a point P, the *antipodal* point to P is the point P' such that PP' is a diameter of the sphere.

We will also refer to *lunas*. A *luna* is the section of a sphere between two half great circles which make an angle θ with each other (like an orange peel, or the lighted portion of the moon visible from the Earth). Recall that the area of a sphere is $4\pi\rho^2$, so the area of a luna of angle θ is

$$\frac{\theta}{2\pi} 4\pi\rho^2 = 2\theta\rho^2.$$

Our first main result is to show that the area of a triangle on the sphere depends only on its angles.

Theorem 10.1.1. *Suppose $\triangle ABC$ is a triangle on the sphere of radius ρ. Then the area of $\triangle ABC$ is given by*

$$|\triangle ABC| = \rho^2(A + B + C - \pi),$$

209

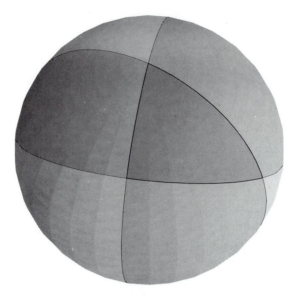

Figure 10.1. A spherical triangle, represented as the intersection of three *lunas*.

where the angles A, B, and C are measured in radians.

Proof. For a triangle $\triangle ABC$ on a sphere, sum up the three lunas associated to the three angles A, B, and C (see Figure 10.1), together with the antipodal lunas. By doing so, we have covered the entire sphere once and covered the triangle and its antipodal an additional two times each. Thus, we get

$$2(2\rho^2)(A + B + C) = 4\pi\rho^2 + 4|\triangle ABC|,$$

from which we get

$$|\triangle ABC| = \rho^2(A + B + C - \pi). \qquad \square$$

This result gives one pause for thought. Recall the similar result (page 164) for triangles in hyperbolic geometry:

$$|\triangle ABC| = k^2(\pi - A - B - C),$$

where the arclength element is

$$ds = k\frac{\sqrt{dx^2 + dy^2}}{y}.$$

This suggests a possible interpretation for hyperbolic space: Might hyperbolic space just be a sphere with imaginary radius ik? How should we interpret that statement?

Exercise 10.1. What is the area of a quadrilateral $ABCD$ on the sphere? What is the area of a polygon on the sphere?

Exercise 10.2. Show that the area A of a circle of radius r on the sphere of radius ρ is

$$A = 4\pi\rho^2 \sin^2\left(\frac{r}{2\rho}\right).$$

[H]

Exercise 10.3. Compare the formula for the area of a circle in spherical geometry with the one in hyperbolic geometry, which is

$$A = 4\pi k^2 \sinh^2\left(\frac{r}{2k}\right).$$

(We derived this for $k = 1$ in Exercise 7.88.) Show that the formula for the area on the sphere gives the formula for the area in $\mathcal{H}$ if we substitute $\rho = ik$. (Hint: Use Euler's formula to show that $i\sin\theta = \sinh i\theta$.)

Exercise 10.4. Substitute the linear approximation for $\sin\theta$ for θ near zero into the formula for the area of a circle on the sphere (see Exercise 10.2). This shows that this formula is consistent with the Euclidean formula for small circles on the sphere. This should be no surprise, since locally, spherical and Euclidean geometry are alike.

10.2   The Geometry of Right Triangles

As mentioned in Section 1.1, the Pythagorean theorem is a theorem in Euclidean geometry, and not in either spherical or hyperbolic geometry. There is, however, a spherical version.

Theorem 10.2.1 (Spherical Pythagorean Theorem). *Let $\triangle ABC$ be a right angle triangle on the unit sphere with right angle at C. As usual, let a, b, and c be the lengths of the sides opposite A, B, and C, respectively. Then*

$$\cos c = \cos a \cos b.$$

There is, not surprisingly, more than one way of proving this. We will use coordinate geometry,[1] which is a decidedly modern approach. The result dates back to Ptolemy (ca. 100 A.D.), whose approach is featured in Exercises 10.13 – 10.15.

Proof. Let the unit sphere **S** be centered at $(0,0,0)$ in $\mathbb{R}^3$. Let $\vec{A}$ and $\vec{B}$ be the vectors in $\mathbb{R}^3$ which represent the points A and B on the sphere **S**. Without loss of generality, we may place C at the North Pole $(0,0,1)$, place A in the xz-plane, and place B in the yz-plane. Then,

$$\vec{A} = (\sin b, 0, \cos b)$$

---

[1]A review of results in coordinate geometry appears in Appendix A.2.

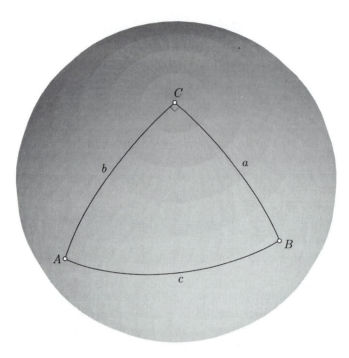

Figure 10.2. A spherical right angle triangle.

$$\vec{B} = (0, \sin a, \cos a).$$

Note that c is the angle between $\vec{A}$ and $\vec{B}$. Thus, taking the dot product of $\vec{A}$ and $\vec{B}$, we get

$$\vec{A} \cdot \vec{B} = \cos a \cos b = ||\vec{A}|| \, ||\vec{B}|| \cos c = \cos c. \qquad \square$$

There is also an analogue of trigonometry on the sphere:

Theorem 10.2.2. *Let $\triangle ABC$ be a right angle triangle on the unit sphere with right angle at C. Then*

$$\sin A = \frac{\sin a}{\sin c}$$

$$\cos A = \frac{\cos a \sin b}{\sin c}.$$

Proof. The angle A is the angle between the xz-plane and the plane spanned by $\vec{A}$ and $\vec{B}$. The vector

$$\vec{A} \times \vec{B} = (-\cos b \sin a, -\sin b \cos a, \sin b \sin a)$$

is perpendicular to the plane spanned by $\vec{A}$ and $\vec{B}$, and points up in Figure 10.2. Its length is the sine of the angle between, which is c. The angle A

is therefore the angle between $(0, -1, 0)$ and $\vec{A} \times \vec{B}$. Taking the dot product, we get

$$(0, -1, 0) \cdot (\vec{A} \times \vec{B}) = ||\vec{A} \times \vec{B}|| \cos A$$
$$\sin b \cos a = \sin c \cos A$$
$$\cos A = \frac{\sin b \cos a}{\sin c}.$$

To get the formula for $\sin A$, we can take the cross product, or use the identity $\sin^2 A + \cos^A = 1$. We opt to use the cross product:

$$||(0, -1, 0) \times (\vec{A} \times \vec{B})|| = ||\vec{A} \times \vec{B}|| \sin A$$
$$||(\sin a \sin b, 0, \sin a \cos b)|| = \sin c \sin A$$
$$\sin A = \frac{\sin a}{\sin c}.$$

We leave verification of the signs up to the reader.   □

In Section 1.1, we pointed out that the Euclidean Pythagorean theorem looks correct when drawn on a piece of paper, even if that paper is thought of as being part of the spherical Earth. That is, the spherical Pythagorean theorem should approximate the Euclidean Pythagorean theorem for very small lengths a, b, and c. This is indeed the case. Recall that the Taylor expansion for $\cos x$ is

$$\cos x = 1 - \frac{x^2}{2} + \frac{x^4}{4!} - \frac{x^6}{6!} + \dots.$$

Thus, the equation

$$\cos c = \cos a \cos b$$

becomes

$$1 - \frac{c^2}{2} + \text{(higher-order terms)} = 1 - \frac{a^2}{2} - \frac{b^2}{2} + \text{(higher-order terms)},$$

and for a, b, and c very small, we may ignore the higher order terms to get

$$c^2 = a^2 + b^2.$$

Exercise 10.5. Let $\triangle ABC$ be a right angle triangle on the unit sphere with right angle at C. Prove that

$$\cos A = \cos a \sin B.$$

What is the corresponding result in Euclidean geometry?   [A]

Exercise 10.6. Let $\triangle ABC$ be a right angle triangle on the unit sphere with right angle at C. Prove that

$$\cot A \cot B = \cos c.$$

Exercise 10.7. Show that the formulas for $\sin A$ and $\cos A$ give the usual definitions for the sine and cosine if the lengths a, b, and c are very small.

Exercise 10.8. What is the Pythagorean theorem on a sphere of radius ρ? What are the trigonometric formulas?

10.3 The Geometry of Spherical Triangles

Note that the side-side-side, side-angle-side, and angle-side-angle theorems can be proved using only Axiom 6. Thus, all sides and angles of a triangle can be found given only three appropriate pieces of information. In Euclidean geometry, the other sides and angles are found using the Law of Sines and Law of Cosines. In spherical geometry, there are similar laws.

Theorem 10.3.1 (Spherical Law of Sines). *Let $\triangle ABC$ be a triangle on the unit sphere with sides a, b, and c opposite the angles A, B, and C, respectively. Then*

$$\frac{\sin a}{\sin A} = \frac{\sin b}{\sin B} = \frac{\sin c}{\sin C}.$$

The proof is essentially the same as one of the proofs in Euclidean geometry.

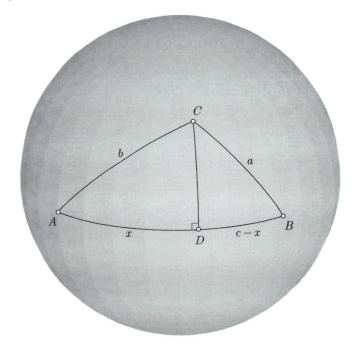

Figure 10.3

Proof. Choose D on AB so that CD is perpendicular to AB, as in Figure 10.3. Let $|CD| = h$. Then,

$$\sin A = \frac{\sin h}{\sin b}$$
$$\sin B = \frac{\sin h}{\sin a}.$$

Thus,

$$\sin A \sin b = \sin B \sin a,$$

which gives us the first equality. By using the altitude at B, we can similarly show

$$\frac{\sin a}{\sin A} = \frac{\sin c}{\sin C}. \qquad \square$$

Theorem 10.3.2 (Spherical Law of Cosines for Sides). *Let $\triangle ABC$ be a triangle on the unit sphere with sides a, b, and c opposite angles A, B, and C, respectively. Then*

$$\cos c = \cos a \cos b + \sin a \sin b \cos C.$$

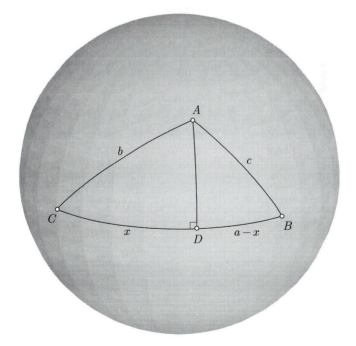

Figure 10.4

Proof. Consider Figure 10.4. From $\triangle ADB$ and the Pythagorean theorem, we get

$$\cos c = \cos h \cos(a - x) = \cos h \cos a \cos x + \cos h \sin a \sin x. \qquad (10.1)$$

From $\triangle ACD$, we get

$$\cos C = \frac{\cos h \sin x}{\sin b}$$

and from the Pythagorean theorem,

$$\cos b = \cos x \cos h.$$

Plugging these into Equation 10.1, we get

$$\cos c = \cos a \cos b + \sin a \sin b \cos C. \qquad \square$$

Note that if $C = \pi/2$, then we just get the spherical Pythagorean theorem.

These two laws allow us to completely solve SSS and SAS problems. In Euclidean geometry, an ASA problem is solved by first finding the third angle and then using the Law of Sines. Since the sum of the angles in a spherical triangle is not constant, we cannot solve ASA problems that way. We need another result – the Law of Cosines for angles:

Theorem 10.3.3 (Spherical Law of Cosines for Angles). *Let $\triangle ABC$ be a triangle with sides a, b, and c opposite angles A, B, and C. Then*

$$\cos C = -\cos A \cos B + \sin A \sin B \cos c.$$

Proof. Let us prove this in a rather uninspired way – let us just expand both sides and compare them.

Consider Figure 10.3. Let D be on AB so that CD is perpendicular to AB. Let $C_1 = \angle ACD$ and $C_2 = \angle DCB$. Let $AD = x$ and $DB = c - x$. Then

$$
\begin{aligned}
\cos C &= \cos(C_1 + C_2) \\
&= \cos C_1 \cos C_2 - \sin C_1 \sin C_2 \\
&= \frac{\cos x \sin h}{\sin b} \frac{\cos(c - x) \sin h}{\sin a} - \frac{\sin x}{\sin b} \frac{\sin(c - x)}{\sin a} \\
&= \frac{(\cos c + \sin x \sin(c - x)) \sin^2 h - \sin x \sin(c - x)}{\sin a \sin b} \\
&= \frac{\cos c \sin^2 h + \sin x \sin(c - x)(\sin^2 h - 1)}{\sin a \sin b} \\
&= \frac{\cos c \sin^2 h - \sin x \sin(c - x) \cos^2 h}{\sin a \sin b}.
\end{aligned}
$$

Expanding the right-hand side of the Law of Cosines, we get

$$- \cos A \cos B + \sin A \sin B \cos c$$
$$= -\frac{\cos h \sin x}{\sin b} \frac{\cos h \sin(c - x)}{\sin a} + \frac{\sin h \sin h}{\sin b \sin a} \cos c,$$

which is the same. □

Exercise 10.9. Show that the spherical Law of Sines is consistent with the Euclidean Law of Sines for very small sides a, b, and c.

Exercise 10.10. Show that the spherical Law of Cosines is consistent with the Euclidean Law of Cosines for very small sides a, b, and c.

Exercise 10.11. What are the Law of Sines and the Laws of Cosines on a sphere of radius ρ?

Exercise 10.12* (The Extended Law of Sines on the Sphere). Let R be the radius of the circumcircle of an arbitrary triangle $\triangle ABC$ on the unit sphere. Prove that

$$\tan R = \frac{\tan(a/2)}{\cos\left(\frac{B+C-A}{2}\right)} = \frac{\tan(b/2)}{\cos\left(\frac{A+C-B}{2}\right)} = \frac{\tan(c/2)}{\cos\left(\frac{A+B-C}{2}\right)}.$$

Verify that, for small lengths a, b, and c, this gives the extended Law of Sines in Euclidean geometry.

10.4   Menelaus' Theorem

Recall the theorem of Menelaus introduced in Section 1.15. There is a similar result in spherical geometry, which was also proved by Menelaus.

Theorem 10.4.1 (Menelaus' Theorem on the Sphere). *Let $\triangle ABC$ be a spherical triangle and let D, E, and F be points on the extended sides BC, CA, and AB, respectively, as in Figure 10.5. Then D, E, and F are collinear if and only if*

$$\frac{\sin|AF|}{\sin|FB|} \frac{\sin|BD|}{\sin|DC|} \frac{\sin|CE|}{\sin|EA|} = -1.$$

Here, the lengths $|AF|$, etc., are taken to be signed lengths, as discussed in Section 1.15. That is, if we take $|AF|$ to be the oriented angle $\angle AOF$, then $|FB|$ is the oriented angle $\angle FOB$.

Lemma 10.4.2. *Let A, B, and F be points on a unit circle centered at O and such that AB is not parallel to OF. Let the line AB intersect OF at F'. Then*

$$\frac{|AF'|}{|F'B|} = \frac{\sin\angle AOF}{\sin\angle FOB}.$$

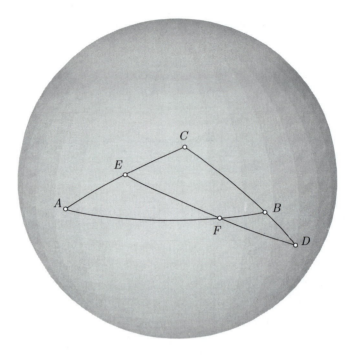

Figure 10.5

If we think of A, B, and F as points on a sphere with center O, then in the language of spherical geometry, this lemma says

$$\frac{|AF'|}{|F'B|} = \frac{\sin|AF|}{\sin|FB|},$$

where the lengths on the left-hand side are (signed) Euclidean lengths, and the lengths inside the trigonometric functions are (signed) spherical lengths.

Proof. Let us first observe that F and its antipodal point F'' generate the same line OF, and therefore the same F'. But $\sin\angle AOF'' = \sin(\pi - \angle AOF) = \sin\angle AOF$, so we may as well assume that F is on the same side of O as the line AB. We will further assume that F is on the shorter arc AB, as in Figure 10.6(a), and leave the other case as an exercise.

Let the bases of the perpendiculars to OF from A and B be labeled A' and B', respectively. Then

$$\triangle AA'F' \sim \triangle BB'F',$$

so

$$\frac{|AF'|}{|F'B|} = \frac{|AA'|}{|BB'|} = \frac{\sin\angle AOF}{\sin\angle FOB}. \qquad \square$$

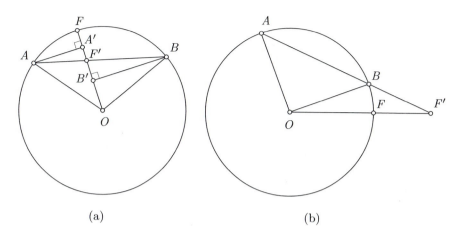

Figure 10.6

Proof of Menelaus' Theorem on the Sphere. Consider the plane through A, B, and C. It is possible that one or more of the lines OD, OE, and OF are parallel to this plane. We leave the proof for these possibilities as an exercise. So, in the case we are considering, the lines OD, OE, and OF intersect this plane at D', E', and F', respectively. The points D, E, and F are collinear on the sphere if and only if the points O, D, E, and F are coplanar. That is, D, E, and F are collinear on the sphere if and only if D', E', and F' are collinear on the plane through A, B, and C. But by Menelaus' theorem in Euclidean geometry, this is the case if and only if

$$\frac{|AF'|}{|F'B|} \frac{|BD'|}{|D'C|} \frac{|CE'|}{|E'A|} = -1. \tag{10.2}$$

By Lemma 10.4.2, Equation 10.2 is equivalent to

$$\frac{\sin|AF|}{\sin|FB|} \frac{\sin|BD|}{\sin|DC|} \frac{\sin|CE|}{\sin|EA|} = -1. \qquad \square$$

Theorem 10.2.2 and the spherical version of the Pythagorean theorem were first proved by Ptolemy. His approach is to use Menelaus' theorem for a particular configuration, which we introduce here. The proofs are left as exercises. Note that our proof of Menelaus' theorem does not use either of these results, so we are not entering a circular argument.

In a right angle triangle $\triangle ABC$ with right angle at C, we pick points D and E on the extended sides AC and AB, respectively, and such that $|AD| = |AE| = \pi/2$, as in Figure 10.7. Note that, if we think of A as a pole of the sphere, then the line EF is the equator. We call the line EF the *polar* of A, and we call A and its antipodal point A' the *poles* of the line EF. Since EF is the polar of A, the angles $\angle AEF$ and $\angle AFE$ are both $\pi/2$. Let us extend EF so that it intersects CB at D. Since both $\angle ACD$

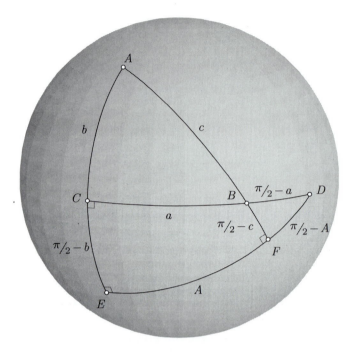

Figure 10.7

and $\angle AED$ are right angles, the point D is a pole of AC. In particular, $|DC|$ and $|DE|$ are both $\pi/2$.

Exercise 10.13. Derive the Pythagorean theorem by applying Menelaus' theorem to the triangle $\triangle ABC$ and the transversal DEF in Figure 10.7.

Exercise 10.14. Apply Menelaus' theorem to the triangle $\triangle AEF$ and the transversal CBD in Figure 10.7. Use the Pythagorean theorem to reduce the obtained expression into one of the formulas of Theorem 10.2.2.

Exercise 10.15. Apply Menelaus' theorem to the triangle $\triangle DEC$ and the transversal ABF in Figure 10.7. Use the results of Exercises 10.13 and 10.14 to reduce the obtained expression into the other formula of Theorem 10.2.2.

Exercise 10.16. Prove Lemma 10.4.2 for the case shown in Figure 10.6(b).

Exercise 10.17. Prove Menelaus' theorem on the sphere in the cases where one or more of OD, OE, and OF are parallel to the plane through A, B, and C.

Exercise 10.18 (Ceva's Theorem in Spherical Geometry). State and prove Ceva's theorem on the sphere.

Exercise 10.19. Suppose $ACBD$ is a quadrilateral with right angles at A, B, and C. Let $a = |BC|$ and $b = |AC|$. Prove

$$\cos D = \sin a \sin b.$$

10.5   Heron's Formula

Though it is clear that there should be some relationship between the different geometries, that relationship is made even more stunning by the vast array of results in Euclidean geometry which have versions in spherical and hyperbolic geometry. We have now seen spherical versions of the Pythagorean theorem, trigonometry, the Law of Sines and Cosines, Menelaus' theorem, and Ceva's theorem. Let us now add to that list a spherical version of Heron's formula.

Since the area of a spherical triangle is determined by its angles and since we can find those angles given the three sides of the triangle, there of course must exist a formula for the area in terms of the three sides. What is very curious is the shape of that formula and how it relates to Heron's formula in Euclidean geometry.

Theorem 10.5.1 (Heron's Formula on the Sphere). *Let $\triangle ABC$ be a spherical triangle with sides a, b, and c, and semiperimeter $s = \dfrac{a+b+c}{2}$. Let $\Delta = |\triangle ABC|$ be a shorthand for the area of $\triangle ABC$. Then*

$$1 - \cos \Delta = \frac{4 \sin s \sin(s-a) \sin(s-b) \sin(s-c)}{(1 + \cos a)(1 + \cos b)(1 + \cos c)}.$$

The proof is left to the reader, with the guidance of the following exercises.

Exercise 10.20. Let us begin with a question of plausibility. Verify that for small lengths a, b, and c, Heron's formula on the sphere is approximated by Heron's formula in Euclidean geometry.

Exercise 10.21. Let $\triangle ABC$ be a right angle triangle with right angle at C and area Δ. Prove that

$$\sin \Delta = \frac{\sin a \sin b}{1 + \cos c}$$

$$\cos \Delta = \frac{\cos a + \cos b}{1 + \cos c}.$$

Exercise 10.22. Let $\triangle ABC$ be a spherical triangle. Let the altitude at A be AD and let $h = |AD|$. Prove that

$$1 - \cos \Delta = \frac{\sin^2 h (1 - \cos c)}{(1 + \cos a)(1 + \cos b)}.$$

What formula in Euclidean geometry is approximated by this formula for small values of a, b, and c?   [H][A]

Exercise 10.23. Using the proof of Heron's formula in Euclidean geometry as a guide (see page 39), use the Law of Cosines and Exercise 10.22 to find a formula for $1 - \cos \Delta$ which depends only on the sides a, b, and c.

Exercise 10.24. Prove the following trig identity, which is found in the inside front or back cover of most calculus texts:

$$\cos \alpha - \cos \beta = -2 \sin \left(\frac{\alpha + \beta}{2} \right) \sin \left(\frac{\alpha - \beta}{2} \right).$$

Exercise 10.25. Finish the proof of Heron's formula on the sphere by reducing the answer found in Exercise 10.23. Use the proof of Heron's formula in Euclidean geometry as a guide, and when that is no longer helpful, use the trig identity in Exercise 10.24.

10.6   Tilings of the Sphere

The angles of a regular n-gon in Euclidean geometry are all equal to

$$\theta_n = \frac{(n - 2)\pi}{n}.$$

In hyperbolic geometry, there exists a regular n-gon whose angles are all equal to θ provided $\theta < \theta_n$. In spherical geometry, there exists a regular n-gon whose angles are all equal to θ provided $\theta_n < \theta < \pi$.

Suppose we have a tiling of the sphere with radius $\rho = 1$. Since there must be at least three tiles to a vertex and since the sum of the angles at each vertex is 2π, the tiles have at most 5 sides. Suppose we have a tiling with pentagons. Then, the angles of the regular pentagons are all $2\pi/3$, so the area of each pentagon is $5(2\pi/3) - 3\pi = \pi/3$ (We prove the area formula for a pentagon in Exercise 10.1.) Since the surface area of the sphere is 4π, there must be twelve pentagons. This, of course, is no surprise. After all, if we connect the vertices of the tiling with line segments in 3-space, then we get a Platonic solid with pentagonal faces. Thus, we get a dodecahedron.

More generally, there is a one-to-one correspondence between regular tilings of the sphere and Platonic solids. Hence, there are only five regular tilings of the sphere.

Similarly, there is a one-to-one correspondence between semiregular tilings of the sphere and semiregular solids.

Exercise 10.26. What is the length of the edges of the triangles and pentagons in the semiregular tiling of the unit sphere which has two triangles and two pentagons at each vertex?   [A][S]

Exercise 10.27. Find the percentage of the area of the sphere that is covered by triangles if the sphere has a semiregular tiling with two triangles and two pentagons at each vertex.   [S]

Exercise 10.28. What percentage of the sphere is covered by squares in the semiregular tiling with representation $(3, 4, 3, 4)$?

Exercise 10.29*. Find the percentage of the area of the soccer ball which is black. That is, find the percentage of the area of the soccer ball which is covered with pentagons.

10.7   The Axioms

Recall that our model is a sphere **S** in 3-space, centered at O and with radius ρ. The distance between two points on the sphere is given by

$$|PQ| = \rho \angle POQ$$

where $\angle POQ \leq \pi$ and is measured in radians.

In this section, we will check which of the axioms and definitions in Chapter 9 are satisfied by this model, and modify those that are not.

We begin with our definition of distance. We must check that it satisfies the three properties of a distance function. The first two are clearly satisfied. The third property, the triangle inequality, is also not too difficult to check. Consider the vertex at O created by PO, QO, and RO. It is clear that we must have $\angle POR + \angle ROQ \geq \angle POQ$.

If $\angle POQ < \pi$, then the set of points R for which we have equality is a section of the great circle that goes through P and Q. These sections satisfy the definition of a line segment, and are unique. However, if P and Q are antipodal, then any half great circle joining P and Q is a line segment joining them. Thus, we must remove the uniqueness statement in Axiom 1:

1. For any two points P and Q in **S**, there exists a line segment joining P and Q.

In Chapter 9, we defined lines to be indefinite continuations of line segments. It is clear that this cannot be done on the sphere. However, there is an alternate definition which works in Euclidean, hyperbolic, and spherical geometry – it is just a little more complicated to describe.

Definition 26. *Line.* A *line* is a set of points l such that if P and Q are in l, then there exists a line segment in l joining P and Q. Furthermore, there exists an $\epsilon > 0$ such that for any $P \in l$ and any real number $0 < r < \epsilon$, there exist exactly two points R and R' in l such that

$$d(P, R) = d(P, R') = r.$$

This definition is not too different than that given on page 197. We can think of continuing a line segment with a ruler of length ϵ. On the plane, this continuation continues indefinitely. On the sphere, it eventually ends.

However, on the sphere, we must be careful that our ruler is not as long as half a great circle.

With this definition of a line, Axiom 2 is satisfied.

The separation axioms are also satisfied on the sphere.

The definition for a circle is the same. Note that the circle with radius $\pi\rho$ has exactly one point, and all circles with radius larger than $\pi\rho$ are empty sets. Note also that if a circle centered at P has radius $r < \pi\rho$, then it is identical to the circle with radius $\pi\rho - r$ centered at P' where P' is the antipodal point to P.

The isometries of a sphere are induced by the isometries of $\mathbb{R}^3$ which fix O. These are rotations about any line through O, and reflections through any plane through O. On $\mathbf{S}$, these maps correspond to rotations about two antipodal points, or reflection through a great circle. Every rotation can be expressed as the product of two reflections.

Finally, we must modify Axiom 5, since every pair of lines on a sphere intersect twice.

5. Every pair of distinct lines intersect in exactly two points.

Thus, to define spherical geometry, we have to modify one definition (of a line) and two axioms (Axioms 1 and 5). The modified definition of a line, though, works well in all geometries, so if we had instead introduced that definition in Chapter 9, then the modification would not have been needed.

Exercise 10.30*. Show that every direct isometry of the sphere has at least two fixed points. Conclude that there are no translations on the sphere.

Exercise 10.31. Find an isometry of the sphere which has no fixed points. Write this isometry as a product of reflections and rotations.

10.8   Elliptic Geometry

The three geometries, Euclidean, hyperbolic, and spherical geometry, are distinguished by the parallel postulate. At one extreme, we have hyperbolic geometry, where parallels are not unique. In Euclidean geometry, the parallels are unique. In spherical geometry, every pair of lines intersects twice. It seems we have missed a possibility. Is it conceivable that there exists a geometry such that every pair of lines intersect exactly once? In this section, we present a model of *elliptic geometry* which satisfies this version of Axiom 5. Our study of elliptic geometry will be rather cursory. However, the next chapter is devoted to projective geometry, which can be thought of as elliptic geometry without a metric.

The model $\mathbf{P}$ of elliptic geometry that we will study is generated by taking the sphere $\mathbf{S}$ and identifying antipodal points. That is, we consider P and P' to be the same point. If P and Q are two points on the sphere, then the distance between them in elliptic geometry is the smaller of $\angle POQ$ and $\angle POQ'$ where Q and Q' are antipodal points. If $\angle POQ = \pi/2$, then $\angle POQ = \angle POQ'$, and the line segment PQ in $\mathbf{P}$ is not unique. However, there are only two line segments – the ones generated by PQ and PQ' on $\mathbf{S}$ (which are the same as those generated by $P'Q'$ and $P'Q$, respectively). Contrast this with spherical geometry, where the number of line segments between P and P' is infinite. Thus, to generate elliptic geometry, we must still modify Axiom 1.

With the definition of a line in Section 10.7, Axiom 2 is satisfied.

The separation axioms are not satisfied in elliptic geometry. The definition of 'same side' can be modified so that Axiom 3 (page 199) will still be satisfied, but with an appropriate modification, there will be only one side of a line in elliptic geometry, and hence, Axiom 4 will not be satisfied. However, the purpose of the separation axioms is to make the geometry two dimensional. This is already guaranteed by the revised Axiom 5, where we demand that every pair of lines intersect.

Exercise 10.32. Do the separation axioms follow from the other axioms of spherical geometry? That is, are they necessary or superfluous?

Exercise 10.33. What is the difference between spherical and elliptic geometry which makes Axiom 3 (with the given definition of 'same side') succeed in one and fail in the other?

The isometries on $\mathbf{P}$ are those induced by the isometries on $\mathbf{S}$.

Exercise 10.34. Find a nontrivial isometry of $\mathbf{S}$ which induces the identity on $\mathbf{P}$.

Exercise 10.35. Suppose Φ is an improper isometry of $\mathbf{S}$ which induces the isometry ϕ of $\mathbf{P}$. Show that there exists a proper isometry Φ' of $\mathbf{S}$ which induces the same isometry ϕ of $\mathbf{P}$.

Chapter 11

Projective Geometry

The inspiration for projective geometry is the human experience and the artist's perspective. Projective geometry describes the way we see, or more precisely, the way a pin-hole camera works. Given some object (say, a cube) in three dimensions, the picture of this object is its projection onto a plane through the pin-hole of the camera (see Figures 11.1 and 11.2).

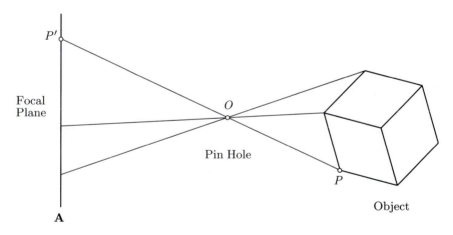

Figure 11.1. The idea behind a pin-hole camera.

We say that the points of the object have been *projected through a point* onto the plane. Such a picture is more realistic than a perpendicular projection since it incorporates perspective.

Mathematically, given a point O and a plane **A**, a point P in three dimensions is sent to the point P' where OP intersects **A** (see Figure 11.1). The point P' exists provided $P \neq O$ and OP is not parallel to **A**. Note that every point on the line OP is sent to P'. Thus, lines through O which are not parallel to **A** have been identified with points in the plane **A**. One

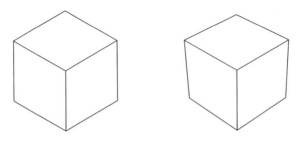

Figure 11.2. A cube, projected onto a plane both perpendicularly and through a point.

of the main ideas of projective geometry is the idea that we should not be constrained to lines not parallel to **A**. So let us extend **A** to include a *line at infinity*, which corresponds to all lines parallel to **A**. The plane **A** together with the line at infinity is called the *projective plane* and we denote it with $\mathbb{P}^2$.

A line in **A** corresponds to a plane through O, so we call the intersection of $\mathbb{P}^2$ with a plane through O a *line* in $\mathbb{P}^2$. Note that every pair of distinct planes through O intersect in a line through O. Thus, every pair of distinct lines in $\mathbb{P}^2$ intersect at exactly one point. We should therefore not be surprised that projective geometry and elliptic geometry are related. Note that the intersection of a sphere **S** centered at O and a line through O is a pair of antipodal points on **S**. In the elliptic geometry **P**, we identified antipodal points on a sphere **S**, so there exists a one-to-one correspondence between points in **P** and points in $\mathbb{P}^2$. A line in **P** corresponds to a great circle on **S**, which can be thought of as the intersection of **S** with a plane through O. Thus, there is also a one-to-one correspondence between lines in **P** and lines in $\mathbb{P}^2$. Since **P** includes a metric, there is a natural way to induce a metric on $\mathbb{P}^2$, but there is actually a lot to be gained by resisting this temptation.

Note that, on **P**, there is no distinguished line. Thus, when we think of $\mathbb{P}^2$ as being the plane **A** together with a line at infinity, we should not give this line any undue importance. It is just like any other line.

The plane **A** may be thought of as a Euclidean plane without a metric. When thought of this way, we call it the *affine plane*.

11.1   Moving a Line to Infinity

Sometimes, carefully chosen definitions can be very powerful because of the way they make us think about things. So far, we have only defined projective geometry, and it appears all we have is Euclidean geometry together with a line at infinity and without a metric. The proof of the following theorem illustrates just how powerful and how much more our definitions actually contain. This theorem is known as Pappus' theorem, and was introduced in Section 4.5 as an example of a result which we were not yet ready to prove but could be nicely demonstrated with Sketchpad. We are now ready to prove this result.

Theorem 11.1.1 (Pappus' Theorem). *Let P_1, P_2, and P_3 be three points on the line l_1, and let Q_1, Q_2, and Q_3 be three points on the line l_2. Let R be the intersection of P_2Q_3 and P_3Q_2; let S be the intersection of P_1Q_3 and P_3Q_1; and let T be the intersection of P_1Q_2 and P_2Q_1. Then R, S, and T are collinear (see Figure 11.3).*

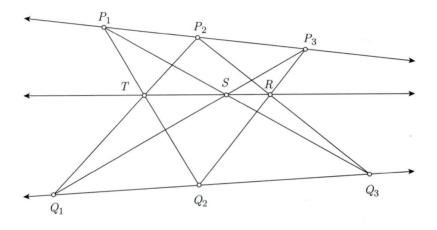

Figure 11.3

Proof. Let us first suppose we have such a diagram on a plane **A** embedded in three dimensions. Pick a point O not on **A**, and think of **A** as a subset of $\mathbb{P}^2$. Let us think of each point and line on **A** as being lines and planes through O respectively. Let l be the line through R and S. We want to show that T lies on l. Let the line l corresponds to a plane L through O, and let T correspond to the line t through O. We therefore want to show that t lies on L.

Let us now consider a different plane **A**$'$ not passing through O. The intersection of **A**$'$ with the lines and planes induced by the original diagram create a different diagram on **A**$'$. Label these new points P_1', P_2', etc. Note that T lies on l if and only if T' lies on l' in the diagram on **A**$'$. Thus, it is enough to prove the result for the diagram on **A**$'$. Our idea is to orient **A**$'$ in such a way that the diagram on **A**$'$ is more convenient to work with.

Let us choose **A**$'$ so that **A**$'$ and L are parallel. Then the line l' is the line at infinity, so R' and S' are at infinity. That is, the lines $P_2'Q_3'$ and $P_3'Q_2'$ are parallel, as are the lines $P_1'Q_3'$ and $P_3'Q_1'$. Hence, we get the convenient diagram pictured in Figure 11.4. The point T' is at infinity if and only if $P_1'Q_2'$ and $P_2'Q_1'$ are parallel, so let us now prove that.

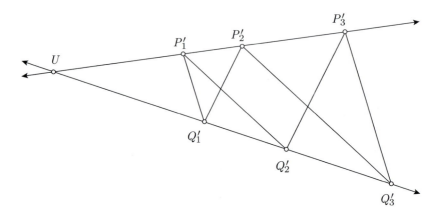

Figure 11.4

Let l_1' and l_2' intersect at U. Let us think of $\mathbf{A}'$ as a Euclidean plane, on which there exists a notion of distance. Since $P_2'Q_3'$ and $P_3'Q_2'$ are parallel, we know

$$\frac{|UP_2'|}{|UQ_3'|} = \frac{|UP_3'|}{|UQ_2'|},$$

and since $P_1'Q_3'$ and $P_3'Q_1'$ are parallel, we get

$$\frac{|UP_1'|}{|UQ_3'|} = \frac{|UP_3'|}{|UQ_1'|}.$$

Combining these two, we get

$$\frac{|UP_1'|}{|UQ_2'|} = \frac{|UP_2'|}{|UQ_1'|},$$

from which it follows that $P_1'Q_2'$ and $P_2'Q_1'$ are parallel. Thus, T' is on the line at infinity, so the line t is on the plane L, so T lies on the line l, as desired.   □

The idea of choosing $\mathbf{A}'$ so that it is parallel to L is called *moving a line to infinity*. This technique can also be used to prove Desargues' theorem, which was introduced in Section 4.5 too.

Theorem 11.1.2 (Desargues' Theorem). *Let P be a point not on $\triangle ABC$. Let A', B', and C' be points on the lines PA, PB, and PC, respectively. Let the (extended) sides BC and $B'C'$ meet at R. Similarly, let AC and $A'C'$ meet at S and let AB and $A'B'$ meet at T. Then R, S, and T are collinear (see Figure 4.5).*

We say the triangles $\triangle ABC$ and $\triangle A'B'C'$ are *perspective from a point* (the point P), or perspective from a line (the line RS).

Exercise 11.1. Prove Desargues' theorem by moving a line to infinity.

Exercise 11.2. Given a convex quadrilateral $ABCD$ in a plane and a point O not in the plane, show that there are points A' on OA, B' on OB, C' on OC, and D' on OD such that $A'B'C'D'$ is a parallelogram.

11.2 Pascal's Theorem

This idea of moving a plane about in three-space to get a different perspective on a diagram can be used in other ways. In this section, we prove Pascal's theorem, which was proved by Blaise Pascal (1623 – 1662) when he was sixteen. We state this result below in a slightly different way than when we first introduced it in Section 4.5. The wording there was chosen to draw attention to its similarity with Pappus' theorem.

Theorem 11.2.1 (Pascal's Theorem). *Let $ABCDEF$ be a hexagon inscribed in a conic $\mathcal{C}$. Let R be the intersection of the opposite sides AB and DE; let S be the intersection of the opposite sides BC and EF; and let T be the intersection of the opposite sides CD and FA. Then the points R, S, and T are collinear (see Figure 11.5).*

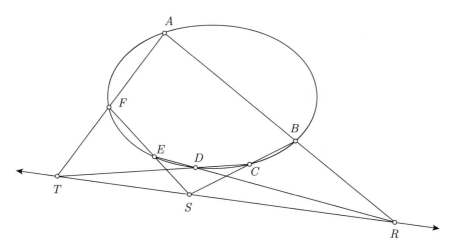

Figure 11.5

The proof is a bit complicated, so we break it down into a few steps. We begin with a lemma, which we were asked to prove in Exercise 1.37.

Lemma 11.2.2. *Let two circles Γ and Γ' intersect at A and B. Let CD be a chord on Γ. Let AC and BD intersect Γ' again at E and F. Then CD and EF are parallel (see Figure 11.6).*

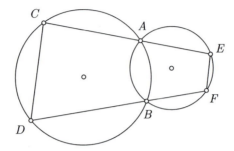

Figure 11.6

Proof. Using the Star Trek lemma twice, we note that

$$\angle ACD = 180° - \angle ABD = \angle ABF = 180° - \angle AEF,$$

so CD and EF are parallel. $\square$

We now prove Pascal's theorem in a special case:

Theorem 11.2.3. *Let $ABCDEF$ be a hexagon inscribed in a **circle** C. Let R be the intersection of the opposite sides AB and DE; let S be the intersection of the opposite sides BC and EF; and let T be the intersection of the opposite sides CD and FA. Then the points R, S, and T are collinear (see Figure 11.7).*

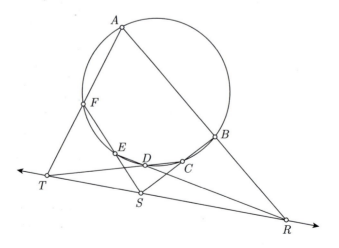

Figure 11.7

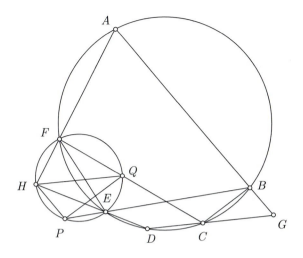

Figure 11.8

This proof is essentially due to van Yzeren [VY].

Proof. As in Figure 11.8, let AB and CD intersect at G, and let DE and FA intersect at H. Let Γ be the circumcircle of $\triangle EFH$. Let BE intersect Γ at P, and let CF intersect Γ at Q. By Lemma 11.2.2, BA is parallel to PH, CD is parallel to QH, and BC is parallel to PQ. Thus, triangles $\triangle BGC$ and $\triangle PHQ$ are similar and have parallel sides. Thus, by Exercise 1.47, BP, CQ, and GH are coincident. That is, the two triangles $\triangle BGC$ and $\triangle PHQ$ are perspective from a point. But then, the triangles $\triangle BGC$ and $\triangle EHF$ are also perspective from a point, so we can apply Desargues' theorem, from which we conclude that R, S, and T are collinear. $\qquad\square$

Proof of Pascal's Theorem. The conics are the ellipses, parabolas, and hyperbolas, and are so named because they can be realized as the intersection of a plane with a right circular cone. Let the conic $\mathcal{C}$ be the intersection of a plane $\mathbf{A}$ with a cone. Let the vertex of the cone be O, and think of $\mathbf{A}$ as a subset of $\mathbb{P}^2$. There exists another plane $\mathbf{A}'$ such that the intersection of the cone and $\mathbf{A}'$ is a circle. Thus, on $\mathbf{A}'$, our diagram is just the circle case of Pascal's theorem, which we just proved. Thus, R, S, and T are collinear. $\qquad\square$

As we just saw, a conic in projective geometry can be thought of as just a cone, and so all conics are essentially the same. The distinction between an ellipse, parabola, or hyperbola in the affine plane $\mathbf{A}$ is dependent on whether the line at infinity (which is the plane through O parallel to $\mathbf{A}$) respectively intersects the cone at exactly O; is tangent to the cone; or intersects the cone in two lines. Though it is possible to choose $\mathbf{A}'$ so that any conic becomes a circle, we can do this for only one conic at a time.

An easy way of seeing this is to note that it is possible for two ellipses to intersect at four points, but a pair of circles intersect in at most two points.

11.3   Projective Coordinates

We represent a point $P \in \mathbb{P}^2$ with the *projective coordinates* or *homogeneous coordinates* $P = (P_0, P_1, P_2)$, a triple of real numbers, not all of which equal zero. Two triples (P_0, P_1, P_2) and (Q_0, Q_1, Q_2) represent the same point if and only if $P_j = \lambda Q_j$ for all j and some $\lambda \neq 0$. The set of points $(\lambda P_0, \lambda P_1, \lambda P_2) \in \mathbb{R}^3$ for fixed P and all $\lambda \in \mathbb{R}$ is the line OP. Thus, a projective coordinate representation for a point P in $\mathbb{P}^2$ is the coordinates of any point on the line through O in $\mathbb{R}^3$ represented by P. We sometimes write $P = (P_0 : P_1 : P_2)$. The colons in this notation are to emphasize that any triple with the same ratios represents the same point.

A polynomial curve in $\mathbb{P}^2$ is the set of solutions to a polynomial equation

$$f(x, y, z) = 0 \tag{11.1}$$

where f is a polynomial with the property that

$$f(\lambda x, \lambda y, \lambda z) = \lambda^d f(x, y, z)$$

for some positive integer d. A polynomial with this property is called a *homogeneous polynomial* of degree d. Note that if $P \in \mathbb{R}^3$ is a solution to Equation 11.1 and $P \neq O$, then so is every point on the line OP. Thus, this equation is well defined on $\mathbb{P}^2$.

The homogeneous polynomial equations of degree one are all equations of the form

$$f(x, y, z) = ax + by + cz = 0.$$

These equations describe the planes through O, and hence the lines in $\mathbb{P}^2$.

Though we started with the projective plane, it is instructive to go down a dimension. The *projective line* is the set of lines through O in $\mathbb{R}^2$. These are the points with homogeneous coordinates (P_0, P_1). There is a natural embedding of the affine line into the projective line: $x \mapsto (x, 1)$. Thus, to create the projective line from an affine line, we must add the single point $(1, 0)$ at infinity.

The affine plane maps naturally into $\mathbb{P}^2$: $(x, y) \mapsto (x, y, 1)$. The points which are missing are all points of the form $(x, y, 0)$, which describe a projective line.

In a similar fashion, we may define projective three-space $\mathbb{P}^3$. This is the set of points with homogeneous coordinates (P_0, P_1, P_2, P_3). Projective three-space is the same as affine three-space together with a copy of $\mathbb{P}^2$ at infinity.

When we write $\mathbb{P}^2$, we usually mean *real projective space*, and if necessary, we emphasize this with the notation $\mathbb{P}^2(\mathbb{R})$. Complex projective space is denoted $\mathbb{P}^2(\mathbb{C})$ and is the set of all points with homogeneous coordinates (z_0, z_1, z_2) where $z_j \in \mathbb{C}$.

We have already investigated the complex projective line $\mathbb{P}^1(\mathbb{C})$, though we did not call it that. This is the set of all points with homogeneous coordinates (z_0, z_1) where $z_j \in \mathbb{C}$. If $z_1 \neq 0$, then

$$(z_0, z_1) = (z_0/z_1, 1),$$

so $\mathbb{P}^1(\mathbb{C})$ is just the complex plane together with the single point $(1, 0)$ at infinity. Recall that (in Chapter 7), we thought of the boundary of the upper half plane $\mathcal{H}$ as the real line together with the point at infinity. The fractional linear transformations can also be thought of in terms of projective coordinates. The point $z \in \mathbb{C}$ can be represented in $\mathbb{P}^1(\mathbb{C})$ with $(z, 1)$. Note that

$$\begin{bmatrix} a & b \\ c & d \end{bmatrix} \begin{bmatrix} z \\ 1 \end{bmatrix} = \begin{bmatrix} az + b \\ cz + d \end{bmatrix},$$

and that $(az + b, cz + d) = \left(\frac{az+b}{cz+d}, 1 \right)$, provided $cz + d \neq 0$. The symbol that we called ∞ in Chapter 7 is the projective point $(1, 0)$. Note that Exercise 7.32 is now trivial, since the composition of functions is just matrix multiplication, so the result follows immediately from the associativity of matrix multiplication.

The cross ratio can be interpreted in projective terms too. The cross ratio of the four points A, B, C and $D \in \mathbb{P}^2$ is

$$(A, B; C, D)$$
$$= ((A_0 C_1 - A_1 C_0)(B_0 D_1 - B_1 D_0), (A_0 D_1 - A_1 D_0)(B_0 C_1 - B_1 C_0)).$$

Thus, the cross ratio of four points in $\mathbb{P}^1$ is a point in $\mathbb{P}^1$. This point is well defined provided at least three of the points A, B, C, and D are distinct. With this definition, we do not need to invent the special symbol ∞ and define its algebra, since ∞ is already represented by $(1, 0)$.

Just as the action of invertible linear maps on $\mathbb{P}^1(\mathbb{C})$ proved interesting, so are the actions of invertible linear maps on $\mathbb{P}^2$. Let $T \in \mathrm{GL}_3(\mathbb{R})$. Since T is linear, it sends lines through O to lines through O, so induces a map on $\mathbb{P}^2$. Because T sends planes through O to planes through O, it sends lines to lines on $\mathbb{P}^2$, and since $T \in \mathrm{GL}_3(\mathbb{R})$, it is invertible. Thus, T sends one model of $\mathbb{P}^2$ to another model of $\mathbb{P}^2$. We call T an *isomorphism* of $\mathbb{P}^2$.

Recall that given any three distinct points A, B, and C in $\mathbb{P}^1(\mathbb{C})$ (or $\mathbb{P}^1(\mathbb{R})$), and their distinct images A', B', and C', there exists a fractional linear transformation γ such that $\gamma(A) = A'$, $\gamma(B) = B'$ and $\gamma(C) = C'$. As one might imagine there exists a similar result in $\mathbb{P}^2$:

Theorem 11.3.1. *Suppose A, B, C, and D are four points in $\mathbb{P}^2(\mathbb{R})$, no three of which are collinear; and suppose A', B', C', and D' are another set of four points in $\mathbb{P}^2(\mathbb{R})$, no three of which are collinear. Then there exists a $T \in \mathrm{GL}_3(\mathbb{R})$ such that $T(A) = A'$, $T(B) = B'$, $T(C) = C'$, and $T(D) = D'$.*

Proof. Let A, B, C, and D be represented with fixed projective coordinates, and think of these representations as points in $\mathbb{R}^3$. Then A, B, and C form a basis of $\mathbb{R}^3$, since they are not collinear, so we can find numbers a, b, and c such that

$$D = aA + bB + cC.$$

Similarly, there exist numbers a', b', and c' such that

$$D' = a'A' + b'B' + c'C'.$$

There exists a linear map T which sends the basis vectors aA, bB, and cC to the basis vectors $a'A'$, $b'B'$, and $c'C'$, respectively. This map T sends D to D', as desired. Thought of as an element of $\mathbb{P}^2(\mathbb{R})$, $A' = (a'/a)A'$, so

$$T(A) = \frac{1}{a}T(aA) = \frac{a'A'}{a} = A'.$$

Similarly, as elements of $\mathbb{P}^2(\mathbb{R})$, $T(B) = B'$ and $T(C) = C'$, as desired. $\square$

This result extends in a couple of obvious ways. We can replace the reals $\mathbb{R}$ with $\mathbb{C}$, or any other field. Also, it is possible to let A, B, and C be noncollinear points, but allow D to be collinear with two of them, provided A', B', and C' are noncollinear, and D' is collinear with the corresponding pair of points.

The consequence of this theorem is that, given some diagram, we can select any four noncollinear points and rearrange them into another set of four noncollinear points, giving a diagram which hopefully is more convenient to work with.

Exercise 11.3. In this exercise, we are asked to render a perspective drawing of a cube. The points (x, y, z) where $x \in \{2, 3\}$, $y \in \{3, 4\}$, and $z \in \{1, 2\}$ form a cube. Project these points on the plane $x = 1$.

Exercise 11.4. Prove that three points P, Q, and $R \in \mathbb{P}^2$ are collinear if and only if

$$\det \begin{bmatrix} P_0 & P_1 & P_2 \\ Q_0 & Q_1 & Q_2 \\ R_0 & R_1 & R_2 \end{bmatrix} = 0.$$

Exercise 11.5. Prove that the equation of the line going through P and Q is

$$\det \begin{bmatrix} x & y & z \\ P_0 & P_1 & P_2 \\ Q_0 & Q_1 & Q_2 \end{bmatrix} = 0.$$

Exercise 11.6. Prove that the intersection of two lines

$$A_0 x + A_1 y + A_2 z = 0$$
$$B_0 x + B_1 y + B_2 z = 0$$

is the point

$$P = \left(\det \begin{bmatrix} A_1 & A_2 \\ B_1 & B_2 \end{bmatrix}, \det \begin{bmatrix} A_2 & A_0 \\ B_2 & B_0 \end{bmatrix}, \det \begin{bmatrix} A_0 & A_1 \\ B_0 & B_1 \end{bmatrix} \right).$$

Exercise 11.7. What is the general form of a homogeneous polynomial equation of degree two in $\mathbb{P}^2$?

Exercise 11.8. The equation

$$f(x, y) = y^2 - x^3 + x + 1$$

is a polynomial equation in the affine plane. Find a homogeneous polynomial $F(x, y, z)$ with the property that $F(x, y, 1) = f(x, y)$. Does this new equation have any points at infinity? Decide how to find the tangent line of $F(x, y, z)$ at some point P on the curve $F(x, y, z) = 0$. What is(are) the tangent line(s) to this curve at the point(s) at infinity?

Exercise 11.9. Prove that the definition for the cross ratio given in this section is consistent with the definition given in Chapter 7.

11.4 Duality

A remarkably intriguing property of the projective plane is the notion of duality. This says that the roles of points and lines are interchangeable. Let us first investigate why this should be possible.

We can think of a point P in $\mathbb{P}^2$ as being a line through O. There exists a unique plane p which passes through O and is perpendicular to P. If $P = (P_0, P_1, P_2)$, then the equation of this plane is $P_0 x + P_1 y + P_2 z = 0$. The plane p represents a line in $\mathbb{P}^2$. Thus, we have a natural way of mapping points in $\mathbb{P}^2$ to lines in $\mathbb{P}^2$. We call the line p (represented by the plane p) the dual to the point P, or the *polar* of P. The point P is the dual or *pole* of the line p.

Suppose P and Q are two distinct points in $\mathbb{P}^2$. They describe a line PQ, which is represented with a plane l through O. Let L be the line perpendicular to l. We call the point L the dual to PQ. Note that the plane p perpendicular to P contains L, and similarly, the plane q perpendicular to Q contains L. Since P and Q are distinct, the planes p and q intersect in a line, which must be L.

Thus, under the dual map, points are sent to lines and lines are sent to points. Furthermore, if two points are on a line, then the intersection of the duals of the two points is the dual of the line through those two points. Similarly, if two lines intersect at a point, then the dual of that point is the line through the duals of the two lines.

For example, let us dualize Desargues' theorem. We begin with triangles $\triangle ABC$ and $\triangle A'B'C'$. The duals of A, B, and C are lines. Their points of intersection are the duals of the lines BC, AC, and AB. Let us call these points D, E, and F, respectively. Thus, the dual of $\triangle ABC$ is $\triangle DEF$. Similarly, let $\triangle D'E'F'$ be the dual of $\triangle A'B'C'$. The duals of the lines AA', BB', and CC' are three points. Let us call them U, V, and W. Since these three lines are coincident, their three dual points must be collinear. But the dual of the line AA' is the intersection of the edges EF and $E'F'$. That is, the edges EF and $E'F'$ intersect at U. Similarly, the pairs of edges DF and $D'F'$ intersect at V, and DE and $D'E'$ intersect at W. We let the intersection of BC and $B'C'$ be R. The dual of BC is D and the dual of $B'C'$ is D', so the dual of R is the line DD'. Similarly, the dual of S is the line EE', and the dual of T is the line FF'. And finally, for the conclusion. The statement that R, S, and T are collinear dualizes to the statement that DD', EE', and FF' are coincident. Thus, by dualizing Desargues' theorem, we get the following theorem:

Theorem 11.4.1. *Let $\triangle DEF$ and $\triangle D'E'F'$ be two triangles. Let EF and $E'F'$ intersect at U; let DF and $D'F'$ intersect at V; and let DE and $D'E'$ intersect at W. If U, V, and W are collinear, then the lines DD', EE', and FF' are coincident.*

That is, the dual of Desargues' theorem is just the converse of Desargues' theorem. Recall, a shortened statement of Desargues' theorem is 'Two triangles which are perspective from a point are perspective from a line.' The notions of *perspective from a point* and *perspective from a line* are dual notions.

Exercise 11.10. Find the dual theorem to Pappus' theorem.

Exercise 11.11. Come up with new definitions of rotations and translations that are duals of each other, are well defined in Euclidean, hyperbolic, and elliptic geometry, and are consistent with our definitions of rotations and translations in Euclidean geometry.   [A]

Exercise 11.12* (Sylvester's Problem). Suppose a set of n points in the plane has the property that a line through any two points goes through a third. Prove that the points are collinear.

11.5   Dual Conics and Brianchon's Theorem

Consider the cone
$$ax^2 + by^2 = cz^2. \tag{11.2}$$

The tangent plane to a point (P_0, P_1, P_2) on this cone is the plane

$$2aP_0 x + 2bP_1 y + 2cP_2 z = 0.$$

The dual of this line is the point $(2aP_0, 2bP_1, 2cP_2)$. Note that this is a solution to the equation

$$\frac{x^2}{a} + \frac{y^2}{b} = \frac{z^2}{c}. \tag{11.3}$$

Thus, the duals of the tangent lines to the cone given by Equation 11.2 are points on the cone given by Equation 11.3, and similarly, the duals of the points on Equation 11.2 are the tangent lines of Equation 11.3. We call Equation 11.3 the *dual conic* to Equation 11.2.

Equation 11.2 is not the general equation of a cone, but the result is true in general. We may think of the dual of a conic as being another conic, and that the duals of the points on the original conic are tangent lines of the dual conic, and vice versa. A consequence of this is the following theorem, which was originally proved this way.

Theorem 11.5.1 (Brianchon's Theorem). *Suppose it is possible to inscribe a conic in the hexagon $ABCDEF$. Then the diagonals AD, BE, and CF are coincident (see Figure 11.9).*

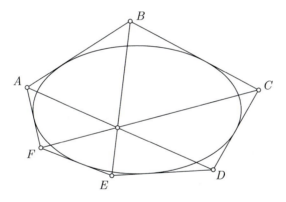

Figure 11.9

Proof. This is the dual of Pascal's theorem. Let us dualize the conditions of the theorem. We can think of the edges of the hexagon as tangent lines of the conic, so the dual scenario is six points on a conic, with edges a, b, c, d, e, and f. The dual of the diagonal AD is the intersection R of the lines a and d; the dual of the diagonal BE is the intersection S of the lines b and e; and the dual of CF is the intersection T of c and f. By Pascal's theorem, these three points R, S, and T are collinear. The dual of the line RS is the point of intersection of AD and BE; and the dual of ST is the intersection of BE and CF. But the lines RS and ST are the same, so their duals are the same. That is, AD, BE, and CF are coincident.   □

Exercise 11.13. This question requires some knowledge of linear algebra. We may write a general polynomial equation of degree two in the form

$$f(x, y, z) = ax^2 + by^2 + cz^2 + 2exy + 2fyz + 2gxz = 0,$$

which we may in turn write in the form

$$\begin{bmatrix} x & y & z \end{bmatrix} \begin{bmatrix} a & e & g \\ e & b & f \\ g & f & c \end{bmatrix} \begin{bmatrix} x \\ y \\ z \end{bmatrix} = 0.$$

Let us define

$$A = \begin{bmatrix} a & e & g \\ e & b & f \\ g & f & c \end{bmatrix}.$$

Let us also write $X = (x, y, z)$, so

$$f(x, y, z) = X^T A X.$$

Since A is symmetric, we know that it is diagonalizable. Prove that the curve $f(X) = 0$ has a solution in $\mathbb{P}^2(\mathbb{R})$ if and only if two of the eigenvalues of A are positive and one is negative, or if one is positive and two are negative. Prove that the tangent line at $P = (P_0, P_1, P_2)$ is the line

$$P^T A X = 0.$$

Finally, prove that the dual conic to $f(X) = 0$ is the conic

$$X^T A^{-1} X = 0.$$

11.6   Areal Coordinates

For the most part, we think of projective geometry as being useful to prove results in Euclidean geometry only if the result does not really involve the metric. In this section, we investigate a coordinate system which garners some of the utility of projective geometry but also keeps some of the consequences of the metric in Euclidean geometry.

We must first generalize our notion of area. The *signed area* of a triangle $\triangle ABC$ has the magnitude of the area of the triangle together with a sign which is positive if the points A, B, and C are oriented counterclockwise, and negative if they are oriented clockwise. We will use the same notation for the signed area as we have for the area of a triangle – namely $|\triangle ABC|$. In this section, all areas of triangles will be signed areas.

A point P in the Euclidean plane is given the *areal coordinates* $P = (P_0, P_1, P_2)$ with respect to a nondegenerate reference triangle $\triangle ABC$ where

$$P = (P_0, P_1, P_2) = \left(\frac{|\triangle PBC|}{|\triangle ABC|}, \frac{|\triangle APC|}{|\triangle ABC|}, \frac{|\triangle ABP|}{|\triangle ABC|} \right).$$

This coordinate system is called *areal* since it involves areas. Note that the vertices of the reference triangle have the coordinates

$$A = (1, 0, 0)$$
$$B = (0, 1, 0)$$
$$C = (0, 0, 1).$$

Note also that, for any point P, we have $P_0 + P_1 + P_2 = 1$, since

$$|\Delta PBC| + |\Delta APC| + |\Delta ABP| = |\Delta ABC|.$$

Remember, these are signed areas.

There is a way to realize areal coordinates in terms of projective coordinates. Let us think of the Euclidean plane as being represented by the plane $z = 1$ in $\mathbb{R}^3$, so a point P is represented with $P = (p_1, p_2, 1)$. Since ΔABC is nondegenerate, there exists a unique $T \in \mathrm{GL}_3(\mathbb{R})$ such that $T(A) = (1, 0, 0)$, $T(B) = (0, 1, 0)$, and $T(C) = (0, 0, 1)$. This is of course the map T such that

$$T^{-1} = \begin{bmatrix} a_1 & b_1 & c_1 \\ a_2 & b_2 & c_2 \\ 1 & 1 & 1 \end{bmatrix}.$$

Though this map sends A, B, and C to the appropriate points, it is not clear that it sends an arbitrary point $P = (p_1, p_2, 1)$ to the areal coordinates of P. Let us establish this.

Lemma 11.6.1. *The signed area of an arbitrary triangle ΔABC in the plane $z = 1$ in $\mathbb{R}^3$ is given by*

$$|\Delta ABC| = \frac{1}{2} \det \begin{bmatrix} a_1 & a_2 & 1 \\ b_1 & b_2 & 1 \\ c_1 & c_2 & 1 \end{bmatrix}.$$

The reader may recall that the absolute value of the determinant of three vectors is the volume of the parallelepiped they subtend. The proof of this result is very similar.

Proof. Let us think of the points A, B, and C as vectors in $\mathbb{R}^3$. Let us emphasize this by writing $\vec{A}$, etc. Recall that

$$\det \begin{bmatrix} a_1 & a_2 & 1 \\ b_1 & b_2 & 1 \\ c_1 & c_2 & 1 \end{bmatrix} = \vec{A} \cdot (\vec{B} \times \vec{C}) = \pm ||\vec{A}|| \, ||\vec{B}|| \, ||\vec{C}|| \sin \theta \cos \phi,$$

where θ is the angle between $\vec{B}$ and $\vec{C}$ and ϕ is the angle between $\vec{A}$ and the normal vector to $\vec{B}$ and $\vec{C}$. The volume V of the tetrahedron with vertices

O, A, B, and C is the area of the triangle with vertices O, B, and C, times one-third the height from this triangle to A. That is,

$$V = \frac{1}{3}(\|\vec{A}\| \cos \phi)\left(\frac{1}{2}\|\vec{B}\|\|\vec{C}\| \sin \theta\right) = \left|\frac{1}{6} \det \begin{bmatrix} a_1 & a_2 & 1 \\ b_1 & b_2 & 1 \\ c_1 & c_2 & 1 \end{bmatrix}\right|.$$

But we can also think of the volume V as the area of $\triangle ABC$ times one-third the distance from the plane $z = 1$ to O. This height is just one, so we also have

$$V = \left|\frac{1}{3}|\triangle ABC|\right|.$$

Finally, we note that the determinant is positive if $\vec{A}$, $\vec{B}$, and $\vec{C}$ obey the right-hand rule, and negative otherwise. That is, the determinant is positive if A, B, and C are ordered counterclockwise, and negative otherwise. Thus, we get

$$|\triangle ABC| = \frac{1}{2} \det \begin{bmatrix} a_1 & a_2 & 1 \\ b_1 & b_2 & 1 \\ c_1 & c_2 & 1 \end{bmatrix},$$

as desired.   □

Corollary 11.6.2. *Let $\triangle PQR$ be a triangle in the plane $z = 1$ in $\mathbb{R}^3$. Let $T \in \mathrm{GL}_3(\mathbb{R})$. Let us use the notation $P = (p_1, p_2, 1)$ and $T(P) = (P_0, P_1, P_2)$, etc. Then*

$$\det \begin{bmatrix} P_0 & Q_0 & R_0 \\ P_1 & Q_1 & R_1 \\ P_2 & Q_2 & R_2 \end{bmatrix} = \frac{1}{2} \det T |\triangle PQR|.$$

Proof. Note that

$$\det \begin{bmatrix} P_0 & Q_0 & R_0 \\ P_1 & Q_1 & R_1 \\ P_2 & Q_2 & R_2 \end{bmatrix} = \det\left(T \begin{bmatrix} p_1 & q_1 & r_1 \\ p_2 & q_2 & r_2 \\ 1 & 1 & 1 \end{bmatrix}\right) = \frac{1}{2} \det T |\triangle PQR|. \quad □$$

Corollary 11.6.3. *Let $\triangle ABC$ be a nondegenerate triangle on the plane $z = 1$ in $\mathbb{R}^3$. Let $T \in \mathrm{GL}_3(\mathbb{R})$ be the map that sends the points A, B, and C to the unit basis vectors. Let $P = (p_1, p_2, 1)$ and let $T(P) = (P_0, P_1, P_2)$. Then (P_0, P_1, P_2) are the areal coordinates of P with respect to the reference triangle $\triangle ABC$.*

Proof. Let $P = A$, $Q = B$, and $R = C$ in Corollary 11.6.2. Then

$$1 = \det \begin{bmatrix} 1 & 0 & 0 \\ 0 & 1 & 0 \\ 0 & 0 & 1 \end{bmatrix} = \frac{1}{2} \det T |\triangle ABC|.$$

If we let $P = P$, $Q = B$, and $R = C$ in Corollary 11.6.2, then we get

$$P_0 = \det \begin{bmatrix} P_0 & 0 & 0 \\ P_1 & 1 & 0 \\ P_2 & 0 & 1 \end{bmatrix} = \frac{1}{2} \det T |\triangle PBC|.$$

Dividing, we get

$$P_0 = \frac{|\triangle PBC|}{|\triangle ABC|}.$$

Similarly,

$$P_1 = \frac{|\triangle APC|}{|\triangle ABC|} \quad \text{and} \quad P_2 = \frac{|\triangle ABP|}{|\triangle ABC|}. \qquad \square$$

Corollary 11.6.4. *Suppose P, Q, and R are expressed in areal coordinates with respect to the reference triangle $\triangle ABC$. Then*

$$\frac{|\triangle PQR|}{|\triangle ABC|} = \det \begin{bmatrix} P_0 & Q_0 & R_0 \\ P_1 & Q_1 & R_1 \\ P_2 & Q_2 & R_2 \end{bmatrix}.$$

Proof. Divide the conclusion of Corollary 11.6.2 by $1 = \frac{1}{2} \det T |\triangle ABC|$.
$\qquad \square$

To help us familiarize ourselves with areal coordinates, let us do a quick exercise.

Exercise 11.14. What are the areal coordinates of I, the incenter of $\triangle ABC$?

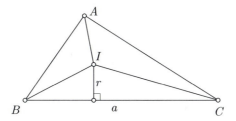

Figure 11.10

Solution. Let r be the inradius. Then, as can be seen in Figure 11.10,

$$I_0 = \frac{|\triangle IBC|}{|\triangle ABC|} = \frac{\frac{1}{2}ra}{\frac{1}{2}r(a+b+c)} = \frac{a}{a+b+c}.$$

We can similarly find I_1 and I_2, giving

$$I = \left(\frac{a}{a+b+c}, \frac{b}{a+b+c}, \frac{c}{a+b+c} \right),$$

or in projective coordinates,

$$I = (a, b, c).$$

Note that $\triangle ABC$ and $\triangle IBC$ will always have the same orientation, so the components of I are all positive. $\qquad\qquad\qquad\qquad\qquad\qquad\square$

Suppose P, Q, and R are collinear. Recall that we defined the *signed ratio of lengths* to be the ratio $\dfrac{|PQ|}{|QR|}$ together with a positive sign if Q is between P and R, and a negative sign otherwise. This definition was introduced in Section 1.15 on Menelaus' and Ceva's theorems.

The following theorem is a generalization of Ceva's theorem.

Theorem 11.6.5 (Routh's Theorem). *Let D, E, and F be points on the sides BC, AC, and AB, respectively, as in Figure 11.11. Let*

$$\lambda = \frac{|BD|}{|DC|}, \quad \mu = \frac{|AE|}{|EC|} \quad and \quad \nu = \frac{|AF|}{|FB|}$$

be signed ratios of lengths. Let BE and CF intersect at P; let AD and CF intersect at Q; and let AD and BE intersect at R. Then

$$|\triangle PQR| = \frac{(\lambda\mu\nu - 1)^2}{(\lambda\mu + \lambda + 1)(\mu\nu + \mu + 1)(\nu\lambda + \nu + 1)}.$$

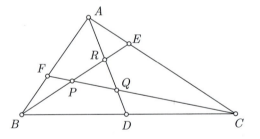

Figure 11.11

Proof. We first note that $|\triangle DBC| = 0$, so $D_0 = 0$. Also,

$$|DC| = \frac{1}{\lambda + 1}|BC|,$$

so

$$|\triangle ADC| = \frac{|BC|h}{2(\lambda + 1)},$$

where h is the altitude of $\triangle ABC$ at A. Since $|\triangle ABC| = \frac{1}{2}|BC|h$, we get

$$D_1 = \frac{1}{\lambda + 1}.$$

Since

$$|BD| = \frac{\lambda}{\lambda + 1}|BC|,$$

we similarly get

$$D_2 = \frac{\lambda}{\lambda + 1}.$$

Thus,

$$D = \left(0, \frac{1}{\lambda + 1}, \frac{\lambda}{\lambda + 1}\right),$$

or in projective coordinates,

$$D = (0, 1, \lambda).$$

Similarly, in projective coordinates, we get

$$E = (\mu, 0, 1)$$
$$F = (1, \nu, 0).$$

We might as well work with projective coordinates for now, since we can convert back to areal coordinates by dividing through by the sum of the components, since in areal coordinates, the sum of the components is one. We now calculate the equation of the line AD. These are the set of points $X = (X_0, X_1, X_2)$ such that $|\Delta ADX| = 0$. That is,

$$\det \begin{bmatrix} 1 & 0 & X_0 \\ 0 & 1 & X_1 \\ 0 & \lambda & X_2 \end{bmatrix} = X_2 - \lambda X_1 = 0.$$

Similarly, BE has the equation $X_0 - \mu X_2 = 0$, and CF has the equation $X_1 - \nu X_0 = 0$. The intersection of BE and CF is therefore

$$P = \left(\det \begin{bmatrix} 0 & -\mu \\ 1 & 0 \end{bmatrix}, \det \begin{bmatrix} -\mu & 1 \\ 0 & -\nu \end{bmatrix}, \det \begin{bmatrix} 1 & 0 \\ -\nu & 1 \end{bmatrix}\right) = (\mu, \mu\nu, 1).$$

Similarly,

$$Q = (1, \nu, \lambda\nu) \qquad \text{and} \qquad R = (\lambda\mu, 1, \lambda).$$

Thus,

$$\frac{|\Delta PQR|}{|\Delta ABC|} = \frac{\det \begin{bmatrix} \mu & 1 & \lambda\mu \\ \mu\nu & \nu & 1 \\ 1 & \lambda\nu & \lambda \end{bmatrix}}{(1 + \mu + \mu\nu)(1 + \nu + \lambda\nu)(1 + \lambda + \lambda\mu)}$$

$$= \frac{(\lambda\mu\nu - 1)^2}{(1 + \mu + \mu\nu)(1 + \nu + \lambda\nu)(1 + \lambda + \lambda\mu)}.$$

The quotient in the above result comes from converting P, Q, and R back to areal coordinates. $\qquad\square$

As mentioned earlier, Routh's theorem is a generalization of Ceva's theorem, since the latter is a corollary of the former:

Corollary 11.6.6 (Ceva's Theorem). *Let D, E, and F divide the segments BC, CA, and AB by the ratios of λ to 1, μ to 1, and ν to 1, respectively. Then AD, BE, and CF are coincident if and only if*

$$\lambda\mu\nu = 1.$$

Proof. The three points P, Q, and R are coincident if and only if the area of $\triangle PQR$ is zero. □

Exercise 11.15. Show that the centroid G has areal coordinates

$$G = (1/3, 1/3, 1/3).$$

Exercise 11.16. Show directly that the orthocenter H has areal coordinates

$$H = (\cot B \cot C, \cot A \cot C, \cot A \cot B).$$

Conclude that, if $A + B + C = 180°$, then

$$\tan A + \tan B + \tan C = \tan A \tan B \tan C.$$

Exercise 11.17. Prove that the areal coordinates of the circumcenter O is

$$O = u(\sin 2A, \sin 2B, \sin 2C),$$

where $1/u = \sin 2A + \sin 2B + \sin 2C$.

Exercise 11.18. What are the areal coordinates of the excenter I_a?   [A]

Exercise 11.19. Use areal coordinates to prove Menelaus' theorem.

Exercise 11.20. In the same way that Routh's theorem is a generalization of Ceva's theorem, there exists a generalization of Menelaus' theorem. Find and prove it.

Exercise 11.21. Let D, E, and F divide each of the sides BC, CA, and AB in the ratio of 2 to 1. Let P, Q, and R be the intersections of BE with CF, CF with AD, and AD with BE, respectively. Prove that the area of $\triangle PQR$ is one-seventh of the area of $\triangle ABC$.

Chapter 12

The Pseudosphere
in Lorentz Space

We have seen on several occasions that hyperbolic geometry might be thought of as a sphere of radius i. There are several interpretations of this analogy, none of which are entirely satisfactory (as analogies). In this chapter, we study one such interpretation, the *pseudosphere* imbedded in three-dimensional *Lorentz space*. Minkowski's study and development of four-dimensional Lorentz space (sometimes called *Minkowski space-time*) was of fundamental importance to the work of his student, Albert Einstein. Like the Poincaré models of hyperbolic geometry, the pseudosphere is also in common use today.

We begin our investigation with the study of a particular model of spherical geometry.

12.1 The Sphere as a Foil

We can think of the sphere **S** as the set of points in Cartesian 3-space a distance 1 away from the origin. This is the set of points $(x, y, z) \in \mathbb{R}^3$ which satisfy

$$x^2 + y^2 + z^2 = 1. \tag{12.1}$$

This model inherits the arclength element of Euclidean three-space:

$$ds = \sqrt{dx^2 + dy^2 + dz^2}. \tag{12.2}$$

Just like our definition for the upper half plane $\mathcal{H}$, this is a complete description of the sphere.

Let us interpret Equations 12.1 and 12.2 in terms of matrix products:

$$\vec{\mathbf{x}}^T I \vec{\mathbf{x}} = 1$$
$$ds^2 = d\vec{\mathbf{x}}^T I d\vec{\mathbf{x}},$$

247

where $\vec{x} = (x, y, z)$ and I is the identity matrix. It is not necessary to include the identity I in either equation above. We include it because we want to think of this model as a foil for the pseudosphere. In our model of the pseudosphere, the matrix I in these equations and the following equations will be substituted with something else.

Suppose a matrix T satisfies

$$T^T I T = I.$$

Then

$$(T d\vec{x})^T I T d\vec{x} = d\vec{x}^T (T^T I T) d\vec{x} = ds^2,$$

so T preserves the arclength element. Furthermore, and not surprisingly, for $\vec{x} \in S$,

$$(T \vec{x})^T I T \vec{x} = \vec{x}^T I \vec{x} = 1.$$

Thus, T is an isometry of **S**.

The group

$$O_3(\mathbb{R}) = \{T \in \mathrm{GL}_3(\mathbb{R}) : T^T I T = I\}$$

is called the *orthogonal group* on $\mathbb{R}^3$. Note that

$$\det(T^T I T) = \det I = 1$$
$$(\det T)^2 = 1,$$

so $\det T = \pm 1$. The group $O_3(\mathbb{R})$ is the full group of isometries on **S** (see Exercise 12.1). The subgroup $SO_3(\mathbb{R})$ is the group of orthogonal matrices with determinant one and is the group of proper isometries on **S** (see Exercise 12.2).

A line on **S** is a great circle, which can be thought of as the intersection of **S** with a plane through the origin O. The distance $|PQ|$ between two points P and Q on **S** is the angle $\angle POQ$. If we think of P and Q as vectors $\vec{P}$ and $\vec{Q}$, then the angle $\angle POQ$ is the angle between them, so

$$\vec{P} \cdot \vec{Q} = ||\vec{P}|| ||\vec{Q}|| \cos |PQ| = \cos |PQ|.$$

Note that the dot product is invariant under the action of $O_3(\mathbb{R})$. To see this, suppose $T \in O_3(\mathbb{R})$. Then

$$T\vec{P} \cdot T\vec{Q} = (T\vec{P})^T I T\vec{Q} = \vec{P}^T (T^T I T)\vec{Q} = \vec{P}^T I \vec{Q} = \vec{P} \cdot \vec{Q}.$$

Exercise 12.1. Let P be a point on **S**. Define

$$R_P(\vec{x}) = \vec{x} - 2(\vec{P} \cdot \vec{x})\vec{P}.$$

Prove that $R_P(\vec{x})$ is reflection through the polar of P. Prove that R_P is in $O_3(\mathbb{R})$ (or more precisely, show that the matrix which represents R_P is in $O_3(\mathbb{R})$). Conclude that $O_3(\mathbb{R})$ is the full group of isometries of **S**.

Exercise 12.2. Suppose $T \in SO_3(\mathbb{R})$. Prove that T has exactly two fixed points, so is a rotation on **S**.

Exercise 12.3. Suppose $T \in O_3(\mathbb{R})$ and $\det T = -1$. Prove that T is not a rotation, so must be an improper isometry.

Exercise 12.4. Show that the arclength element in polar coordinates is given by

$$ds^2 = \frac{dr^2}{1 - r^2} + r^2 d\theta^2$$

(for $r \neq 1$).

Exercise 12.5. Recall that the location of a point P in $\mathbb{R}^3$ can be described using spherical coordinates. The spherical coordinates of P are (ρ, ϕ, θ), where $|OP| = \rho$, ϕ is the angle OP makes with the z-axis, and θ is the angle that the projection of OP onto the xy-plane makes with the x-axis. Recall also that the Cartesian coordinates (x, y, z) of P are given by

$$(x, y, z) = (\rho \sin \phi \cos \theta, \rho \sin \phi \sin \theta, \rho \cos \phi).$$

Prove that the arclength element in three-space is

$$ds = \sqrt{\rho^2 \sin^2 \phi d\theta^2 + \rho^2 d\phi^2 + d\rho^2}.$$

Conclude that the arclength element on the sphere of radius one is

$$ds = \sqrt{\sin^2 \phi d\theta^2 + d\phi^2}.$$

Exercise 12.6. The map $-I$ is an isometry of S. Write it as a product of reflections in **S**.

Exercise 12.7. Prove the following properties of the cross product:

$$\vec{u} \times (\vec{v} \times \vec{w}) = (\vec{u} \cdot \vec{v})\vec{w} - (\vec{u} \cdot \vec{w})\vec{v}$$

$$(\vec{u} \times \vec{v}) \cdot (\vec{w} \times \vec{x}) = \det \begin{bmatrix} \vec{u} \cdot \vec{x} & \vec{u} \cdot \vec{w} \\ \vec{v} \cdot \vec{x} & \vec{v} \cdot \vec{w} \end{bmatrix}.$$

In the following sections, we will define analogues of the dot and cross products.

12.2 The Pseudosphere

Let us consider the surface $\mathcal{V}$ described by the equation

$$x^2 + y^2 - z^2 = -1,$$

which is a hyperboloid of two sheets (see Figure 12.1). Using the previous

Figure 12.1. A hyperboloid of two sheets.

section as a guide, let us write

$$\vec{x}^T J \vec{x} = -1,$$

where

$$J = \begin{bmatrix} 1 & 0 & 0 \\ 0 & 1 & 0 \\ 0 & 0 & -1 \end{bmatrix}.$$

This surface is imbedded in $\mathbb{R}^3$, which we usually think of as having a metric. Let us abandon that metric and instead define

$$ds^2 = d\vec{x}^T J d\vec{x} = dx^2 + dy^2 - dz^2. \qquad (12.3)$$

Unlike the arclength element on **S**, it is not so clear that this arclength element is well defined. That is, it is conceivable that the right-hand side of Equation 12.3 might sometimes be negative. This is in fact not possible, as seen in Exercise 12.8.

Again, we note that if

$$T^T J T = J,$$

then T preserves the arclength element, and furthermore, sends $\mathcal{V}$ to itself, so is an isometry of $\mathcal{V}$.

Unlike the sphere, $\mathcal{V}$ has two components, so there are no lines (on $\mathcal{V}$) joining points on one sheet with points on the other sheet. We will therefore restrict our attention to one sheet, the sheet $\mathcal{V}^+$ with $z > 0$. We call $\mathcal{V}^+$ the *pseudosphere*. The surface $\mathcal{V}^+$ can equivalently be thought of as $\mathcal{V}$ where every point P is identified with its antipodal point P' (the point P' represented by the vector $-\vec{P}$). In this respect, the surface $\mathcal{V}^+$ is more of an analogue of elliptic geometry.

The group of isometries on $\mathcal{V}$ can be classified as follows:

$$O_J = \{T \in \mathrm{GL}_3(\mathbb{R}) : T^T J T = J\}$$
$$O_J^+ = \{T \in O_J : T(\mathcal{V}^+) = \mathcal{V}^+\}$$
$$\mathrm{SO}_J^+ = \{T \in O_J^+ : \det T = 1\}.$$

The group O_J is called the *Lorentz group*, O_J^+ is the group of isometries on $\mathcal{V}^+$, and SO_J^+ is the group of proper isometries on $\mathcal{V}^+$ (see Exercises 12.13 and 12.14).

We expect that lines on $\mathcal{V}^+$ are the intersection of $\mathcal{V}^+$ with planes through the origin O. We verify this in Exercise 12.12.

Let us consider a particular line on $\mathcal{V}^+$, the line created by the intersection of $\mathcal{V}^+$ with the plane $y = 0$. This is the curve

$$x^2 - z^2 = -1,$$

which can be parameterized as

$$\vec{x}(\phi) = (\sinh \phi, 0, \cosh \phi).$$

The arclength of this curve for $\phi \in [0, \phi_0]$ is

$$s = \int_0^{\phi_0} \sqrt{\left(\frac{dx}{d\phi}\right)^2 + \left(\frac{dy}{d\phi}\right)^2 - \left(\frac{dz}{d\phi}\right)^2}\, d\phi$$
$$= \int_0^{\phi_0} \sqrt{\cosh^2 \phi - \sinh^2 \phi}\, d\phi$$
$$= \phi_0.$$

Thus, the distance between the 'North Pole' $N = (0, 0, 1)$ and the point $(\sinh \phi_0, 0, \cosh \phi_0)$ is ϕ_0. More generally, since both $\mathcal{V}$ and ds are radially symmetric with respect to the z-axis, the distance from N to an arbitrary point $P = (P_1, P_2, P_3)$ is $|NP| = \operatorname{arccosh} P_3$.

The use of the parameter ϕ hints at another important analogy. For any values θ and ϕ, the point

$$\vec{x} = (\cos \theta \sinh \phi, \sin \theta \sinh \phi, \cosh \phi)$$

lies on the surface $\mathcal{V}^+$.

The *Lorentz inner product* is our analogue of the dot product and is given by

$$\vec{x} \circ \vec{y} = \vec{x}^T J \vec{y}.$$

If $T \in O_J^+$, then

$$T\vec{x} \circ T\vec{y} = (T\vec{x})^T J (T\vec{y}) = \vec{x}^T (T^T J T) \vec{y} = \vec{x} \circ \vec{y},$$

so the Lorentz inner product is invariant under the action of O_J^+. In particular, taking the dot product of $N = (0, 0, 1)$ and $P = (P_1, P_2, P_3)$, we get

$$\vec{N} \circ \vec{P} = -P_3 = -\cosh|NP|.$$

For any points P and Q, there exists an isometry $T \in O^+$ which sends Q to N. Let $P' = T(P)$. Since T preserves the Lorentz product,

$$\vec{P} \circ \vec{Q} = \vec{P}' \circ \vec{N} = -\cosh|P'N|,$$

and since T is an isometry, it preserves lengths, so

$$|P'N| = |PQ|.$$

Thus, for any two points P and Q on $\mathcal{V}^+$, we have

$$\cosh|PQ| = -\vec{P} \circ \vec{Q}.$$

The space $\mathbb{R}^3$ equipped with the Lorentz inner product is called a *Lorentz space* and is sometimes denoted $\mathbb{R}^{2,1}$. The 'length' of a vector $\vec{x}$ in this space is defined to be

$$\|\vec{x}\| = \sqrt{\vec{x} \circ \vec{x}}.$$

The equation for the surface $\mathcal{V}$ can be written as

$$\vec{x} \circ \vec{x} = -1,$$

so it is the set of points a distance i away from the origin. That is, $\mathcal{V}$ is a sphere of radius i.

Exercise 12.8. Prove that the arclength element on $\mathcal{V}$ is given by

$$ds^2 = \frac{1}{z^2} \left(dx^2 + dy^2 + (y dx - x dy)^2 \right).$$ [S]

Exercise 12.9. Let $\vec{x}(t)$ be a piecewise smooth curve on $\mathcal{V}^+$ such that $\vec{x}(0) = N = (0, 0, 1)$ and $\vec{x}(t_0) = P = (x_0, 0, z_0)$. Prove that the arclength of $\vec{x}(t)$ for $t \in [0, t_0]$ is at least $\operatorname{arccosh}(z_0)$. Conclude that the line through N and P is the intersection of $\mathcal{V}^+$ with the plane $y = 0$.

Exercise 12.10. Let R_θ rotate $\mathcal{V}$ an angle θ about the z-axis. What is the matrix representation of R_θ? Prove that $R_\theta \in O_J^+$.   [A]

Exercise 12.11. Prove that the map

$$T_\phi = \begin{bmatrix} \cosh \phi & 0 & \sinh \phi \\ 0 & 1 & 0 \\ \sinh \phi & 0 & \cosh \phi \end{bmatrix}$$

is in O_J^+.

Exercise 12.12. Let P and Q be points on $\mathcal{V}^+$. Use Exercises 12.10 and 12.11 to prove that there exists a $T \in O_J^+$ such that $TQ = N$ and $TP = P'$ where the y component of P' is zero. Use Exercise 12.9 to conclude that the line NP' is the intersection of $\mathcal{V}^+$ with the plane $y = 0$. Conclude that the line PQ is the intersection of $\mathcal{V}^+$ with the plane through P, Q and the origin O.

Exercise 12.13†. Let $\vec{a} \circ \vec{a} = 1$. Prove that the plane

$$\vec{a} \circ \vec{x} = 0$$

intersects $\mathcal{V}$. Define the map

$$R_{\vec{a}}(\vec{x}) = \vec{x} - 2(\vec{a} \circ \vec{x})\vec{a}.$$

Prove that the matrix which represents $R_{\vec{a}}$ is in O_J^+, and that it is reflection through the line described by the intersection of $\mathcal{V}^+$ with the plane $\vec{a} \circ \vec{x} = 0$. Conclude that, if $\mathcal{V}^+$ is a model of hyperbolic geometry, then O_J^+ is the full group of isometries of $\mathcal{V}^+$.

Exercise 12.14. Suppose $T \in O_J^+$. Prove that T is a proper isometry if and only if $\det T = 1$. [H]

Exercise 12.15. Show that the arclength element on $\mathcal{V}$ expressed in polar coordinates is given by

$$ds^2 = \frac{dr^2}{1 + r^2} + r^2 d\theta^2.$$

Exercise 12.16. Recall, the surface $\mathcal{V}$ can be parameterized by

$$\vec{x}(\theta, \phi) = (\cos \theta \sinh \phi, \sin \theta \sinh \phi, \cosh \phi).$$

What is the arclength element on $\mathcal{V}$ expressed in terms of θ and ϕ?

Exercise 12.17. Suppose $\vec{u} = (u_1, u_2, u_3)$ and $\vec{v} = (v_1, v_2, v_3)$ are two vectors with $u_3, v_3 > 0$ and $\vec{u} \circ \vec{u} < 0$, $\vec{v} \circ \vec{v} < 0$. Let us define ϕ such that

$$\vec{u} \circ \vec{v} = ||\vec{u}|| ||\vec{v}|| \cosh \phi.$$

Prove that ϕ is well defined and depends only on the directions of $\vec{u}$ and $\vec{v}$, and not on their magnitudes.

Exercise 12.18. We have discovered that the distance between two points P and Q on $\mathcal{V}^+$ is given by

$$|PQ| = \text{arccosh}(-\vec{P} \circ \vec{Q}).$$

Prove directly that this distance function satisfies the triangle inequality.

12.3   Angles and the Lorentz Cross Product

An angle between two lines on the sphere **S** is the angle between the two planes which define the lines. We would therefore expect something similar on the pseudosphere $\mathcal{V}^+$.

In Exercise 12.13, we saw that a plane which intersects $\mathcal{V}^+$ can be written as

$$\vec{a} \circ \vec{x} = 0,$$

where $\vec{a} \circ \vec{a} = 1$. More generally, a plane is defined by a vector $\vec{a}$ where $\vec{a} \circ \vec{a} > 0$. We therefore expect that the angle between two lines defined by $\vec{a} \circ \vec{x} = 0$ and $\vec{b} \circ \vec{x} = 0$ should be some quantity which depends only on $\vec{a}$ and $\vec{b}$; is independent of the magnitudes of these vectors; and is invariant under the action of O_J^+.

Let us begin with $\vec{a}$ and $\vec{b}$ such that $||\vec{a}|| = ||\vec{b}|| = 1$. There exists a $T \in O_J^+$ such that $T\vec{a} = (1, 0, 0)$ and $T\vec{b} = \vec{b}' = (b_1', b_2', 0)$ (see Exercise 12.19). Both points lie on the surface

$$x^2 + y^2 - z^2 = 1$$

and have the z component equal to zero, so lie on the curve

$$x^2 + y^2 = 1.$$

This is a circle, so it can be parameterized by θ via $(x, y, z) = (\cos\theta, \sin\theta, 0)$. In particular, there exists a θ such that $\vec{b}' = (\cos\theta, \sin\theta, 0)$. Then,

$$\vec{a} \circ \vec{b} = (1, 0, 0) \circ (\cos\theta, \sin\theta, 0) = \cos\theta.$$

Now, let us suppose that $\vec{a}$ and $\vec{b}$ both have positive lengths. By dividing through by their lengths, we get vectors with length one, and since the Lorentz product is linear, we get

$$\frac{\vec{a} \circ \vec{b}}{||\vec{a}|| \, ||\vec{b}||} = \left(\frac{\vec{a}}{||\vec{a}||}\right) \circ \left(\frac{\vec{b}}{||\vec{b}||}\right) = \cos\theta$$

for some angle θ which is independent of the length of the vectors $\vec{a}$ and $\vec{b}$, and is independent of the action of O_J^+. Thus, we define the angle between two lines on $\mathcal{V}^+$ described by the equations $\vec{a} \circ \vec{x} = 0$ and $\vec{b} \circ \vec{x} = 0$ to be the angle θ such that

$$\vec{a} \circ \vec{b} = ||\vec{a}|| \, ||\vec{b}|| \circ \cos\theta.$$

Note that, if we restrict θ to the range $[0, \pi]$, then for any two planes there are two possible angles, depending on whether one of the planes is described by $\vec{a}$ or $-\vec{a}$. These two angles are, of course, supplementary angles.

Let us now turn our attention to finding $\vec{a}$. Given points P and Q on $\mathcal{V}^+$, let the line through P and Q be defined by the plane $\vec{a} \circ \vec{x} = 0$. Note that

$$\vec{a} \circ \vec{x} = \vec{a}^T J \vec{x} = J\vec{a} \cdot \vec{x}.$$

Thus, we can choose $\vec{a}$ so that

$$J\vec{a} = \vec{P} \times \vec{Q}.$$

We therefore define the *Lorentz cross product* to be

$$\vec{P} \otimes \vec{Q} = J(\vec{P} \times \vec{Q}).$$

Exercise 12.19. Suppose $||\vec{a}|| = ||\vec{b}|| = 1$. Prove that there exists a $T \in O_J^+$ such that $T\vec{a} = (1, 0, 0)$ and $T\vec{b} = \vec{b}' = (b_1', b_2', 0)$. [H]

Exercise 12.20. Prove the following properties of the Lorentz cross product:

$$\vec{u} \otimes \vec{v} = J\vec{v} \times J\vec{u}$$
$$\vec{u} \otimes \vec{v} = -\vec{v} \otimes \vec{u}$$
$$\vec{u} \circ (\vec{v} \otimes \vec{w}) = \det \begin{bmatrix} u_1 & u_2 & u_3 \\ v_1 & v_2 & v_3 \\ w_1 & w_2 & w_3 \end{bmatrix}$$
$$\vec{u} \otimes (\vec{v} \otimes \vec{w}) = (\vec{u} \circ \vec{v})\vec{w} - (\vec{u} \circ \vec{w})\vec{v}$$
$$(\vec{u} \otimes \vec{v}) \circ (\vec{w} \otimes \vec{x}) = \det \begin{bmatrix} \vec{u} \circ \vec{x} & \vec{u} \circ \vec{w} \\ \vec{v} \circ \vec{x} & \vec{v} \circ \vec{w} \end{bmatrix}.$$

Exercise 12.21. Suppose $\vec{a}$ and $\vec{b}$ have positive lengths. Prove that

$$||\vec{a} \otimes \vec{b}|| = i||\vec{a}||||\vec{b}|| \sin \theta,$$

where θ is the angle defined by $\vec{a} \circ \vec{b} = ||\vec{a}||||\vec{b}|| \cos \theta$.

Exercise 12.22. Construct a proof of the hyperbolic Pythagorean theorem and Theorem 7.16.2 which is in the spirit of the proofs of the spherical Pythagorean theorem and Theorem 10.2.2 given in Section 10.2. [H]

Exercise 12.23. We saw that the surface $\mathcal{V}^+$ can be parameterized by ϕ and θ so that a point $P(\phi, \theta)$ has the coordinates

$$P(\phi, \theta) = (\sinh \phi \cos \theta, \sinh \phi \sin \theta, \cosh \phi).$$

The appropriate analogue on the sphere **S** is

$$P(\phi, \theta) = (\sin \phi \cos \theta, \sin \phi \sin \theta, \cos \phi).$$

What is the analogue of this parameterization in Euclidean geometry? [A]

12.4   A Different Perspective

Though we have frequently referred to the pseudosphere as a model of hyperbolic geometry, we have not yet proved this assertion. We will do so in this section.

We begin by taking a second look at the Poincaré disc model. In this model, the disc represents the entire plane, and distances get longer as we approach the boundary of the disc. We might imagine that we are looking down at the hyperbolic plane, and that the plane is curving away from us.

Now, let us take the pseudosphere and rotate it away from us, until we are looking up along the z-axis, as in Figure 12.3. What we see is a disc. Thus, it appears as though the Poincaré disc model might just be the pseudosphere viewed from a different perspective. This is indeed the case, as we shall see.

We project $\mathcal{V}^+$ onto the plane $z = 0$ using stereographic projection through the point $(0, 0, -1)$. If we use polar coordinates, then the angle θ remains unchanged under the projection, so we really have a two-dimensional problem. That is, we can think of $\mathcal{V}^+$ as the surface of revolution found by rotating the top half of the hyperbola $r^2 - z^2 = -1$.

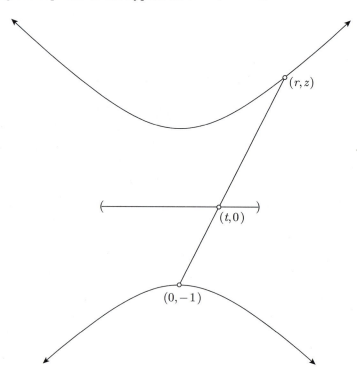

Figure 12.2

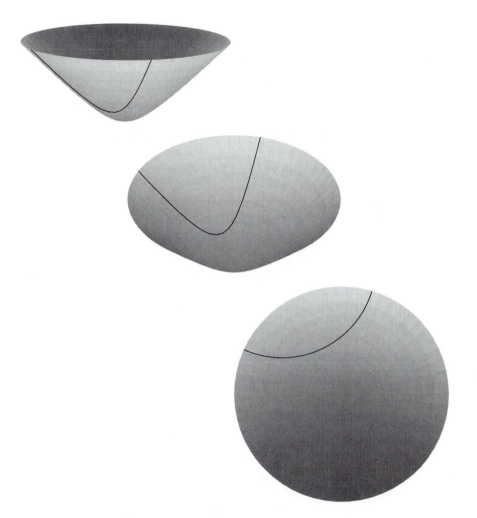

Figure 12.3. We rotate the pseudosphere $\mathcal{V}^+$ away from us until we are looking down the z-axis. In the last figure, which looks like a disc, we choose our point of view to be a particular point on the z-axis. From this vantage, a hyperbolic line looks like an arc of a circle. Furthermore, in this last figure we are looking at all of $\mathcal{V}^+$, not just a portion of it. The boundary of the disc is the 'horizon' of the plane.

In Figure 12.2, let us denote a point on the hyperbola with (r, z) and a point on the line $z = 0$ with $(t, 0)$. Then, the vector from $(0, -1)$ to (r, z) is a multiple of the vector from $(0, -1)$ to $(t, 0)$, so

$$(r, z + 1) = \lambda(t, 1)$$

$$r = \lambda t \tag{12.4}$$

$$z = \lambda - 1. \tag{12.5}$$

The set of points on the top half of the hyperbola are mapped to the interval $(-1, 1)$, and the bottom half is mapped to $(-\infty, -1) \cup (1, \infty)$. No points are mapped to ± 1. Thus, $\mathcal{V}^+$ is mapped to the unit disc, not including its boundary. This gives us a disc model, but not necessarily the Poincaré disc model. To see that this is in fact the case, it is enough to check that the arclength elements are the same.

In polar coordinates, the arclength element on $\mathcal{V}^+$ is given by (see Exercise 12.15)

$$ds^2 = \frac{dr^2}{1 + r^2} + r^2 d\theta^2.$$

The arclength element on the Poincaré disc $\mathcal{D}$ (in polar coordinates using t to represent the distance from the origin) is given by (see Exercise 7.81)

$$ds^2 = \frac{4dt^2 + 4t^2 d\theta^2}{(1 - t^2)^2}.$$

Let us plug Equations 12.4 and 12.5 into the equation of the hyperbola to get

$$\lambda(\lambda t^2 - \lambda + 2) = 0.$$

It is no coincident that $\lambda = 0$ is a solution. This gives the point $(0, -1)$, which we carefully chose to be on the hyperbola. We are interested in the solutions with $\lambda \neq 0$, so we get

$$\lambda = \frac{2}{1 - t^2},$$

and hence

$$r = \frac{2t}{1 - t^2} \quad \text{and} \quad z = \frac{1 + t^2}{1 - t^2}.$$

Hence,

$$dr = \frac{2(1 - t^2)dt + 4t^2 dt}{(1 - t^2)^2} = \frac{2(1 + t^2)dt}{(1 - t^2)^2},$$

so

$$ds^2 = \frac{4(1 + t^2)^2 dt^2 (1 - t^2)^2}{(1 - t^2)^4 ((1 - t^2)^2 + 4t^2)} + \frac{4t^2 d\theta^2}{(1 - t^2)^2}$$

$$= \frac{4dt^2 + 4t^2 d\theta^2}{(1 - t^2)^2},$$

as desired. Thus, the geometry of the surface $\mathcal{V}^+$ is the same as the geometry of the Poincaré disc $\mathcal{D}$. In particular, we now know that the pseudosphere $\mathcal{V}^+$ is a model of hyperbolic geometry.

Exercise 12.24. Suppose the intersection of $\mathcal{V}^+$ with $\vec{a} \circ \vec{x} = d$ is bounded and nonempty. Prove that this intersection is a (hyperbolic) circle. What are the necessary and sufficient conditions on $\vec{a}$ and d for this to happen? What is the center of this circle?   [H]

12.5   The Beltrami-Klein Model

The success of the projection done in the previous section depends very much on the vantage point of our perspective. That is, the choice of the point $(0, 0, -1)$ through which we project is very important. In this section, we will instead project onto the plane $z = 1$ and through the point $(0, 0, 0)$. The surface $\mathcal{V}^+$ again projects to a unit disc, not including its boundary. This time, rather than investigate the arclength element, let us just ask what happens to lines under this projection. Since lines on $\mathcal{V}^+$ are the intersection of $\mathcal{V}^+$ with planes through the origin, the projection of a line is just the intersection of the plane that describes the line together with the unit disc on the plane $z = 1$. That is, lines in this new model are just chords of the unit disc (see Figure 12.4).

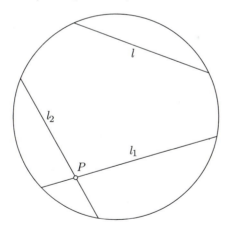

Figure 12.4. The Klein model of hyperbolic geometry together with some lines and a point which illustrate the fifth axiom.

This model was first developed by Eugenio Beltrami (1835 – 1900). The model is called the Beltrami-Klein model or Klein model, after Beltrami and Felix Klein (1849 – 1925). The formula for distance in this model was developed by Klein. This model predates both the Poincaré models and the pseudosphere.

12.6   Menelaus' Theorem

By now, the similarities between the sphere and the pseudosphere should be quite apparent, so it should not be any surprise that there is a version of Menelaus' theorem in hyperbolic geometry.

Theorem 12.6.1 (Menelaus' Theorem in Hyperbolic Geometry). *Let $\triangle ABC$ be a triangle in hyperbolic geometry. Let D, E, and F be points on the extended sides BC, AC, and AB, respectively. Then D, E, and F are collinear if and only if*

$$\frac{\sinh|AF|}{\sinh|FB|}\frac{\sinh|BD|}{\sinh|DC|}\frac{\sinh|CE|}{\sinh|EA|} = -1.$$

Again, in keeping with the convention concerning signed lengths introduced in Section 1.15, we give the ratio

$$\frac{\sinh|AF|}{\sinh|FB|}$$

a positive sign if F is between A and B, and a negative sign otherwise.

To prove Menelaus' theorem, we need an analogue of Lemma 10.4.2:

Lemma 12.6.2. *Let A, B, and F be collinear points on $\mathcal{V}^+$. Let F' be the intersection of the (Euclidean) lines OF and AB. Then*

$$\frac{|AF'|}{|F'B|} = \frac{\sinh|AF|}{\sinh|FB|},$$

where the lengths on the left are Euclidean lengths, the lengths inside the hyperbolic trigonometric functions are hyperbolic lengths, and the ratios are signed ratios.

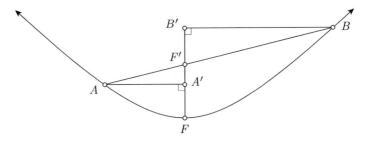

Figure 12.5

Proof. Let us first consider a special case – the case where F is the 'North Pole' $(0, 0, 1)$ and A and B lie on the plane $y = 0$, as in Figure 12.5. Then, as we saw in Section 12.2, the points A and B have the coordinates

$(\pm \sinh |AF|, 0, \cosh |AF|)$ and $(\pm \sinh |BF|, 0, \cosh |BF|)$, respectively. Let the bases of the perpendiculars to OF from A and B be A' and B', respectively. Then, $|AA'| = \sinh |AF|$ and $|BB'| = \sinh |BF|$, where the lengths on the left are Euclidean, and the lengths inside the trig functions are hyperbolic. We note that $\triangle AFA' \sim \triangle BFB'$, so we get the following equality of unsigned ratios:

$$\frac{|AF|}{|FB|} = \frac{|AA'|}{|B'B|} = \frac{\sinh |AF|}{\sinh |FB|}.$$

Applying the correct signs to the ratios at either end, we note that F' is between A and B if and only if F is between A and B, so the signed ratios are equal too.

Now, suppose F is not the point $(0, 0, 1)$, or that either A or B is not on the plane $y = 0$. Then, there exists an isometry $T \in O_J^+$ such that the $T(F) = (0, 0, 1)$ and $T(A)$ lies on the plane $y = 0$. Since A, B, and F are collinear, the point $T(B)$ also lies on the plane $y = 0$. Thus,

$$\frac{|T(A)T(F)'|}{|T(F)'T(B)|} = \frac{\sinh |T(A)T(F)|}{\sinh |T(F)T(B)|},$$

where $T(F)'$ is the intersection of $T(F)O$ with $T(A)T(B)$. Since T is an isometry,

$$\frac{\sinh |T(A)T(F)'|}{\sinh |T(F)'T(B)|} = \frac{\sinh |AF|}{\sinh |FB|}.$$

Since T is linear, it sends lines to lines, so $T(F)' = T(F')$. Also, if we set

$$\lambda = \frac{|AF'|}{|AB|},$$

(together with the appropriate sign), then, as vectors,

$$\vec{F}' = \vec{A} + \lambda \vec{B} - \vec{A} = \lambda \vec{A} + (1 - \lambda)\vec{B},$$

and

$$\frac{|AF'|}{|F'B|} = \frac{\lambda}{1 - \lambda}.$$

But since T is linear,

$$\vec{TF'} = T(\lambda \vec{A} + (1 - \lambda)\vec{B}) = \lambda T\vec{A} + (1 - \lambda)T\vec{B},$$

so

$$\frac{|T(A)T(F')|}{|T(F')T(B)|} = \frac{\lambda}{1 - \lambda} = \frac{|AF'|}{|F'B|}.$$

Thus,

$$\frac{|AF'|}{|F'B|} = \frac{\sinh |AF|}{\sinh |FB|},$$

as claimed. □

Menelaus' theorem in hyperbolic geometry now follows in the same way that it does in spherical geometry, so we leave the proof as an exercise.

Exercise 12.25. Prove Menelaus' theorem in hyperbolic geometry.

Exercise 12.26 (Ceva's Theorem in Hyperbolic Geometry). State and prove Ceva's theorem in hyperbolic geometry.

Exercise 12.27. Despite the similarities, not everything in spherical geometry translates nicely to the pseudosphere. Explain why Ptolemy's proof of the spherical Pythagorean theorem (see Exercise 10.13) cannot be adapted to the pseudosphere.

Chapter 13

Finite Geometries

Geometry has an inherent algebraic structure. We first saw this in Chapter 3 when we studied the algebra of constructions. In this chapter, we present another stunning example of the relationship between these two apparently diverse subjects. We will present both algebraic and geometric definitions of finite affine and projective planes. Though the two definitions look very different, they in fact define the same objects. The core arguments of this amazing result are due to David Hilbert (1862 – 1943) who used a geometrically defined algebra discovered by Karl Georg Christian von Staudt (1798 – 1867).

13.1   Algebraic Affine Planes

As was stressed in Chapter 9, the choice of axioms one uses to define a particular theory can be a matter of taste. In this section, we present an algebraic set of axioms to define affine planes. Our model is the real affine plane $\mathbb{R}^2$, on which lines are defined via the equations

$$ax + by = c,$$

where a and b are not both zero. To come up with new geometries, we merely replace $\mathbb{R}$ with an arbitrary *field*.

Definition 27. *Field.* A set F together with two binary operations $+$ and $\cdot$ is called a *field* if for any a, b, and c in F, we have

1.	$a + b \in F$	(Closure under addition.)
2.	$a \cdot b \in F$	(Closure under multiplication.)
3.	$(a + b) + c = a + (b + c)$	(Associativity of addition.)
4.	$a + b = b + a$	(Commutativity of addition.)

5. There exists an element $0 \in F$ such that $a + 0 = a$ for all $a \in F$.
   (Existence of an additive identity.)

6. For any element $a \in F$, there exists an element $-a \in F$ such that
$a + (-a) = 0$.   (Existence of additive inverses.)

7. $(a \cdot b) \cdot c = a \cdot (b \cdot c)$ (Associativity of multiplication.)

8. $a \cdot b = b \cdot a$ (Commutativity of multiplication.)

9. There exists an element $1 \in F$ such that $a \cdot 1 = a$ for any $a \in F$.
(Existence of a multiplicative identity.)

10. For any $a \in F$ with $a \neq 0$, there exists an element $a^{-1} \in F$ such that
$a \cdot a^{-1} = 1$.   (Existence of multiplicative inverses.)

11. $a \cdot (b + c) = a \cdot b + a \cdot c$ (The distributive law.)

The notation $\cdot$ for multiplication is often omitted.

Given a field F, we define the *algebraic affine plane* F^2 to be the set *points* denoted by ordered pairs of elements of F:

$$F^2 = \{(a, b) : a, b \in F\}.$$

We define a *line* in F^2 to be a subset l of F^2 given by

$$l = \{(x, y) \in F^2 : ax + by = c\},$$

for any elements a, b, and c in F such that a and b are not both zero. Note that two lines are equal if they are equal as sets. Thus, two distinct equations $ax + by = c$ and $a'x + b'y = c'$ might define the same line.

The reader might note that this definition apparently contains no axioms. This is because there is little distinction between definitions and axioms. For example, consider a set F together with two operations $+$ and $\cdot$. In the above, we *define* F to be a field *if* it satisfies the listed properties. We could have instead taken the point of view that F *is* a field, and that we *assume* it satisfies these properties. Because of this point of view, the properties of a field are sometimes called the *field axioms*.

There are as many different algebraic affine planes as there are fields. The smallest finite field is $\mathbb{Z}/2\mathbb{Z}$, the integers modulo two.[1] The geometry $(\mathbb{Z}/2\mathbb{Z})^2$ can be modeled abstractly as in Figure 13.1. Note that the 'diagonals' do not intersect.

This model has a couple of familiar geometric properties: (1) For any two distinct points P and Q, there exists a unique line l through P and Q. (2) For any line l and any point P not on l, there exists a line l' through P which does not intersect l.

These two properties are, in fact, satisfied by any algebraic affine plane F^2:

Theorem 13.1.1. *Given any two points P and Q in F^2, there exists a unique line l which contains both P and Q.*

---

[1]See Section A.4 in the Appendix A for a quick review of modular arithmetic.

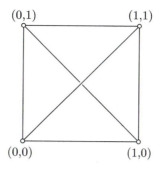

Figure 13.1. An abstract rendering of the four point geometry $(\mathbb{Z}/2\mathbb{Z})^2$.

Theorem 13.1.2. *Given a line l and a point P not on l, there exists a unique line l' which contains P and does not intersect l.*

We leave the proofs as exercises.

As is shown in most (if not all) undergraduate texts in modern algebra, for every prime power $q = p^r$ with $r \geq 1$, there exists exactly one field with q elements. This field is usually denoted with $\mathbb{F}_q$. There are no fields with n elements for n not a power of a prime. For $r = 1$, the field $\mathbb{F}_p$ is just $\mathbb{Z}/p\mathbb{Z}$. In Exercise 13.8, we develop the field with four elements.

There are therefore algebraic affine planes $\mathbb{F}_q^2$ with $q^2 = p^{2r}$ elements for any prime p and integer $r \geq 1$.

Exercise 13.1. Prove Theorem 13.1.1.

Exercise 13.2. Prove Theorem 13.1.2.

Exercise 13.3. Let P and Q be distinct points in $(\mathbb{Z}/n\mathbb{Z})^2$. Prove that there exists a line l which goes through P and Q.

Exercise 13.4. Find two distinct points P and Q in $(\mathbb{Z}/6\mathbb{Z})^2$ such that there exist two distinct lines through both of them. [H]

Exercise 13.5. Find lines l, l_2, and l_3 in $(\mathbb{Z}/6\mathbb{Z})^2$ such that l_2 and l_3 intersect, but neither intersect l.

Exercise 13.6. Prove that every line in $(\mathbb{Z}/p\mathbb{Z})^2$ has p points on it.

Exercise 13.7. Draw an abstract rendering of the nine point affine geometry $(\mathbb{Z}/3\mathbb{Z})^2$.

Exercise 13.8. Though $(\mathbb{Z}/4\mathbb{Z})^2$ is not an affine plane, there does exist a sixteen point affine plane. This is constructed by first finding a four point field. Let α satisfy

$$\alpha^2 + \alpha + 1 = 0$$

and define

$$\mathbb{F}_4 = \{a + b\alpha : a, b \in \mathbb{Z}/2\mathbb{Z}\}.$$

Verify that every nonzero element (there are only three) has a multiplicative inverse. Verify that $\mathbb{F}_4$ is a field.

13.2   Algebraic Projective Planes

Recall that we define the set of points in $\mathbb{P}^2(\mathbb{R})$ to be the set of lines which go through the origin in $\mathbb{R}^3$. A line in $\mathbb{P}^2(\mathbb{R})$ is represented by a plane in $\mathbb{R}^3$ which goes through the origin. For an arbitrary field F, we define $\mathbb{P}^2(F)$ in a similar fashion. A point in $\mathbb{P}^2(F)$ is represented by a line through the origin in F^3, and a line in $\mathbb{P}^2(F)$ is represented by a plane in F^3 which goes through the origin.

Just as in $\mathbb{P}^2(\mathbb{R})$, we represent a point $P \in \mathbb{P}^2(F)$ with the homogeneous coordinates (P_1, P_2, P_3), a triple of elements in F, not all of which are zero. The triple (P_1', P_2', P_3') represents the same point if $P_i' = \lambda P_i$ for all i and some nonzero $\lambda \in F$. A line in $\mathbb{P}^2(F)$ is the set of solutions to the equation

$$a_1 x + a_2 y + a_3 z = 0.$$

Again, using the smallest field $F = \mathbb{Z}/2\mathbb{Z}$, we get the smallest algebraic projective plane $\mathbb{P}^2(\mathbb{Z}/2\mathbb{Z})$, the seven point geometry shown in Figure 13.2. The points are labeled with homogeneous coordinates, though in this case, there is no distinction between homogeneous coordinates and coordinates in $(\mathbb{Z}/2\mathbb{Z})^3$.

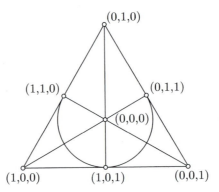

Figure 13.2. An abstract rendering of the seven point projective geometry.

Finite algebraic projective planes also satisfy several familiar geometric properties:

Theorem 13.2.1. *For every pair of points P and Q in $\mathbb{P}^2(F)$, there exists a unique line l through both.*

Theorem 13.2.2. *Every pair of distinct lines intersect in exactly one point.*

The proofs of these two we leave as exercises.

Theorem 13.2.3 (Desargues' Theorem). *Let F be a field. Two triangles $\triangle ABC$ and $\triangle A'B'C'$ in $\mathbb{P}^2(F)$ are perspective from a point if and only if they are perspective from a line.*

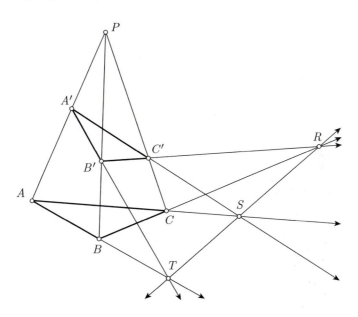

Figure 13.3

We gave a proof of Desargues' theorem in Euclidean geometry on page 95. That proof involves visualization in three-dimensional Euclidean geometry. In Exercise 11.1, we asked for another proof, but the proof sought for there involves a comparison of similar triangles, in the fashion of the proof of Pappus' theorem given on page 229. Thus, neither proof is applicable to our current situation.

Proof. Let $\vec{A}$ be a vector in F^3 which represents the point A in $\mathbb{P}^2(F)$, and so on.

Suppose first that $\triangle ABC$ and $\triangle A'B'C'$ are perspective from the point P, as in Figure 13.3. This figure is meant to be helpful, but keep in mind that it is a figure in Euclidean geometry and only models our finite geometry. Since A' is on the line PA, $\vec{A}'$ is in the plane spanned by $\vec{P}$ and $\vec{A}$. By Exercise 13.9, we can write $\vec{A}'$ as a linear combination of $\vec{P}$ and $\vec{A}$, so

$$\vec{A}' = r\vec{P} + s\vec{A}.$$

If we multiply $\vec{A}'$ by r^{-1}, we get another vector which also represents A'. Thus, without loss of generality, we may write

$$\vec{A}' = \vec{P} + a\vec{A}$$

for some $a \in F$. We can similarly write

$$\vec{B}' = \vec{P} + b\vec{B} \quad \text{and} \quad \vec{C}' = \vec{P} + c\vec{C}.$$

Let R be the point represented by

$$\vec{R} = b\vec{B} - c\vec{C}.$$

Note that $\vec{R}$ is a linear combination of $\vec{B}$ and $\vec{C}$, so R is on the line BC. Note also that

$$\vec{R} = (\vec{P} + b\vec{B}) - (\vec{P} + c\vec{C}) = \vec{B}' - \vec{C}',$$

so R is on the line $B'C'$. Hence, R is the point of intersection of the lines BC and $B'C'$. Similarly, the points S and T represented by

$$\vec{S} = a\vec{A} - c\vec{C} \quad \text{and} \quad \vec{T} = a\vec{A} - b\vec{B}$$

are the points of intersection of the pairs of lines AC and $A'C'$, and BC and $B'C'$, respectively. But then,

$$\vec{T} = a\vec{A} - b\vec{B} = (a\vec{A} - c\vec{C}) - (b\vec{B} - c\vec{C}) = \vec{S} - \vec{R},$$

so $\vec{T}$ is in the plane spanned by $\vec{R}$ and $\vec{S}$. That is, T is on the line RS, so $\triangle ABC$ and $\triangle A'B'C'$ are perspective from a line.

Finally, recall that the dual of the above result is its converse, so follows by Exercise 13.12.   □

Exercise 13.9. Let A, B, and C be points on the plane

$$ax + by + cz = 0$$

in F^3. Let $\vec{A}$ be the vector OA, and so on. Prove that there exist elements r and s in F such that

$$\vec{C} = r\vec{A} + s\vec{B}.$$

Exercise 13.10. How many points are in $\mathbb{P}^2(\mathbb{Z}/3\mathbb{Z})$? Draw an abstract rendering of this geometry.

Exercise 13.11. How many points are in $\mathbb{P}^2(\mathbb{F}_q)$?

Exercise 13.12. Define the duals of points and lines in $\mathbb{P}^2(F)$. Suppose that if P and Q are points whose duals are p and q, then the point of intersection of p and q is the dual of the line PQ. Conclude that the dual of any theorem in $\mathbb{P}^2(F)$ is also a result in $\mathbb{P}^2(F)$.

Exercise 13.13. A *linear automorphism* of an affine or projective geometry $\mathcal{G}$ is an invertible map f from $\mathcal{G}$ to itself which sends lines to lines. Prove that every linear automorphism of a projective geometry $\mathbb{P}^2(F)$ is given by a linear transformation $T \in \mathrm{SL}_3(F)$.

Exercise 13.14. How many linear automorphisms are there in $\mathbb{P}^2(\mathbb{F})$? [H]

Exercise 13.15*. We define a conic in $\mathbb{P}^2(\mathbb{F}_q)$ to be the set of solutions to an equation of the form

$$X^T A X = 0,$$

where $X = (X_0, X_1, X_2)$ and A is a symmetric invertible 3×3 matrix with entries in $\mathbb{F}_q$. How many points are on such a conic? Hint: Read up on stereographic projection in Chapter 15.

13.3   Weak Incidence Geometry

Though the algebraically defined finite planes of the previous two sections satisfy several familiar geometric properties, it does not seem possible that they should be the only finite geometries, so we should probably take the point of view that they are only models of certain finite planes. In general, finite geometries should be defined geometrically. In this section, we define *weak incidence geometries* via a set of geometric axioms.

Let $\mathcal{G}$ be a set of points on which there are subsets called lines. We call $\mathcal{G}$ a *weak affine plane* if it satisfies the following axioms:

1. Given any two distinct points P and Q in $\mathcal{G}$, there exists a unique line l through P and Q.

2. Every line l contains at least two points.

3. There exist at least two lines.

4a. Given a line l and a point P not on l, there exists a unique line l' through P which does not intersect l.

In Section 13.1, we verified that all algebraic affine planes satisfy these properties, so we have a large number of examples. The smallest example has four points. It is not hard to see that there are no weak affine planes with three points (see Exercise 13.16). This is perhaps our first indication that this set of axioms is somewhat restrictive. In fact, every weak affine plane contains n^2 points for some n (see Corollary 13.3.4).

Let us consider more sets $\mathcal{G}$. We call $\mathcal{G}$ a *weak projective plane* if it satisfies Axioms 1 and 3, and

2p. Every line l contains at least three points.

4p. Every pair of lines intersect.

Again, in Section 13.2, we verified that all algebraic projective planes satisfy these axioms.

There is a strong relationship between affine and projective planes:

Theorem 13.3.1. *Suppose $\mathcal{G}$ is a weak projective plane and L is a line on $\mathcal{G}$. Let $\mathcal{G}' = \mathcal{G} \setminus L$ and call a subset l' of $\mathcal{G}'$ a line if there exists a line l in $\mathcal{G}$ such that $l' = l \cap \mathcal{G}'$. Then $\mathcal{G}'$ is a weak affine geometry.*

Proof. Axioms 1 and 3 are trivially satisfied by $\mathcal{G}'$. To see that Axiom 2 is satisfied, we note that every line l intersects L exactly once (by Axioms 1 and 4p), so $l' = l \cap \mathcal{G}'$ contains one less point then l, and hence at least two points. Now, suppose l' is a line in $\mathcal{G}'$ and $P \in \mathcal{G}'$ is a point not on l'. Let l intersect the line L at $Q \in \mathcal{G}$. Then there exists a unique line $l_1 = PQ$. The line l_1' does not intersect l. Suppose there exists another line l_2' which goes through P and does not intersect l'. But l_2 intersects l, so that point of intersection must be on L. That is, l_2 intersects l at Q. Hence, $l_2 = PQ = l_1$. Thus, there is a unique line l_1' through P which does not intersect l. $\qquad\square$

Given a weak affine plane $\mathcal{G}'$, one can also construct a weak projective plane (see Exercise 13.19).

There are several properties that we would like all geometries to satisfy. For example, we would like all lines and points to *look* the same. In geometries for which we have a metric, we express this notion via isometries. In incidence geometries, we have no metric so cannot talk about isometries. The following theorem shows that all lines in finite affine and projective geometries potentially look alike.

Theorem 13.3.2. *Suppose $\mathcal{G}$ is a finite weak projective geometry, and suppose l is a line on $\mathcal{G}$ which contains n points. Then every line on $\mathcal{G}$ contains n points.*

Proof. Let l' be another line and let l and l' intersect at P. By Axiom 2p, there exist points Q and Q' on l and l', respectively, neither of which is P. Also by Axiom 2p, there exists a third point R on QQ'. This point R does not lie on either l or l'. We now use stereographic projection from l to l' through R. For every point S on l, let S' be the point on l' where RS and l' intersect (note that S' exists by Axiom 4p, and is uniquely defined by Axiom 1). This defines an invertible map from l to l'. Hence, l and l' contain the same number of points. That is, l' contains n points. Since l' was arbitrary, every line in $\mathcal{G}$ contains n points. $\qquad\square$

Corollary 13.3.3. *Suppose $\mathcal{G}$ is a finite weak projective geometry and suppose l is a line on $\mathcal{G}$ which contains n points. Then $\mathcal{G}$ contains $n^2 - n + 1$ points.*

Corollary 13.3.4. *Suppose $\mathcal{G}$ is a finite weak affine geometry. Then $\mathcal{G}$ contains n^2 points for some integer $n \geq 2$.*

We leave the proofs of these as exercises. Note that there are algebraic affine planes with 2^2, 3^3, 4^2, 5^2, and 7^2 points, but none with 6^2 points. This raises the question: *Are there any weak affine planes with 36 points?*

In 1782, Euler posed a similar problem, the *thirty-six officer problem*. Suppose there are six regiments, each with six officers. Suppose that each officer has one of six different ranks and that no two officers in the same regiment have the same rank. Is there a way to arrange these officers in six rows and columns such that no two officers in the same row or column are from the same regiment or have the same rank?

This question is related to weak affine geometries in the following way:

Theorem 13.3.5. *Suppose there exists a weak affine geometry with n^2 points where $n > 2$. Then there is a way of organizing n regiments of n officers with n different ranks in n rows and n columns such that no two officers in any row or column are from the same regiment or have the same rank.*

The answer to Euler's problem is *no*, which was shown by exhaustive search in 1900. Consequently, there are no weak affine planes with thirty-six points. This is our second indication that the set of axioms which define weak affine planes is rather restrictive.

Before we continue, let us say a little about *finite hyperbolic geometries*. The weakest form of such a geometry $\mathcal{G}$ satisfies Axioms 1 – 3 and the following:

4h. For any line l and P not on l, there exist at least two lines through P which do not intersect l.

The smallest such geometry has five points. Since there is no analogue of Theorem 13.3.2, we often also require that Axiom 2 be modified so that it reads,

2h. Every line contains exactly n points.

Even with this modification, it is clear that we are not yet defining satisfactory analogues of hyperbolic geometry. For example, the geometries $(\mathbb{Z}/3\mathbb{Z})^3$ and $\mathbb{P}^3(\mathbb{Z}/2\mathbb{Z})$ both satisfy Axioms 1, 2h with $n = 3$, 3, and 4h, but are more properly analogues of three-dimensional affine and projective geometries. Let us add an axiom which will ensure that our geometry is two dimensional. For any three points A, B, and C in $\mathcal{G}$, let $\mathfrak{p}(A, B, C)$ be the smallest subset of $\mathcal{G}$ which contains A, B, and C, and has the property that for every pair of points P and Q in $\mathfrak{p}(A, B, C)$, the line PQ lies entirely within $\mathfrak{p}(A, B, C)$. For noncollinear points A, B, and C, this set may be thought of as the plane that contains them. Let us now assume

5h. For any three noncollinear points A, B, and C in $\mathcal{G}$, the set $\mathfrak{p}(A, B, C)$ is all of $\mathcal{G}$.

This axiom eliminates the three-dimensional examples given above. It is not too hard to show that the smallest geometry which satisfies these axioms (Axioms 1, 2h, 3, 4h, and 5h) with $n = 3$ must have at least thirteen points (see Exercise 13.23). Such a geometry exists and is developed in Exercise 13.27.

Exercise 13.16. Prove that there is only one three point geometry which satisfies Axioms 1 – 3.

Exercise 13.17. Find all four point geometries which satisfy Axioms 1 – 3. [H]

Exercise 13.18. How many lines are there in an affine geometry with n^2 points?

Exercise 13.19*. Suppose $\mathcal{G}$ is a weak affine plane. Describe how to add a *line at infinity* to $\mathcal{G}$ to get a weak projective geometry $\mathcal{G}'$. Prove that $\mathcal{G}'$ is a weak projective geometry.

Exercise 13.20. Prove Corollary 13.3.3. [S]

Exercise 13.21. Prove Corollary 13.3.4.

Exercise 13.22*. Prove Theorem 13.3.5. [H]

Exercise 13.23. Prove that any geometry which satisfies Axioms 1, 2h, 3, and 4h with $n = 3$ has at least thirteen points. [S]

Exercise 13.24. Let $\mathcal{G}$ be a geometry which satisfies Axioms 1, 2h with $n = 3$, and 3. Let A, B, and C be three noncollinear points in $\mathcal{G}$. Prove that $\mathfrak{p}(A, B, C)$ contains at least seven points.

Exercise 13.25. Suppose $\mathcal{G}$ is a weak projective geometry and that A, B, and C are three noncollinear points in $\mathcal{G}$. Prove that $\mathfrak{p}(A, B, C) = \mathcal{G}$. That is, prove that Axiom 5h is a theorem in weak projective geometry.

Exercise 13.26. Suppose $\mathcal{G}$ is a weak affine geometry with at least three points on a line. Let A, B, and C be three noncollinear points in $\mathcal{G}$. Prove that $\mathfrak{p}(A, B, C) = \mathcal{G}$. What happens if there are only two points on a line?

Exercise 13.27*. In this exercise, we develop the thirteen point hyperbolic geometry. Let $\mathcal{G} = \{P_0, ..., P_{12}\}$. Let the twenty-six lines on $\mathcal{G}$ be the sets $l_k = \{P_{0+k}, P_{1+k}, P_{4+k}\}$ and $l'_k = \{P_{0+k}, P_{2+k}, P_{7+k}\}$ for $k = 0, 1, ..., 12$ (and indexing modulo 13). Prove that $\mathcal{G}$ satisfies Axioms 1, 2h, 3, 4h, and 5h. [H]

13.4 Geometric Projective Planes

In a finite weak projective plane, every line contains the same number of points. There exist algebraic projective planes for which every line contains $q + 1$ points where q is a power of a prime. There does not exist a weak projective plane with 7 points on each line. A natural question to ask is, *Are there any weak projective geometries with $n + 1$ points on each line where n is not a power of a prime?* Though we will not answer this question, we will prove something a little more impressive but on a smaller set of geometries.

A *geometric projective plane* is a weak projective plane which satisfies one more axiom:

5. Desargues' theorem holds.

Since we proved Desargues' theorem for $\mathbb{P}^2(F)$, every algebraic projective plane is a geometric projective plane. In the rest of this chapter, we will show that every *finite* geometric projective plane is an algebraic projective plane. Thus, our algebraic and geometric descriptions of finite projective planes are equivalent.

We prove this in several steps. Let $\mathcal{G}$ be a geometric projective plane and let l be a line in $\mathcal{G}$. Let P be a point on l and let $l_* = l \setminus \{P\}$. We first define von Staudt's addition and multiplication on l_*. Following Hilbert, we prove that l_* is a *division ring*, which is an object that satisfies all the properties of a field except possibly commutativity of multiplication (that is, properties $1 - 7$ and $9 - 11$ on page 263). If $\mathcal{G}$ is finite, then l_* is finite, so by Wedderburn's theorem (a rather deep result in abstract algebra), l_* must in fact be a field. We will then build a coordinate system on $\mathcal{G}$ to show that $\mathcal{G}$ is equivalent to $\mathbb{P}^2(l_*)$.

Exercise 13.28. The principle of duality is an important property of projective the projective plane. Dualize the statements of these axioms. Prove each of these dual statements from this set of axioms. Conclude that the dual statement of any theorem in projective the projective plane is in fact a theorem.

13.5 Addition

Let $\mathcal{G}$ be a geometric projective plane. Let us choose three distinct lines l, l', and l'' in $\mathcal{G}$. Let l and l'' intersect at P, let l' and l'' intersect at P', and let l and l' intersect at O. Let $l_* = l \setminus \{P\}$ and let $l'_* = l' \setminus \{P'\}$. Let us think of l'' as a 'line at infinity' so that when we talk about parallel lines, we mean lines which intersect at a point on l''. We will construct an algebra on l_* (and in passing, on l'_* too). The point O will be our zero element.

Let A and B be points on l_*, and let $A' \neq 0$ be an arbitrary point on l'_* (see Figure 13.4). Construct the line through A' and parallel to l. That is, construct the line $A'P$. Construct also the line through B and parallel to l' (i.e., BP'). Let these two lines intersect at D. The point C where the line through D and parallel to AA' intersects l is the point we call $A + B$.

Let us first check that this addition is well defined. Suppose A'' is another nonzero point on l'_*. Let $A''P$ intersect the line BD at D'. Our definition of addition is well defined if $D'C$ is parallel to AA''. To see this, we apply Desargues' theorem to the triangles $\triangle AA'A''$ and $\triangle CDD'$. The lines AC, $A'D$, and $A''D'$ are parallel, so are coincident at infinity. Thus, the two triangles are perspective from a point, and are therefore perspective from a line. Since AA' is parallel to CD and $A'A''$ is parallel to DD', the line of perspectivity is the line at infinity. Thus, AA'' and CD' intersect at infinity. That is, they are parallel, as desired.

Lemma 13.5.1. *Addition on l_* is commutative.*

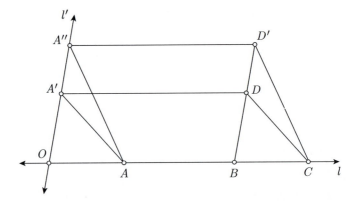

Figure 13.4

Proof. Let A and B be points on l_*. Let A' be an arbitrary point on l'_* and let B' be the point on l'_* such that BB' is parallel to AA'. In Figure 13.5, the points D and D' are constructed so that $A'D$ and $B'D'$ are parallel to l, and AD' and BD are parallel to l'. The point $A + B$ is the intersection of l with the line through D and parallel to AA'. The point $B + A$ is the intersection of l with the line through D' which is parallel to BB'. Thus, $A + B = B + A$ if DD' is parallel to AA'. Let $A'D$ and AD' intersect at A'', and let BD and $B'D'$ intersect at B''. We apply Desargues' theorem to $\Delta AA'A''$ and $\Delta BB'B''$. These two triangles are perspective from the

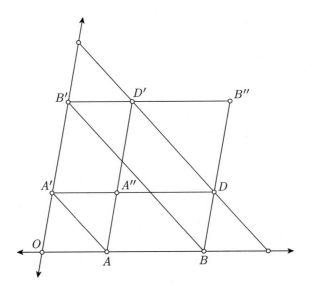

Figure 13.5

line at infinity, so are perspective from a point. That is, O, A'', and B'' are collinear. We now apply Desargues' theorem to $\triangle AA'A''$ and $DD'B''$, which are perspective from the point O, so are perspective from a line. That line is the line at infinity, so DD' is parallel to AA', as desired. □

Lemma 13.5.2. *Addition on l_* is assosciative.*

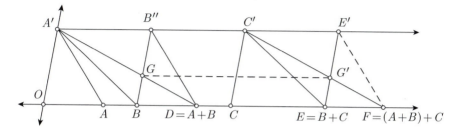

Figure 13.6

Proof. Let A, B, and C be points on l_*, and consult Figure 13.6. Choose A' on l'_* and let B' be the intersection of $A'P$ and BP'. Let D be the intersection of l with the line through B' and parallel to AA'. Then $D = A + B$. Let C' be the intersection of $A'B'$ and CP', and let E be the intersection of l with the line through C' and parallel to BA'. Then $E = B+C$. Let F be the intersection of l with the line through C' and parallel to DA'. Then $F = D+C = (A+B)+C$. Finally, let E' be the intersection of $A'B'$ and EP'. If $E'F$ is parallel to AA', then $F = A+E = A+(B+C)$, as desired. So let us show $E'F$ is parallel to AA'. We do this using Desargues' theorem a couple of times. First, let G be the intersection of BB' and DA', and let G' be the intersection of EE' and FC'. The triangles $\triangle BGA'$ and $\triangle EG'C'$ are perspective from the line at infinity, so are perspective from a point, the point P at infinity. Thus, GG' is parallel to l. The triangles $\triangle DB'G$ and $\triangle FE'G'$ are perspective from the point P, so are perspective from a line, the line at infinity. Thus, FE' is parallel to DB', which is parallel to AA', as desired. □

Exercise 13.29. Verify that $O + A = A$ for all A.

Exercise 13.30. Describe how to find $-A$.

Exercise 13.31. Let I and I' be fixed nonzero points on l_* and l'_*, respectively. For any Q on l_*, define Q' to be the point on l'_* such that QQ' is parallel to II'. Similarly, for a point Q'_* on l'_*, define $(Q')'$ to be the point Q on l_* such that QQ' is parallel to II'. For any A and B on l_*, prove that

$$(A' + B')' = A + B.$$

13.6   Multiplication

Recall the method of multiplying constructible lengths (see page 74). Our definition of · is inspired by that construction. We first pick nonzero points I and I' on l_* and l'_*, respectively. The point I will be our multiplicative identity. Let A and B be points in l_*. Let B' be the point on l'_* such that BB' is parallel to II' (see Figure 13.7). The point C where the line through B' and parallel to AI' intersects l is the point we call $A \cdot B$.

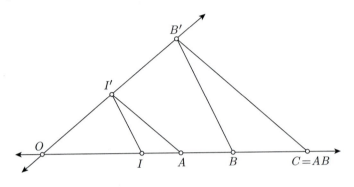

Figure 13.7

Lemma 13.6.1. *Multiplication is assosciative.*

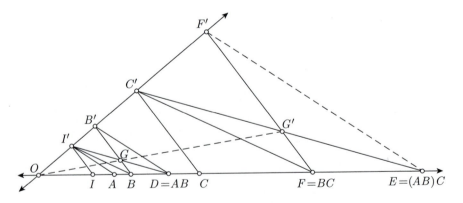

Figure 13.8

Proof. Given I, I', A, B, and C, as in Figure 13.8, we find the following points: B' is the point on l' such that BB' is parallel to II'; D is the point on l such that DB' is parallel to AI'; C' is the point on l' such that CC' is parallel to II'; and E is the point on l such that EC' is parallel to DI'. By construction, $D = AB$ and $E = DC = (AB)C$. The point F is the point on l such that FC' is parallel to BI'; F' is the point on l' such that FF' is parallel to II'. By construction, $F = BC$, and if EF' is parallel to AI', then $E = AF = A(BC)$. We show this by applying Desargues' theorem twice. Let G be the intersection of DI' and BB'; and let G' be the the intersection of EC' and FF'. Note that $\triangle BI'G$ and $\triangle FC'G'$ are perspective from the line at infinity, so they are perspective from a point. Hence, O, G, and G' are collinear. Thus, $\triangle DGB'$ and $EG'F'$ are perspective from a point, so they are perspective from a line. Hence EF' is parallel to DB', which is parallel to AI', as desired. $\qquad\square$

Exercise 13.32. Show that $IA = AI = A$ for all A.

Exercise 13.33. Describe how to find A^{-1}.

Exercise 13.34. Let I and I' be fixed nonzero points on l_* and l'_*, respectively. For any Q on l_*, define Q' to be the point on l'_* such that QQ' is parallel to II'. Similarly, for a point Q' on l'_*, define $(Q')'$ to be the point Q on l_* such that QQ' is parallel to II'. For any A and B on l_*, prove that

$$(A'B')' = AB.$$

13.7 The Distributive Law

We will prove only one direction of the distributive law and leave the other case as an exercise (Exercise 13.35).

Lemma 13.7.1. *Let A, B, and C be points on l_*. Then*

$$A(B + C) = AB + AC.$$

Proof. Let $I \neq 0$, A, B, and C be points on l_*. Let $I' \neq 0$ be a point on l'_*. Let A' and B' be points on l' such that AA' and BB' are parallel to II' (see Figure 13.9). Let D' and E' be the points on l' such that $D'B$ and $E'C$ are parallel to $A'I$. Then $D' = A'B'$ and $E' = A'C'$. Let F be the intersection of the lines $D'P$ and CP'. (Recall that P and P' are the points at infinity on l and l', respectively, so $D'P$ and CP' are just shorthand notations for lines parallel to l and l'.) Let G and H' be the points on l and l', respectively, such that GH' goes through F and is parallel to IA'. Note that $B + C = G$, and hence, $H' = A'(B' + C')$ (using Exercises 13.31 and 13.34). Thus, H (not shown) is $A(B + C)$. Note also that $E' + D' = H'$, so $H' = A'C' + A'B'$. Hence, we also have $H = AC + AB = AB + AC$. Thus,

$$H = A(B + C) = AB + AC. \qquad\square$$

Exercise 13.35*. Prove the other direction of the distributive law. That is, prove that

$$(A + B)C = AC + BC.$$

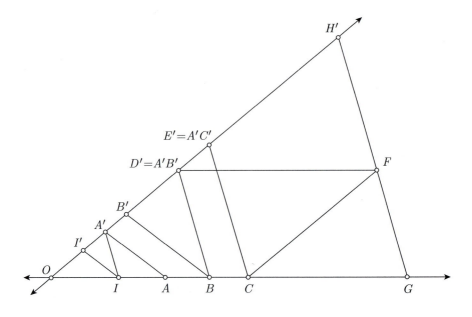

Figure 13.9

13.8 Commutativity, Coordinates, and Pappus' Theorem

In the last three sections, we have defined an algebra on l_* and shown that all the properties of a field are satisfied except one – we have not shown that multiplication is commutative. That is, we have shown that l_* is a *division ring*. If $\mathcal{G}$ is finite, then this is in fact enough to show that l_* is a field. This follows from Wedderburn's theorem:

Theorem 13.8.1 (Wedderburn's Theorem). *Every finite division ring is a field.*

See any good textbook in algebra for a proof (e.g., [Her]).

As suggested by the title of this section, the commutativity of multiplication and Pappus' theorem are related. They are, in fact, equivalent. Let us first recall Pappus' theorem, which is more correctly called a property, since we do not know whether it holds or not.

Property (Pappus' Theorem). Let P_1, P_2, and P_3 be three points on the line l_1, and let Q_1, Q_2, and Q_3 be three points on the line l_2. Let R be the intersection of P_2Q_3 and P_3Q_2; let S be the intersection of P_1Q_3 and P_3Q_1; and let T be the intersection of P_1Q_2 and P_2Q_1. Then R, S, and T are collinear.

Theorem 13.8.2. *Multiplication in the division ring l_* is commutative if and only if Pappus' theorem is true.*

In our proof that the commutativity of multiplication in l_* implies Pappus' theorem, we use a result in algebraic projective planes. So let us first establish that if l_* is a field, then $\mathcal{G}$ is equivalent to $\mathbb{P}^2(l_*)$.

Let us begin by building a coordinate system on the affine subset $\mathcal{G}' = \mathcal{G} \setminus l''$. Let P be an arbitrary point in $\mathcal{G}'$. Let the line through P and parallel to l' intersect l at R. Similarly, let the line through P and parallel to l intersect l' at S', as in Figure 13.10. We give the point P the coordinates (R, S) (where, as before, S is the point on l such that SS' is parallel to II'). To show that this gives an affine coordinate system, we must show that lines in $\mathcal{G}'$ are described by the appropriate equations.

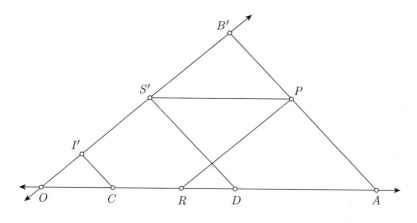

Figure 13.10

Let A be a point on l and let B' be a point on l'. Note that the coordinates of A and B' are $(A, 0)$ and $(0, B)$, respectively. Thus, both points lie on the curve described by

$$BX + AY = AB.$$

Let P be on the line AB' and let P have coordinates (R, S). We must show that

$$BR + AS = AB.$$

To see this, consider Figure 13.10. Let I' be the identity on l'. Let C and D be the points on l such that CI' and DS' are parallel to AB'. Then $CB = A$, $CS = D$, and $D + R = A$. Thus, $C = AB^{-1}$ and hence,

$$AB^{-1}S + R = A$$
$$AS + BR = AB.$$

Thus, our choice of coordinates defines an affine coordinate system for $\mathcal{G}'$. This coordinate system can be extended to a projective coordinate system on $\mathcal{G}$ (see Exercise 13.36). Thus $\mathcal{G}$ and $\mathbb{P}^2(l_*)$ define the same geometry.

Let us now return to the proof of Theorem 13.8.2.

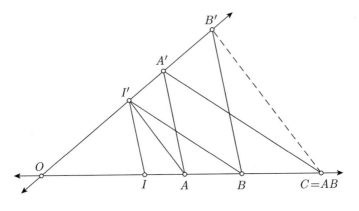

Figure 13.11

Proof of Theorem 13.8.2. Let us first suppose that Pappus' theorem is true. Let the points I, I', A, and B be as in Figure 13.11. We construct A' and B' so that AA' and BB' are parallel to II'. Let C be the point on l such that CB' is parallel to AI'. Then, by construction, $C = AB$. The point BA is the intersection of l with the line through A' which is parallel to BI'. Hence, to show $AB = BA$, we must show that CA' is parallel to BI'. Let us apply Pappus' theorem to the points A, B, and C on l, and the points B', A', and I' on l' (and in that order). Note that AA' and BB' are parallel, and that AI' and CB' are parallel, so the points we call T and S are on the line at infinity. Thus, the point R, which is the intersection of BI' and CA', is also at infinity. That is, BI' and CA' are parallel, as desired.

Let us now prove the other direction in the special case when the lines l_1 and l_2 (as stated in the conditions of Pappus' theorem) are the coordinate axis l and l', and the points R and T are on the line l'' at infinity. Then, we have exactly the situation in Figure 13.11, with $P_1 = A$, $P_2 = B$, $P_3 = C = AB$, $Q_1 = B'$, $Q_2 = A'$, and $Q_3 = I'$. Let S be the intersection of P_1Q_3 with $RT = l''$. Then by definition, RB' intersects l at the product BA. Since multiplication is commutative, $AB = BA$, so RB' intersects l at C, as desired.

Finally, let us consider an arbitrary case. Let l_1 and l_2 intersect at D. Let the line RT intersect l_1 and l_2 at E and E', respectively. Since $\mathcal{G}$ is equivalent to $\mathbb{P}^2(l_*)$, by the comments following Theorem 11.3.1, there exists a linear transformation $T \in \mathrm{GL}_3(l_*)$ which sends D to O, E to P, E' to P', and Q_3 to I'. This map sends lines to lines, so in particular, preserves the constructions of our addition and multiplication. Thus, if we were to develop an algebra on l_1 and l_2 with the line RT at infinity, then we will get the same algebra as l_*. Hence, we may view l_1 and l_2 as coordinate axes, and proceed as before.   □

Remarks. It is not necessary for l_* to be a field in order to develop a coordinate system, but more care must be taken in this case. Similarly, using a little more care, one can define algebraic projective geometries for division rings. In particular, *Hamilton's quaternion algebra* $\mathbb{H}$ (which is developed in Exercise 13.37) is a noncommutative division ring, and in the geometry $\mathbb{P}^2(\mathbb{H})$, Pappus' theorem is not true.

Exercise 13.36. Extend the affine coordinate system on $\mathcal{G}'$ to a projective coordinate system on $\mathcal{G}$.

Exercise 13.37. Hamilton's quaternion algebra is the four-dimensional vector space over $\mathbb{R}$ given by

$$\mathbb{H} = \{a + bi + cj + dk : a, b, c, d \in \mathbb{R}\},$$

and in which we define products using the rules $i^2 = j^2 = k^2 = -1$, $ij = k$, $jk = i$, $ki = j$, and using the associative and distributive laws. Prove that $ji = -k$, $kj = -i$, and $ik = -j$. Evaluate

$$(a + bi + cj + dk)(a - bi - cj - dk).$$

For any nonzero element $r = a + bi + cj + dk$, find the inverse of r and express r^{-1} as a linear combination of 1, i, j, and k. Prove that $\mathbb{H}$ is a division ring.

13.9   Weak Projective Space and Desargues' Theorem

When we treat Desargues' theorem as an axiom, it might seem as if we are accepting rather a lot, so there might be the inclination to search for a weaker set of axioms from which Desargues' theorem follows. We find a clue as to how to do this in the proof of Desargues' theorem given on page 95. This proof only involves intersections of lines and planes in three-space, so can potentially be adapted to planes imbedded in a *weak projective space*. Let us begin with its definition.

A *weak projective space* is a set $\mathcal{G}$ whose elements are called points and which has subsets called lines and planes. The set $\mathcal{G}$ further satisfies the following axioms:

1. Given any two distinct points P and Q in $\mathcal{G}$, there exists a unique line l through P and Q.

2s. Every line l contains at least three points.

3s. Given any three noncollinear points P, Q, and R in $\mathcal{G}$, there exists a unique plane α which contains all three.

4s. Every plane α contains at least two lines.

5s. If two points P and Q are in a plane α, then the line PQ is in α.

6s. Every pair of distinct lines in a plane α intersect.

7s. Every pair of distinct planes intersect in at least two points.

8s. There exist at least two planes.

In weak projective space, we have the following three results which we leave as exercises.

Theorem 13.9.1. *Let l be a line and let P be a point not on l. Prove that there exists a plane which contains l and P.*

Theorem 13.9.2. *Prove that the intersection of two distinct planes is exactly a line.*

Theorem 13.9.3. *Prove that three distinct planes intersect at exactly one point.*

Theorem 13.9.4 (Desargues' Theorem). *Let α be a plane in a weak projective space $\mathcal{G}$. Let $\triangle ABC$ and $\triangle A'B'C'$ be two nondegenerate triangles in α. Let R be the intersection of BC and $B'C'$; let S be the intersection of AC and $A'C'$; and let T be the intersection of AB and $A'B'$. Then the lines AA', BB', and CC' are coincident at a point P if and only if the points R, S, and T are collinear.*

Proof. This proof is inspired by the proof given on page 95. Our idea is to *lift* a line out of the plane to create a three-dimensional object. We begin by choosing a point O in $\mathcal{G}$ which is not on the plane α. Think of this point as our point of view – the point from which we are viewing the plane α and Figure 13.3.

Suppose the lines AA', BB', and CC' are coincident at a point P. Let B_* be a point on OB which is not on α. Such a point exists by Axiom 2s. Let β be the plane which contains P, O, and B. That is, in the notation given on page 271, $\beta = \mathfrak{p}(P, O, B)$. Note that the line PB is on β, so B' is on β, by Axiom 5s. The lines OB' and PB_* are both on β, so by Axiom 6s, intersect at a point which we label B'_*. Let $\gamma = \mathfrak{p}(A, B_*, C)$ and $\gamma' = \mathfrak{p}(A', B'_*, C')$. By Axiom 7s, these two planes intersect at a line l_*. Note that the line AC is on γ and the line $A'C'$ is on γ', so their point of intersection S lies on l_*. Since B'_* lies on PB_* and A' lies on PA, the lines AB_* and $A'B'_*$ both lie on the plane $\mathfrak{p}(P, A, B_*)$, so they intersect. Let this point be T_*. Since AB_* lies on γ and $A'B'_*$ lies on γ', the point T_* lies on

l_*. Similarly, the lines CB_* and $C'B'_*$ intersect at a point R_* on l_*. Now, consider the planes $\mathfrak{p}(A, B, O)$ and $\mathfrak{p}(A', B', O)$. The points O and T are on both of them, so the intersection of these planes is the line OT. But B_* is in $\mathfrak{p}(A, B, O)$ and B'_* is in $\mathfrak{p}(A', B', O)$, so T_* lies on OT. Similarly, R_* lies on OR. Thus, the plane through O and l_* intersects the plane α at the line l through R and T. Since S is on $l*$, S is in this plane, so is also on l. That is, S is on the line RT, as desired.

Suppose now that R, S, and T are collinear. Let P be the intersection of BB' and CC'. Note that $\triangle BB'T$ and $CC'S$ are perspective from a point. Thus, by the first part of this proof, the points A, A', and P are collinear. That is, $\triangle ABC$ and $\triangle A'B'C'$ are perspective from a point, as desired. □

In light of Theorem 13.9.4, Desargues' theorem no longer seems so difficult to accept. But consider this fact: There exist weak projective planes for which Desargues' theorem is not true (see [Hi] §23). That is, there exist weak projective planes which cannot be imbedded in a weak projective space. Such a geometry is called a *non-Desarguesian geometry*.

Exercise 13.38. Prove Theorem 13.9.1.

Exercise 13.39. Prove Theorem 13.9.2.

Exercise 13.40. Prove Theorem 13.9.3.

Chapter 14

Nonconstructibility

The algebra we developed for constructible lengths is adequate to prove that the regular pentagon and even the regular 17-gon are both constructible. In principle, it is powerful enough to show that the regular 257-gon is constructible, though such a proof would be tedious. However, to prove that a point is not constructible, we will need to identify more structure in the set of constructible points.[1]

The major failing of the algebra of constructible lengths is that we cannot construct negative lengths, and consequently, this algebra does not have the structure of a *field*. In this chapter, we direct our attention to the set of *constructible numbers*, which do have the structure of a field.

14.1   The Field of Constructible Numbers

Recall that a complex number $z \in \mathbb{C}$ can be uniquely represented as

$$z = x + iy$$

where $i^2 = -1$ and both x and y are real numbers. We can therefore represent z with the unique pair (x, y), which represents a point in the plane. This representation of $\mathbb{C}$ is called the Argand plane.

In our rules for constructible points, we first select two distinct points. Let us call these points 0 and 1 in $\mathbb{C}$. Then, every constructible point P can be thought of as a complex number in the Argand plane. If a point $z \in \mathbb{C}$ represents a constructible point, then we call z a *constructible number*. The set $\mathcal{C}$ of constructible numbers is a subset of $\mathbb{C}$, and so we can talk about the

---

[1]**Note to the reader:** In this chapter, we briefly touch on the notions of fields, vector spaces, the dimension of vector spaces, irreducible polynomials, and groups. These topics are usually covered in courses on modern algebra and linear algebra. We expect the reader to be at least familiar with vector spaces. The reader should expect this chapter to be more difficult than other chapters, if the reader is not familiar with at least some of the other topics.

sums and products of constructible numbers. In this section, we will show that the constructible numbers, together with the operations of addition and multiplication, form a field (see page 263 for the definition of a field).

Lemma 14.1.1. *Suppose A and B are constructible numbers. Then so are $A + B$ and $A - B$.*

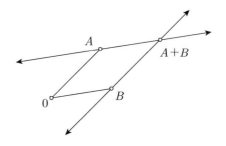

Figure 14.1

Proof. Recall, from Lemma 3.3.4, that it is possible to construct the line through B which is parallel to OA, and similarly, it is possible to construct the line through A which is parallel to OB. The point of intersection of these two lines is the point $A + B$ (see Figure 14.1). To construct $A - B$, first construct $-B$ by finding the intersection of the line $0B$ with the circle $C_0(|0B|)$. Now think of $A - B$ as $A + (-B)$. $\qquad\square$

Lemma 14.1.2. *Suppose A and B are constructible numbers. Then the product AB is a constructible number. If $B \neq 0$, then the quotient A/B is also a constructible number.*

Proof. Let us write $A = ae^{i\alpha}$ and $B = be^{i\beta}$. Then $AB = abe^{i(\alpha+\beta)}$, and $A/B = (a/b)e^{i(\alpha-\beta)}$. We saw in Chapter 3 that the lengths ab and a/b are constructible. The angles $\alpha \pm \beta$ are constructible by Lemma 3.3.5. $\qquad\square$

Theorem 14.1.3. *Suppose $K \subset F$ where F is a field. Then K is a field if and only if K is closed under addition, subtraction, multiplication, and division.*

If F is a field, $K \subset F$, and K is a field, then we call K a *subfield* of F (with the addition and multiplication of F).

Corollary 14.1.4. *The set of constructible numbers $\mathcal{C}$ forms a field.*

The field of constructible numbers $\mathcal{C}$ is sometimes called the *surd* field. Let us also include the analogue of Lemma 3.4.5 here.

Lemma 14.1.5. *Suppose A is a constructible number. Then so is $\sqrt{A}$.*

Proof. We again write $A = ae^{i\alpha}$. Then $\sqrt{A} = \sqrt{a}e^{i\alpha/2}$. By Lemma 3.4.5, $\sqrt{a}$ is a constructible length, and of course, we know how to bisect the angle α (Lemma 3.3.1). □

Exercise 14.1. Prove Theorem 14.1.3.

Exercise 14.2. Suppose K is a subfield of $\mathbb{C}$. Show that $\mathbb{Q}$ is a subfield of K.

14.2   Fields as Vector Spaces

In the Argand representation of $\mathbb{C}$, we think of $z = x + iy$ as representing a point (x, y) in $\mathbb{R}^2$. But note that if $z_1 = x_1 + iy_1$ and $z_2 = x_2 + iy_2$, then

$$z_1 + z_2 = (x_1 + x_2) + i(y_1 + y_2).$$

The addition in $\mathbb{C}$ therefore gives us an addition in $\mathbb{R}^2$, which is defined by

$$(x_1, y_1) + (x_2, y_2) = (x_1 + x_2, y_1 + y_2).$$

This, of course, is just vector addition. Hence, $\mathbb{C}$ may be thought of as a vector space over $\mathbb{R}$. This is in fact true of any subfield of a field.

Theorem 14.2.1. *Suppose K is a subfield of a field L. Then L is a vector space over K.*

Before we prove this, let us recall the definition of a vector space:

Definition 28.   *Vector space.* A *vector space* V over a field F is a set with two operations, *vector addition* and *scalar multiplication*, and which satisfies the following properties. For any $\vec{u}$, $\vec{v}$, and $\vec{w} \in V$ and any a and $b \in F$, we have

    1. $\vec{u} + \vec{v} \in V$.   (Closure of vector addition.)

    2. $(\vec{u} + \vec{v}) + \vec{w} = \vec{u} + (\vec{v} + \vec{w})$.

    3. $\vec{u} + \vec{v} = \vec{v} + \vec{u}$.

4. There exists a vector $\vec{0}$ such that $\vec{0} + \vec{v} = \vec{v}$ for any $\vec{v} \in V$.

5. For any $\vec{v} \in V$, there exists an element $-\vec{v} \in V$ such that $\vec{v} + (-\vec{v}) = \vec{0}$.

    6. $a\vec{v} \in V$.   (Closure of scalar multiplication.)

    7. $1\vec{v} = \vec{v}$.

    8. $a(\vec{v} + \vec{w}) = a\vec{v} + a\vec{w}$.   (Distribution of scalar multiplication over vector addition.)

    9. $(a + b)\vec{v} = a\vec{v} + b\vec{v}$.

10. $(ab)\vec{v} = a(b\vec{v})$.

Note that in property 9, the first $+$ is addition in the field, while the second $+$ is vector addition. Also, in property 10, the first product is in the field, while both products on the right-hand side are scalar products.

Proof of Theorem 14.2.1. We usually think of a vector as an ordered set of numbers. This is sometimes a limited view. We are now proposing, in this theorem, that we think of some elements of a field as vectors, and a few, at times also as scalars. Let us denote elements in L with Greek letters, and when we wish to think of elements in K as scalars, let us denote them with lowercase letters.

Let us check that L satisfies the properties of a vector space. Property 1, suitably translated, reads $\alpha + \beta \in L$. This is true since L is a field and is therefore closed under addition. Similarly, Properties $2 - 5$ follow from the properties of the field L. Property 6, again translated, reads $a\alpha \in L$. This follows since $a \in K \subset L$, so $a \in L$, and L is closed under multiplication. Similarly, Property 7 is true since the multiplicative identity in K is the multiplicative identity in L, and Properties $8 - 10$ follow since elements of K are elements of L and L is a field. $\square$

The reason we want to think of fields as vector spaces is that we understand the concept of dimension for vector spaces, and we would like to apply this concept to fields. So let us recall some more linear algebra.

We call a vector space V over a field F *finite dimensional* if there exists a set of vectors $\{\vec{v}_1, ..., \vec{v}_n\}$ such that for any $\vec{v} \in V$, there exist elements $a_1, ..., a_n \in F$ such that

$$\vec{v} = a_1\vec{v}_1 + ... + a_n\vec{v}_n.$$

If we also have

$$a_1\vec{v}_1 + ... + a_n\vec{v}_n = \vec{0} \qquad \text{if and only if} \qquad a_1 = a_2 = ... = a_n = 0,$$

then we call the set $\{\vec{v}_1, ..., \vec{v}_n\}$ a *basis* of V, and we call n the *dimension* of V.

Exercise 14.3. Prove that the definition of dimension is well defined. That is, prove that if $\{\vec{v}_1, ..., \vec{v}_n\}$ and $\{\vec{w}_1, ..., \vec{w}_m\}$ are both bases for V, then $m = n$.

If K is a subfield of L, then L is a vector space over K. We use the symbol $[L : K]$ to denote the dimension of L over K as a vector space, and call $[L : K]$ the *degree* of L over K. If L is not a finite dimensional vector space over K, then we write $[L : K] = \infty$. We call $[L : K]$ the *degree* of L over K.

The main result of this section is the following:

Theorem 14.2.2. *Suppose K, L, and M are fields such that $K \subset L \subset M$, and both $[M : L]$ and $[L : K]$ are finite. Then*

$$[M : K] = [M : L][L : K].$$

This theorem is true even if $[M : L]$ or $[L : K]$ is infinite, but we will not need such a result.

Proof. Let us write elements of K with lowercase letters, elements of L with Greek letters, and elements of M with uppercase letters. Let $\{\alpha_1, ..., \alpha_n\}$ be a basis of L over K, and let $\{A_1, ..., A_m\}$ be a basis of M over L. Then, $[L : K] = n$ and $[M : L] = m$.

Let B be an element of M. Then there exist $\beta_1, ..., \beta_m \in L$ such that

$$B = \beta_1 A_1 + ... + \beta_m A_m.$$

For each j, we can find elements $b_{j1}, ..., b_{jn}$ in K such that

$$\beta_j = b_{j1}\alpha_1 + ... + b_{jn}\alpha_n.$$

Thus, we can write

$$B = \sum_{j=1}^{m} \sum_{k=1}^{n} b_{jk}\alpha_k A_j.$$

But α_k can also be thought of as an element of M, so B can be written as a linear combination of the elements $\{\alpha_k A_j\} \subset M$ over the field K. This set has mn elements.

Finally, suppose

$$\sum_{j=1}^{m} \sum_{k=1}^{n} b_{jk}\alpha_k A_j = 0.$$

Then, since $\{A_1, ..., A_m\}$ is a basis of M, we know

$$b_{j1}\alpha_1 + ... + b_{jn}\alpha_n = 0$$

for $j = 1, ..., m$. But since $\{\alpha_1, ..., \alpha_n\}$ is a basis of L, we know $b_{jk} = 0$ for every j and k. Thus, the set $\{\alpha_k A_j\}$ is a basis of M over K. Hence,

$$[M : K] = mn = [M : L][L : K]$$

as desired. $\qquad\qquad\qquad\qquad\qquad\qquad\qquad\qquad\qquad\qquad\qquad\qquad\qquad$ $\square$

Exercise 14.4. Let

$$\mathbb{Q}[\sqrt{2}] = \{a + b\sqrt{2} : a, b \in \mathbb{Q}\}.$$

Prove that $\mathbb{Q}[\sqrt{2}]$ is a field and is a vector space over $\mathbb{Q}$.

Exercise 14.5. Let

$$\mathbb{Q}[\sqrt[3]{2}] = \{a + b\sqrt[3]{2} + c(\sqrt[3]{2})^2 : a, b, c \in \mathbb{Q}\}.$$

Prove that $\mathbb{Q}[\sqrt[3]{2}]$ is a field and that $[\mathbb{Q}[\sqrt[3]{2}], \mathbb{Q}] = 3$.

Let $f(x) = a_n x^n + \ldots + a_1 x + a_0$ be a polynomial with coefficients a_i in a field F. We say $f(x)$ factors over F if there exist polynomials $g(x)$ and $h(x)$ with degree greater than 0 and coefficients in F such that $f(x) = g(x)h(x)$. If there does not exist any such factorization, we say $f(x)$ is *irreducible* over F.

Exercise 14.6. Suppose α is a root of $f(x)$ and $g(x)$. Show that there exists a polynomial $h(x)$ such that $h(\alpha) = 0$ and $h(x)$ divides both $f(x)$ and $g(x)$.

Exercise 14.7. Let α satisfy the equation $\alpha^3 + \alpha^2 - 2\alpha - 1 = 0$. Let

$$\mathbb{Q}[\alpha] = \{a + b\alpha + c\alpha^2 : a, b, c \in \mathbb{Q}\}.$$

Prove that $\mathbb{Q}[\alpha]$ is a field and that $[\mathbb{Q}[\alpha] : \mathbb{Q}] = 3$.

Solution. Let us first establish that the polynomial $f(x) = x^3 + x^2 - 2x - 1$ is irreducible over $\mathbb{Q}$. Suppose it is not. Then $f(x) = g(x)h(x)$ for some polynomials g and h with rational coefficients and degree greater than or equal to one. Since the sum of the degrees of g and h is three, one of these factors must have degree one. Thus, $f(x)$ must have a rational root. Let this root be r/s where r and s are relatively prime numbers. Then

$$s^3 f(r/s) = r^3 + r^2 s - 2rs^2 - s^3 = 0.$$

Note that r divides the first three terms, so must also divide the last. Since r and s are relatively prime, this is possible only if $r = \pm 1$. Similarly, $s = \pm 1$. Hence the rational root must be ± 1. But $f(1) = -1$ and $f(-1) = 1$, so neither is a root. Thus, $f(x)$ must be irreducible over the rationals.

To show $\mathbb{Q}[\alpha]$ is a field, we must show it is closed under addition and multiplication, and that the additive and multiplicative inverses are in $\mathbb{Q}[\alpha]$.

It is clear that $\mathbb{Q}[\alpha]$ is closed under addition, and that the additive inverses are in $\mathbb{Q}[\alpha]$.

If we multiply two elements in $\mathbb{Q}[\alpha]$, we *a priori* get a fourth degree polynomial in α. However, let us note that

$$\alpha^3 = -\alpha^2 + 2\alpha + 1$$

and

$$\alpha^4 = -\alpha^3 + 2\alpha^2 + \alpha$$
$$= \alpha^2 - 2\alpha - 1 + 2\alpha^2 + \alpha$$
$$= 3\alpha^2 - \alpha + 1.$$

Hence, we can rewrite the product as a polynomial in α of degree two. Hence, $\mathbb{Q}[\alpha]$ is closed under multiplication.

To see that multiplicative inverses exist, we fix a_0, a_1, and a_2, which gives us a system of three linear equations in the three variables b_1, b_2, and b_3. This system of linear equations is solvable provided the matrix is invertible. If this matrix is not invertible, then there exists some nonzero vector (b_0, b_1, b_2) which is sent to $(0, 0, 0)$. But this means

$$(a_0 + a_1\alpha + a_2\alpha^2)(b_0 + b_1\alpha + b_2\alpha^2) = 0.$$

Since both numbers are numbers in $\mathbb{C}$, this can happen only if one of them is 0. If $b_0 + b_1\alpha + b_2\alpha^2 = 0$, then α is the root of a nonzero polynomial of degree less than three, which contradicts the fact that $f(x)$ is irreducible. Thus, we must have $a_0 + a_1\alpha + a_2\alpha^2$ is zero. Hence, every nonzero element of $\mathbb{Q}[\alpha]$ has an inverse in $\mathbb{Q}[\alpha]$.

To see that $[\mathbb{Q}[\alpha] : \mathbb{Q}] = 3$, we show that $\{1, \alpha, \alpha^2\}$ forms a basis. Of course, every element is a linear combination of these three, so all we have to do is show that

$$a_0 + a_1\alpha + a_2\alpha^2 = 0$$

implies $a_0 = a_1 = a_2 = 0$. But, if the coefficients are not all zero, then we have α as a root of a nonzero polynomial of degree less than three, which again contradicts the irreducibility of $f(x)$. Hence, $\{1, \alpha, \alpha^2\}$ forms a basis of $\mathbb{Q}[\alpha]$, and hence $[\mathbb{Q}[\alpha] : \mathbb{Q}] = 3$. $\square$

Exercise 14.8. Suppose α is a root of a polynomial $f(x)$ which is irreducible over $\mathbb{Q}$ and has degree n. Let

$$\mathbb{Q}[\alpha] = \{a_0 + a_1\alpha + \ldots + a_{n-1}\alpha^{n-1} : a_i \in \mathbb{Q}\}.$$

Show that $\mathbb{Q}[\alpha]$ is a field, and that $[\mathbb{Q}[\alpha] : \mathbb{Q}] = n$.

Exercise 14.9 (Eisenstein's Criterion). Suppose $f(x) = a_n x^n + \ldots + a_0$ is a polynomial with coefficients in $\mathbb{Z}$. Suppose there exists a prime p such that p does not divide a_n, p divides a_k for all $k \neq n$, and p^2 does not divide a_0. Prove that $f(x)$ is irreducible over $\mathbb{Q}$.

Exercise 14.10. Use Eisenstein's criterion and the substitution $x = u + 1$ to show that the polynomial

$$f(x) = x^{p-1} + x^{p-2} + \ldots + x + 1$$

is irreducible over $\mathbb{Q}$ for any prime p.

14.3 The Field of Definition for a Construction

To construct a point P, we start with two points $P_0 = 0$ and $P_1 = 1$. We then construct some lines and/or circles, and construct a point P_2 at the intersection of two of these. As we proceed through our construction, we create a sequence of points P_3, P_4, ... and so on, until we construct $P_n = P$. We call the sequence $\mathcal{P} = \{P_0, P_1, ..., P_n\}$ a *construction* for P. Let $\mathbb{Q}[\mathcal{P}]$ be the smallest field which contains $\mathcal{P}$. We call $\mathbb{Q}[\mathcal{P}]$ the *field of definition* for the construction $\mathcal{P}$.

The main result of this section is the following:

Theorem 14.3.1. *Let $\mathcal{P} = \{P_0, ..., P_n\}$ be a construction. Then*

$$[\mathbb{Q}[\mathcal{P}] : \mathbb{Q}] = 2^r$$

for some integer $r \leq n$.

We will prove this using a sequence of lemmas. First, for a field $K \subset \mathbb{C}$, let us define

$$\overline{K} = \{\overline{z} = x - iy : z = x + iy \in K\}.$$

Lemma 14.3.2. *Suppose $K = \overline{K}$. Let R, S, T, and U be points in K. Let P be the intersection of the line RS and the line TU. Then P is in K.*

Proof. Let $R = R_1 + iR_2$, etc. Since $K = \overline{K}$, we know $R_1 - iR_2 \in K$. Thus, $R_1 = \dfrac{R + \overline{R}}{2}$ and $R_2 = \dfrac{R - \overline{R}}{2}$ are both in K. If $R_1 \neq S_1$, then the line RS is given by

$$y = \frac{S_2 - R_2}{S_1 - R_1}(x - R_1) + R_2 = m_1 x + b_1,$$

where both m_1 and b_1 are in K. Similarly, the line TU is given by $y = m_2 x + b_2$, where both m_2 and b_2 are in K. We solve for the point of intersection:

$$m_1 x + b_1 = m_2 x + b_2$$
$$x = \frac{b_2 - b_1}{m_1 - m_2},$$

so $x \in K$. Since $x = P_1$, and $P_2 = m_1 P_1 + b_1 \in K$, we have $P \in K$, as desired.

The case when $R_1 = S_1$ is treated in a similar way. $\qquad\square$

Exercise 14.11. Let K be a subfield of $\mathbb{C}$. Suppose $D \in K$ and $\sqrt{D} \notin K$. Let

$$K[\sqrt{D}] = \{a + b\sqrt{D} : a, b \in K\}.$$

Show that $K[\sqrt{D}]$ is a field and that $[K[\sqrt{D}] : K] = 2$.

Lemma 14.3.3. *Suppose $K = \overline{K}$. Let R, S, T, and U be points in K. Let P and Q be the points of intersection of the line RS and the circle $C_T(|TU|)$. Then there exists a D such that both P and Q are in $K[\sqrt{D}]$. Furthermore, $\overline{K[\sqrt{D}]} = K[\sqrt{D}]$.*

Proof. The equation of the circle $C_T(|TU|)$ is

$$(x - T_1)^2 + (y - T_2)^2 = (T_1 - U_1)^2 + (T_2 - U_2)^2.$$

Let the line have the equation $y = mx + b$ where $m, b \in K$, as in the previous proof (assuming $R_1 \neq S_1$). Plugging this into the above equation of the circle, we get a quadratic equation in x, which again has coefficients in K, say

$$Ax^2 + Bx + C = 0.$$

Let $D = B^2 - 4AC$ be the discriminant of the quadratic. Then the roots are

$$x = \frac{-B \pm \sqrt{D}}{2A} \in K[\sqrt{D}].$$

These roots are P_1 and Q_1. Plugging these values into $y = mx + b$, we get P_2 and Q_2, which are both in $K[\sqrt{D}]$. Thus, P and Q are in $K[\sqrt{D}]$, as desired.

Again, the case when $R_1 = S_1$ can be treated in a similar fashion. Now, suppose $a + b\sqrt{D} \in K[\sqrt{D}]$. Then

$$\overline{a + b\sqrt{D}} = \bar{a} + \bar{b}\sqrt{D}.$$

Note that D is real. Thus, $\overline{\sqrt{D}} = \sqrt{D}$, and $\overline{a + b\sqrt{D}} \in K[\sqrt{D}]$. Hence, $\overline{K[\sqrt{D}]} = K[\sqrt{D}]$. $\qquad\square$

Lemma 14.3.4. *Suppose $K = \overline{K}$. Let R, S, T, and U be points in K. Let P and Q be the points of intersection (if there are any) of the two circles $C_R(|RS|)$ and $C_T(|TU|)$. Then there exists a D such that both P and Q are in $K[\sqrt{D}]$. Furthermore, $\overline{K[\sqrt{D}]} = K[\sqrt{D}]$.*

Proof. The equation of the two circles are

$$(x - R_1)^2 + (y - R_2)^2 = (R_1 - S_1)^2 + (R_2 - S_2)^2$$
$$(x - T_1)^2 + (y - T_2)^2 = (T_1 - U_1)^2 + (T_2 - U_2)^2.$$

Subtracting, we get the equation of a line with coefficients in K. Thus, the intersection of these two circles is the same as the intersection of a circle and a line, which we covered in the previous case. $\qquad\square$

We are now ready to prove our main result.

Proof of Theorem 14.3.1. For a construction $\mathcal{P} = \{P_0, P_1, ..., P_n\}$, let $\mathcal{P}_k = \{P_0, ..., P_k\}$. We proceed using induction on the following statement: $\overline{\mathbb{Q}[\mathcal{P}_k]} = \mathbb{Q}[\mathcal{P}_k]$ and $[\mathbb{Q}[\mathcal{P}_k] : \mathbb{Q}] = 2^{r_k}$ where $r_k \leq k$.

For the base case, note that $\mathbb{Q}[\mathcal{P}_1] = \mathbb{Q}$, that $\overline{\mathbb{Q}} = \mathbb{Q}$, and that $[\mathbb{Q}[\mathcal{P}_1] : \mathbb{Q}] = 1$.

Now, assume the statement is true for k. If the point $P_{k+1} \in \mathbb{Q}[\mathcal{P}_k]$, then we are finished. Otherwise, the point P_{k+1} is a point of intersection of either a line and a circle, or of two circles, where this line and/or circle(s) are constructed using points in $\mathbb{Q}[\mathcal{P}_k]$. Thus, by the previous lemmas, $\mathcal{P}_{k+1} \in \mathbb{Q}[\mathcal{P}_k][\sqrt{D}]$ for some real D. Hence, $[\mathbb{Q}[\mathcal{P}_k][\sqrt{D}] : \mathbb{Q}[\mathcal{P}_k]] = 2$. Clearly, $\mathbb{Q}[\mathcal{P}_k] \subset \mathbb{Q}[\mathcal{P}_{k+1}] \subset \mathbb{Q}[\mathcal{P}_k][\sqrt{D}]$. But, by Theorem 14.2.2, there cannot be any fields between these two. Hence, since $\mathcal{P}_{k+1} \notin \mathbb{Q}[\mathcal{P}_k]$, we have $\mathbb{Q}[\mathcal{P}_{k+1}] = \mathbb{Q}[\mathcal{P}_k][\sqrt{D}]$. Thus, $\overline{\mathbb{Q}[\mathcal{P}_{k+1}]} = \mathbb{Q}[\mathcal{P}_{k+1}]$, and $[\mathbb{Q}[\mathcal{P}_{k+1}] : \mathbb{Q}] = 2^{r_k+1} = 2^{r_{k+1}}$, so $r_{k+1} = r_k + 1 \leq k + 1$. $\qquad\square$

Corollary 14.3.5. *Suppose α satisfies a polynomial $f(x)$ of degree n which is irreducible over $\mathbb{Q}$, and suppose n has an odd prime factor. Then α is not constructible.*

Proof. Suppose α is constructible. Then there exists a construction $\mathcal{P}$ such that $\alpha \in \mathbb{Q}[\mathcal{P}]$. Hence, $\mathbb{Q}[\alpha] \subset \mathbb{Q}[\mathcal{P}]$. Thus, by Theorem 14.2.2, $[\mathbb{Q}[\alpha] : \mathbb{Q}]$ divides 2^r for some integer r. But by Exercise 14.8, $[\mathbb{Q}[\alpha] : \mathbb{Q}] = n$. Hence n divides 2^r, which is not possible, since n has an odd prime factor. $\qquad\square$

Remark. The converse of Corollary 14.3.5 is not true. It is possible to find a number α that is the root of a polynomial of degree four which is irreducible over $\mathbb{Q}$, but for which there is no field K such that $\mathbb{Q} \subset K \subset \mathbb{Q}[\alpha]$ and $[\mathbb{Q}[\alpha] : K] = [K : \mathbb{Q}] = 2$. Such an α is not constructible.

14.4   The Regular 7-gon

We can now prove that the regular 7-gon is not constructible, as well as the impossibility of doubling a square or trisecting an arbitrary angle.

Theorem 14.4.1. *We cannot construct the regular 7-gon.*

Proof. The proof is very similar to the proof that the regular pentagon is constructible. Suppose we can construct the regular 7-gon. Then we can construct the point $\omega = e^{i2\pi/7}$. But

$$\omega^7 = e^{i2\pi} = 1,$$

so

$$\omega^7 - 1 = 0$$
$$(\omega - 1)(\omega^6 + \omega^5 + \omega^4 + \omega^3 + \omega^2 + \omega + 1) = 0.$$

Since $\omega \neq 1$, the latter factor must be zero. Let

$$x = \omega + \omega^{-1}.$$

Then

$$x^2 = \omega^2 + 2 + \omega^{-2}$$
$$x^3 = \omega^3 + 3\omega + 3\omega^{-1} + \omega^{-3}.$$

Thus,
$$x^3 + x^2 - 2x - 1 = 0.$$
This polynomial is irreducible (see Exercise 14.7), so $[\mathbb{Q}[x] : \mathbb{Q}] = 3$. Thus, x is not constructible, and hence ω is not constructible. □

Exercise 14.12. Prove that it is impossible to double the cube.

Exercise 14.13. Prove that it is impossible to construct the regular 9-gon. Conclude that it is impossible to trisect $60°$, and hence, impossible to trisect an arbitrary angle.

Exercise 14.14. Prove that the p^r-gon is not constructible for any odd prime p and $r \geq 2$.

Exercise 14.15. Let p be a prime, and let $\omega = e^{2\pi i/p}$. By Exercise 14.10, we know that ω is the root of an irreducible polynomial of degree $p - 1$. Thus, the regular p-gon is constructible only if $p - 1 = 2^r$ for some r. Suppose r is composite. Show that $2^r + 1$ is not a prime. Conclude that the regular p-gon is constructible only if

$$p = 2^{2^k} + 1$$

for some integer k.

Exercise 14.16. In 1882, Lindemann proved that π is transcendental. That is, π is not the root of any polynomial. Use this to prove $\mathbb{Q}[\pi]$ is an infinite dimensional vector space over $\mathbb{Q}$. Conclude that it is impossible to square the circle.

14.5   The Regular 17-gon

In 1796, at the age of eighteen, Gauss showed that the regular 17-gon is constructible. Thirty-five years later, Évariste Galois developed the revolutionary theory which bears his name. Galois theory is a generalization of the tools used by Abel to prove that it is impossible to solve an arbitrary quintic using radicals, and the tools used by Gauss to decide which regular polygons are not constructible. We will, in this section, use this problem to very briefly introduce the reader to Galois theory.

By Exercise 14.15, we know that the 17-gon might be constructible, since the field $\mathbb{Q}[\omega]$ where $\omega = e^{2\pi i/17}$ has degree 16, which is a power of 2. To show that it is constructible, it is enough to find fields K_1, K_2, and K_3 such that $\mathbb{Q}[\omega] \supset K_1 \supset K_2 \supset K_3 \supset \mathbb{Q}$ and $[\mathbb{Q}[\omega] : K_1] = [K_3 : K_2] = [K_2 : K_3] = [K_3 : \mathbb{Q}] = 2$. We will find these fields using Galois theory.

Exercise 14.17. Suppose K and L are fields, and that $K \subset L \subset \mathbb{C}$. Suppose also that $[L : K] = 2$. Show that there exists an element $D \in K$ such that $L = K[\sqrt{D}]$.

We call a map σ a *homomorphism* (of fields) if σ maps the elements of a field K into a field L and if for all a and $b \in K$, we have

$$\sigma(a + b) = \sigma(a) + \sigma(b)$$
$$\sigma(ab) = \sigma(a)\sigma(b).$$

That is, σ preserves the additive and multiplicative structure of the field.

We call σ an *isomorphism* if σ is a homomorphism and it is invertible. We say the two fields K and L are *isomorphic*.

If $L = K$, and σ is an isomorphism, then we call σ an *automorphism* of K.

Exercise 14.18. Suppose K is a field which contains $\mathbb{Q}$. Suppose σ is an automorphism of K. Show that $\sigma(x) = x$ for all $x \in \mathbb{Q}$.

Suppose α is a root of an irreducible polynomial $f(x)$ with coefficients in a field K. As we saw in Exercise 14.7, the structure of the field $K[\alpha]$ depends only on the polynomial $f(x)$, and not on which root α is chosen. Hence, if α and β are two roots of an irreducible polynomial $f(x)$, then the map σ from $K[\alpha]$ to $K[\beta]$ defined by

$$\sigma(a_0 + a_1\alpha + ... + a_{n-1}\alpha^{n-1}) = a_0 + a_1\beta + ... + a_{n-1}\beta^{n-1}$$

is an isomorphism. In particular, if $\beta \in K[\alpha]$, then σ is an automorphism.

Exercise 14.19. Suppose α is a root of an irreducible polynomial $f(x)$ over $\mathbb{Q}$. Suppose σ is an isomorphism which sends $\mathbb{Q}[\alpha]$ to some field $K \subset \mathbb{C}$. Show that $\sigma(\alpha)$ is a root of $f(x)$. Conclude that $K = \mathbb{Q}[\beta]$ for some root β of $f(x)$.

Exercise 14.20. Suppose α is a root of an irreducible polynomial $f(x)$, and suppose both σ and τ are automorphisms on $\mathbb{Q}[\alpha]$. Show that $\sigma(\alpha) = \tau(\alpha)$ if and only if $\sigma = \tau$. Conclude that there are at most n automorphisms of $\mathbb{Q}[\alpha]$ where n is the degree of $f(x)$.

The regular 17-gon is constructible if and only if $\omega = e^{2\pi i/17}$ is a constructible number. We know ω satisfies an irreducible polynomial $f(x)$ of degree 16 (by Exercise 14.10). We also know

$$\omega^{17} - 1 = (\omega - 1)f(\omega) = 0.$$

Consider the number $\omega^k = e^{2\pi i k/17}$ for $k = 1, 2, ..., 16$. Note that

$$f(\omega^k)(\omega^k - 1) = (\omega^k)^{17} - 1 = \omega^{17k} - 1 = 0.$$

Since $\omega^k \neq 1$, we must have $f(\omega^k) = 0$. Thus, the sixteen roots of $f(x)$ are $\{\omega, \omega^2, ..., \omega^{16}\}$, all of which are in $\mathbb{Q}[\omega]$. Consider the automorphism σ where $\sigma(\omega) = \omega^3$. Since σ is an automorphism of $\mathbb{Q}[\omega]$, the map σ^2 defined by $\sigma(\sigma(\omega)) = \sigma(\omega^3) = \omega^9 = \omega^{-8}$ is also an automorphism of $\mathbb{Q}[\omega]$.

Similarly, the maps σ^0, σ^1, σ^2, ..., are all automorphisms of $\mathbb{Q}[\omega]$, and the image of ω under each of these maps are respectively,

$$\omega, \omega^3, \omega^{-8}, \omega^{-7}, \omega^{-4}, \omega^5, \omega^{-2}, \omega^{-6}, \omega^{-1}, \omega^{-3}, \omega^8, \omega^7, \omega^4, \omega^{-5}, \omega^2, \omega^6, \omega, ...$$

Thus, σ^{16} is the identity, and the set $G = \{\sigma^0, \sigma, \sigma^2, ..., \sigma^{15}\}$ is the full set of automorphisms of $\mathbb{Q}[\omega]$. This set is in fact a group, and is called the *Galois group* for $\mathbb{Q}[\omega]$ over $\mathbb{Q}$.

Exercise 14.21. Check that G is a group.

The group G has the following subgroups:

$$G_1 = \{\sigma^0, \sigma^8\}$$
$$G_2 = \{\sigma^0, \sigma^4, \sigma^8, \sigma^{12}\}$$
$$G_3 = \{\sigma^0, \sigma^2, \sigma^4, \sigma^6, \sigma^8, \sigma^{10}, \sigma^{12}, \sigma^{14}\}.$$

It also has the subgroups $G_0 = \{\sigma^0\}$ and $G_4 = G$. We now prove a special case of an important result in Galois theory:

Theorem 14.5.1. *Let $\mathbb{Q}[\omega]$ and G be as they are defined above. Let G_j be a subgroup of G. Let*

$$K_j = \{a \in \mathbb{Q}[\omega] : \sigma(a) = a \text{ for all } \sigma \in G_j\}.$$

Then K_j is a field.

We call K_j the *fixed field* for the group G_j.

Proof. Since K_j is a subset of a field $\mathbb{Q}[\omega]$, we need only show that K_j is closed under addition, multiplication, subtraction, and division.

Note that, for any $x \in \mathbb{Q}$ and any automorphism σ, we have $\sigma(x) = x$. In particular, $\sigma(-1) = -1$.

Let $\sigma \in G_j$, and a, $b \in K_j$. Then

$$\sigma(a + b) = \sigma(a) + \sigma(b) = a + b$$
$$\sigma(a - b) = \sigma(a) + \sigma(-1)\sigma(b) = a - b$$
$$\sigma(ab) = \sigma(a)\sigma(b) = ab.$$

Finally,

$$a = \sigma(a) = \sigma((a/b)b) = \sigma(a/b)\sigma(b) = \sigma(a/b)b,$$

so $\sigma(a/b) = a/b$. Thus, K_j is a field. $\square$

Remark. Note that we nowhere used that G_j is a group. This result is in fact true for any set $S \subset G$. However, the fixed field for S is the same as the fixed field for the smallest group which contains S, so we may as well just talk about the fixed field for subgroups.

We now have five fields: $K_0 = \mathbb{Q}[\omega]$, K_1, K_2, K_3, and $K_4 = \mathbb{Q}$. Note that if $G_j \leq G_k$ (read 'G_j is a subgroup of G_k'), then $K_j \supset K_k$. This is clear, since anything fixed by G_k must be fixed by G_j, so every element of K_k is in K_j. We therefore have

$$\mathbb{Q} = K_4 \subset K_3 \subset K_2 \subset K_1 \subset K_0 = \mathbb{Q}[\omega].$$

Furthermore, these five fields are distinct. To see that $K_1 \neq K_0$, consider the element $\omega \in \mathbb{Q}[\omega]$. This element is not in K_1 since $\sigma^8(\omega) = \omega^{-1} \neq \omega$. To see that $K_2 \neq K_1$, consider $x = \omega + \omega^{-1}$. Since $\sigma^8(x) = \sigma^8(\omega + \omega^{-1}) = \omega^{-1} + \omega = x$, this element is in K_1. But, $\sigma^4(x) = \sigma^4(\omega + \omega^{-1}) = \omega^4 + \omega^{-4} \neq x$, so $x \notin K_2$. To see that $K_3 \neq K_2$, consider the element $y = \omega + \omega^{-1} + \omega^4 + \omega^{-4}$, and to show $\mathbb{Q} \neq K_3$, consider the element $z = \omega + \omega^{-1} + \omega^4 + \omega^{-4} + \omega^8 + \omega^{-8} + \omega^2 + \omega^{-2}$.

Since none of these fields are equal, we have $[K_j : K_{j+1}] \geq 2$. However, since $[K_0 : K_4] = 16$, we get $[K_j : K_{j+1}] = 2$, as desired. Hence, the regular 17-gon is constructible.

This argument also gives us a way to construct the fields K_1, K_2, and K_3. We do so in the following set of exercises.

Exercise 14.22. Let

$$z = \omega + \omega^2 + \omega^4 + \omega^8 + \omega^{-1} + \omega^{-2} + \omega^{-4} + \omega^{-8} = \sum_{\tau \in G_3} \tau(\omega).$$

Show that

$$z^2 + z - 4 = 0.$$

Hence, $K_3 = \mathbb{Q}[z] = \mathbb{Q}[\sqrt{17}]$.

Exercise 14.23. Let

$$y = \omega + \omega^4 + \omega^{-1} + \omega^{-4} = \sum_{\tau \in G_2} \tau(\omega).$$

Show that

$$y^2 - zy + 1 = 0.$$

Find a $D \in K_3$ such that $K_2 = K_3[\sqrt{D}]$.

Exercise 14.24. Let

$$x = \omega + \omega^{-1} = \sum_{\tau \in G_1} \tau(\omega).$$

Show that

$$2x^2 - 2yx + (yz - z + y - 3) = 0.$$

Find a $D \in K_2$ such that $K_1 = K_2[\sqrt{D}]$.

Chapter 15

Modern Research in Geometry

A false but often said opinion is that geometry is dead. This false perception most probably arises because we label modern results in geometry as being part of the fields of differential geometry, algebraic geometry, arithmetic geometry, or Diophantine geometry. In this chapter, we investigate a few ideas in modern geometry. The topics I have chosen are a reflection of my background, which is in number theory. Hence, these topics are for the most part in algebraic, arithmetic, and Diophantine geometry. There are many texts in geometry which slowly drift into differential geometry, so I do not feel too guilty for leaving the subject out.

15.1   Pythagorean Triples

Diophantine geometry is an intersection of geometry and number theory. Though the subject is very active in modern research, its roots are among the ancient Greeks, most notably Diophantus of Alexandria (ca. 250 A.D.) after whom the subject is named. The problem of finding Pythagorean triples is a classical example of a Diophantine problem which arises in geometry.

A Pythagorean triple is an integer triple (a, b, c) such that

$$a^2 + b^2 = c^2.$$

We are probably all familiar with the examples $(3, 4, 5)$ and $(5, 12, 13)$. There are in fact an infinite number of Pythagorean triples, as we will show in this section. This result may be derived via a number of algebraic manipulations and repeated observations that certain numbers are squares, but in this section, we will take a more geometric approach. Our main tool is stereographic projection, which was introduced in Chapter 6.

Note that if (a, b, c) is a Pythagorean triple, then $(a/c, b/c)$ is a point on the circle

$$x^2 + y^2 = 1.$$

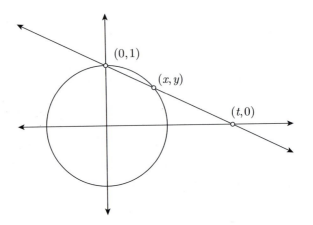

Figure 15.1

Let us consider the line through $(0,1)$ and a point $(t,0)$ on the x-axis. This line intersects the circle at two points – at $(0,1)$ and (say) at (x,y) (see Figure 15.1). Comparing slopes, we get

$$\frac{y-1}{x} = \frac{-1}{t}$$

so

$$x = t(1-y).$$

Since (x,y) is a point on the circle, this gives

$$t^2(y-1)^2 + y^2 - 1 = 0$$
$$(y-1)(t^2(y-1) + (y+1)) = 0.$$

The solution $y = 1$ gives the North Pole, so we are interested in the other solution, which is given by

$$y(t^2+1) + (1-t^2) = 0$$
$$y = \frac{t^2-1}{t^2+1}$$
$$x = t\left(1 - \frac{t^2-1}{t^2+1}\right)$$
$$= \frac{2t}{t^2+1}.$$

Thus, in terms of t, the points on the circle are $\left(\frac{2t}{t^2+1}, \frac{t^2-1}{t^2+1}\right)$. Note that if t is rational, then so are x and y, and if x and y are rational, then so is t. Now, let us write $t = p/q$. Then

$$(x,y) = \left(\frac{2pq}{p^2+q^2}, \frac{p^2-q^2}{p^2+q^2}\right).$$

If $p > q \geq 0$ are relatively prime, positive, and of different parity, then

$$(a, b, c) = (2pq, p^2 - q^2, p^2 + q^2).$$

This gives us all Pythagorean triples with a even. Those with a odd are given by the pairs (p, q) both odd, and $(a, b, c) = (pq, (p^2 - q^2)/2, (p^2 + q^2)/2)$. One can also get those with a odd by finding the solutions with a even and switching a and b.

One advantage of this method is that it works for all quadratic curves.

Though there are infinitely many integer solutions to the Pythagorean equation, there are no integer solutions to the equation

$$a^n + b^n = c^n$$

with $abc \neq 0$ and $n \geq 3$. This result is known as Fermat's Last Theorem, and was recently solved by Andrew Wiles (1995). Fermat first made the conjecture in a famous marginal note, in which he claimed to have a proof, but that the margin was too small to include it there. The book was his copy of *Arithmetica* by Diophantus. Fermat is only credited with proving this result for $n = 4$. Euler had the main ideas that lead to a solution for $n = 3$. Before Wiles' result, we knew it was correct for all $n \leq 4000000$[Ro].

Exercise 15.1. Use stereographic projection to find the integer solutions to

$$a^2 - 2b^2 = c^2.$$

Exercise 15.2. Show that there are no integer solutions to

$$a^2 + b^2 = 2c^2.$$

Exercise 15.3. Let $\triangle ABC$ be a right angle triangle with sides of integer length. Prove that the inradius r and the exradii r_a, r_b, and r_c are all integers.

Exercise 15.4. In general, stereographic projection will not work on cubics or curves of higher degree. (This is a fundamental concept in *algebraic geometry*.) An exception is the curve

$$y^2 = x^3 - x^2.$$

Apply stereographic projection to this curve using the point $(0, 0)$ and the line $x = 1$.

Exercise 15.5. How many solutions are there to the equation

$$x^2 + y^2 = z^2$$

in $\mathbb{P}^2(\mathbb{F}_q)$?

Exercise 15.6. In this text, we have developed another method of generating Pythagorean triples. Recall that the map

$$\phi = \begin{bmatrix} 1 & -i \\ -i & 1 \end{bmatrix}$$

sends the Poincaré upper half plane to the Poincaré disc. In particular, it sends the real line to the unit circle. Show that ϕ sends rational numbers to rational points on the unit circle, and vise versa. Use this to derive a formula for Pythagorean triples.

15.2　Bezout's Theorem

Bezout's theorem is a fundamental theorem in algebraic geometry. Though we will not prove it, we will show that it is a generalization of the Fundamental Theorem of Algebra, and use it to give an alternate proof of Pascal's theorem, thereby showing its connection to both geometry and algebra.

Theorem 15.2.1 (Bezout's Theorem). *Let $f(X_0, X_1, X_2) = 0$ and $g(X_0, X_1, X_2) = 0$ be two homogeneous polynomials of degree m and n which define two curves in $\mathbb{P}^2(\mathbb{C})$. If these two curves do not have a component in common (that is, if $\gcd(f,g) = 1$), then they intersect in exactly mn points (counting multiplicity).*

A special case of this theorem is the Fundamental Theorem of Algebra.

Theorem 15.2.2 (The Fundamental Theorem of Algebra). *A polynomial $f(x)$ of degree n has exactly n roots over $\mathbb{C}$ (counting multiplicity). That is, $f(x)$ factors into linear terms.*

Proof. (That is, when thought of as a corollary of Bezout's theorem.) For one of the curves, let us take

$$X_1 X_2^{n-1} - X_2^n f(X_0/X_2) = 0.$$

If we set $(X_0, X_1, X_2) = (x, y, 1)$, then this is just the curve $y = f(x)$. The other curve we take is
$$X_1 = 0,$$

which describes the x-axis. By Bezout's theorem, these two curves intersect n times, counting multiplicity. But this is just the number of times f intersects the x-axis, which is the number of zeros of f. Note that there is no point of intersection at infinity. □

The consequence of Bezout's theorem which will be of most interest to us is the Cayley-Bacharach theorem (in the cubic case). Let $f(x, y, z)$ be a homogeneous polynomial of degree three. We call the curve defined by $f(x, y, z) = 0$ a *cubic curve*.

Theorem 15.2.3 (Cubic Cayley-Bacharach Theorem). *Let f and g be two cubic curves in $\mathbb{P}^2(\mathbb{C})$ which intersect in exactly nine points $P_1,...,P_9$ (that is, they do not have a component in common). If h is another cubic curve which goes through $P_1,...,P_8$, then it also goes through P_9.*

Proof. Let

$$\vec{V}(x, y, z) = (x^3, y^3, z^3, x^2y, x^2z, y^2x, y^2z, z^2x, z^2y, xyz).$$

Then, the equation of a general cubic curve has the form

$$\vec{a} \cdot \vec{V}(x, y, z) = 0.$$

In particular, we can write $f = \vec{a} \cdot \vec{V}(x, y, z)$ and $g = \vec{b} \cdot \vec{V}(x, y, z)$ for some ten-dimensional vectors $\vec{a}$ and $\vec{b}$. Let M be the matrix with coefficients M_{ij}, where M_{ij} is the jth component of $\vec{V}(P_i)$. Then, since $f(P_i) = g(P_i) = 0$ for all i, we know

$$M\vec{a} = M\vec{b} = 0.$$

Since $\vec{a}$ and $\vec{b}$ are distinct, we therefore know that M has a kernel that is at least two dimensional, and hence, M has rank of at most eight. Thus, the vectors $\vec{V}(P_i)$ for $i = 1, ..., 9$ are dependent.

Suppose that the vectors $\vec{V}(P_i)$ for $i = 1, ..., 8$ are also dependent. Then there exists a point P_i such that $\vec{V}(P_i)$ can be written as a linear combination of the other seven $\vec{V}(P_j)$. Without loss of generality, we may assume that $i = 8$. Then, we can write

$$\vec{V}(P_8) = r_1\vec{V}(P_1) + ... + r_7\vec{V}(P_7).$$

Let $\vec{c}$ represent a cubic curve which goes through $P_1, ..., P_7$. Then

$$\vec{c} \cdot \vec{V}(P_i) = 0, \qquad i = 1, ..., 7,$$

and hence

$$\vec{c} \cdot \vec{V}(P_8) = r_1\vec{c} \cdot \vec{V}(P_1) + ... + r_7\vec{c} \cdot \vec{V}(P_7) = 0.$$

That is, the curve goes through P_8 too.

We will now show that there exists a cubic curve which goes through $P_1, ..., P_7$ but does not go through P_8. To see this, first note that no four of these points lie on a line, for if they did, then there would be a line which intersects the cubic f in four points. By Bezout's theorem, that means the line is a component of f. Similarly, the line is a component of g, so f and g have a component in common. Thus, there must exist two points, say P_2 and P_3, such that P_8 is on neither P_1P_2 nor P_1P_3. Consider the degenerate cubic curve which consists of the line P_1P_2 and the quadratic through the points $P_3, ..., P_7$ (by Exercise 15.7, there exists a quadratic through any five points). If P_8 does not lie on this quadratic, then we are finished. If it does lie on this quadratic, then consider the cubic which consists of the line $P_1 P_3$ and the quadratic through $P_2, P_4, ..., P_7$. If P_8 does not lie on this quadratic, then we are finished. If it does lie on this quadratic, then we have found two quadratics which intersect in the five points $P_4, ..., P_8$, so by Bezout's theorem, they must be the same quadratic. But then we have seven points, $P_2, ..., P_8$, which lie on this quadratic. That is, the cubic f and this quadratic intersect in seven points, so the quadratic is a component of f. Similarly, the quadratic is a component of g, and we have a contradiction. Thus, there must exist a cubic which passes through $P_1, ..., P_7$ but not P_8. Hence, the vectors $\vec{V}(P_i)$ for $i = 1, ..., 8$ must be independent.

On the other hand, the vectors $\vec{V}(P_i)$ for $i = 1, ..., 9$ are dependent, so we must be able to write

$$\vec{V}(P_9) = r_1 \vec{V}(P_1) + ... + r_8 \vec{V}(P_8),$$

and arguing as before, any cubic that goes through $P_1, ..., P_8$ must also go through P_9. □

Note that Pascal's theorem is a corollary of the Cayley-Bacharach theorem.

Corollary 15.2.4 (Pascal's Theorem). *Let $ABCDEF$ be a hexagon inscribed in a conic $\mathcal{C}$. Let R be the intersection of the opposite sides AB and DE; let S be the intersection of the opposite sides BC and EF; and let T be the intersection of the opposite sides CD and FA. Then the points R, S, and T are collinear.*

Proof. The line AB can be described with a polynomial of degree one. Let us call this polynomial F_{AB}. Let us choose our two polynomials of degree three to be

$$f = F_{AB} F_{CD} F_{EF}$$
$$g = F_{BC} F_{DE} F_{FA}.$$

These two degree three polynomials intersect at the points A, B, C, D, E, F, R, S, and T. The conic $\mathcal{C}$ can be described by a degree two polynomial $F_\mathcal{C}$. Let h be the degree three polynomial

$$h = F_\mathcal{C} F_{RS}.$$

Then h intersects eight of the nine points of intersection, so it must also go through the ninth point, namely T. If T lies on $\mathcal{C}$, then f and $F_\mathcal{C}$ intersect in seven points, which contradicts Bezout's theorem, so T must lie on RS, as desired. □

Exercise 15.7. Show that, given any five points $P_1, ..., P_5$, there exists a quadratic that goes through all five points.

Exercise 15.8. Suppose two homogeneous degree two polynomials f and g in $\mathbb{P}^2(\mathbb{C})$ intersect in exactly four points P_1, P_2, P_3, and P_4. Suppose another degree two polynomial h goes through P_1, P_2, and P_3. Prove that h goes through P_4.

15.3   Elliptic Curves

An *elliptic curve* is a cubic curve together with a group structure. It is not at all obvious that a cubic curve should include any sort of group structure, and this section is devoted primarily to describing the group. The study of elliptic curves has evolved into a very important branch of mathematics. The intent of this section is to merely give a taste of the subject. Two excellent undergraduate texts on the subject are the books *A Friendly Introduction to Number Theory* by J. H. Silverman [Sil] and *Rational Points on Elliptic Curves* by J. H. Silverman and J. Tate [S-T].

Let E be a cubic curve in $\mathbb{P}^2(\mathbb{C})$. That is, let E be the curve described by a homogeneous polynomial of degree three. Let P and Q be two points on E. By Bezout's theorem, the line PQ and E intersect in exactly three points, namely P, Q, and another point R. Let us define a binary operation on E by

$$P * Q = R,$$

where P, Q, and R are collinear. If $P = Q$, then we take the tangent line to E at P. This line intersects E at P with multiplicity two or three, and in either case, there exists a unique third point of intersection, so the operation is well defined. Note that if $P * Q = R$, then $P * R = Q$ and $Q * R = P$.

Now, let us fix a point O on E. We define an addition on E by

$$P + Q = (P * Q) * O. \tag{15.1}$$

See Figure 15.2.

Theorem 15.3.1. *The set of points on the curve E together with the addition defined in (15.1) form a commutative group.*[1]

Proof. To show that this forms a group, we must show it satisfies the following properties.

1. Associativity: $(P + Q) + R = P + (Q + R)$

2. Commutativity: $P + Q = Q + P$

3. There exists an element 0 such that $P + 0 = P$ for all P.

4. For every P, there exists an element $-P$ such that $P + (-P) = 0$.

---

[1]See Appendix A for a reminder of the definition of a group.

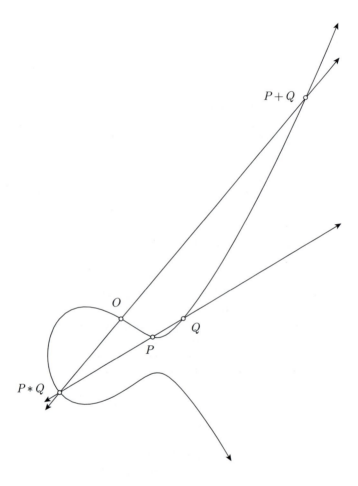

Figure 15.2. The elliptic curve $y^2 = x^3 - 4x + 4$ (or, in homogeneous coordinates, $y^2 z = x^3 - 4xz^2 + 4z^3$). The addition of the point $P = [1, 1]$ and $Q = [2, 2]$ is illustrated, using the zero element $O = [0, 2]$. The result is $P + Q = [6, 4]$.

The first property is the difficult one to establish, so let us first dispense with the others.

It is clear that $P * Q = Q * P$, so $P + Q = Q + P$.

The zero element is the point O. To see this, let $R = P * O$. That is, R is the point so that P, O, and R are collinear. Then $R * O = P$, so $P + O = (P * O) * O = R * O = P$.

Let the tangent at O intersect E again at O'. That is, let $O' = O * O$. Then, $-P = P * O'$. To see this, note that $P * (-P) = O'$, so

$$P + (-P) = (P * (-P)) * O = O' * O = O.$$

Finally, let us prove associativity. Let $S = P * Q$, $T = O * S$, $X = T * R$, $U = Q * R$, $V = O * U$, and $Y = V * P$. Then,

$$(P + Q) + R = ((P * Q) * O) + R = (((P * Q) * O) * R) * O$$
$$= ((S * O) * R) * O = (T * R) * O = X * O,$$

and

$$P + (Q + R) = P + ((Q * R) * O) = (P * ((Q * R) * O)) * O$$
$$= (P * (U * O)) * O = (P * V) * O = Y * O.$$

Thus, we must show $X = Y$. Consider the cubic C_1 which is the union of the three lines PQ, TR, and UV. Then C_1 intersects E at P, Q, $P * Q = S$, T, R, $T * R = X$, U, V, and $U * V = O$. Consider also the cubic C_2 which is the union of the three lines QR, ST, and VP, which intersects the cubic E at Q, R, $Q * R = U$, S, T, $S * T = O$, V, P, and $V * P = Y$. Thus, the cubic C_2 intersects at eight of the nine points of intersection between the cubics E and C_1. By the Cayley-Bacharach theorem, it must therefore go through the ninth point. That is, we must have $Y = X$, as desired. $\square$

By making a change of basis, it is possible to write any cubic curve in the form

$$y^2 z = x^3 + axz^2 + bz^3.$$

If $z = 0$, then $x = 0$, so $y = 1$. Thus, by setting $z = 1$, we get the equation

$$y^2 = x^3 + ax + b,$$

which represents every point on E except the point at infinity (the solution with $z = 0$). This is the standard form for an elliptic curve. We usually choose the zero element O to be the point at infinity. Lines through O are vertical lines. The point O at infinity has the nice property that it is a point of inflection. That is, $O * O = O$.

If we think of x as being a function of t, and let $y = x'$, the derivative of x with respect to t, then the equation

$$(x')^2 = x^3 + ax + b$$

is a differential equation whose solution is

$$t = \int \frac{dx}{\sqrt{x^3 + ax + b}},$$

which is (in general[2]) an *elliptic integral*. That is, this integral is related to the problem of finding the arclength of an ellipse. Thus, an elliptic curve does after all have something to do with an ellipse. The solution $x(t)$ to this differential equation is known as the Weirstrass $\wp$-function, which is often introduced in courses in complex analysis. The Weirstrass $\wp$-function, when thought of as a function over complex values, is a doubly periodic function. Since the elliptic curve is defined in $\mathbb{P}^2(\mathbb{C})$, it is a surface, if thought of as a real object (in the same way that the complex line $\mathbb{C}$ can be thought of as a real plane). The Weirstrass $\wp$-function gives us a covering of the elliptic curve. Since it is doubly periodic, an elliptic curve is topologically a torus.

Exercise 15.9. Let E be defined by $y^2 = x^3 + ax + b$, and suppose that $a, b \in \mathbb{Q}$. We say a point P on E is a *rational point* if it has rational components. Suppose P and Q are two rational points on E. Show that $P + Q$ is a rational point. Thus, the subset of rational points on E forms a subgroup of E. We call this subgroup $E(\mathbb{Q})$.

Exercise 15.10. Suppose O is a point of inflection, so that $O * O = O$. Show that P has the property that

$$P + P + P = O$$

if and only if P is a point of inflection. Show that if P and Q are points of inflection, then so is $P + Q$.

Exercise 15.11. Let E be defined by the equation $y^2 = x^3 + ax + b$. Show that the point at infinity is a point of inflection. (We can do this by first writing this equation is projective coordinates and then setting $y = 1$. Now, the point at infinity is the point $(x, z) = (0, 0)$. Show that this point is a point of inflection.)

Exercise 15.12. Let E be defined by the equation $y^2 = x^3 + ax + b$. Show that E has exactly nine points of inflection, counting the point at infinity. Conclude that there exist nine noncollinear points in $\mathbb{P}^2(\mathbb{C})$ such that a line through any two points goes through a third. Why does this not contradict Sylvester's problem (Exercise 11.12)?

Exercise 15.13. Show that if P is on the line through the point $(0, 1, 0)$ and the point $(a, 0, 1)$, then $P = (a, y, 1)$ for some y. That is, the lines through $(0, 1, 0)$ are vertical lines.

---

[2]The integral is an elliptic integral if the curve is nonsingular. We will not define what it means to be singular or nonsingular, but point out that the elliptic curve in Exercise 15.4 is singular with the singularity $(0, 0)$. This is why stereographic projection through this point works.

Exercise 15.14. Let E be described by the equation $y^2 = x^3 + ax + b$. Let O be the point at infinity. Find a formula for $P + Q$ if P and Q are distinct points on E. Find another formula for $P + P$.

Exercise 15.15*. Let E be a degenerate elliptic curve consisting of a line and a circle that are disjoint. Let O be a point on the circle. Show that if P and Q are on the circle, then $P + Q$ is on the circle. Describe the group action on this circle.

Exercise 15.16*. Let E be the cubic $y^2 = x^3 - x^2$. We saw in Exercise 15.4 that we can stereographically project the points on E to the line $x = 1$. Let $(1, s)$ and $(1, t)$ be two points on this line. Let them project to P and Q on E. Find $P + Q$ and project this point back to the line. What do we get?

Exercise 15.17**. Consider the curve C defined by

$$y^2 = x^4 + a. \tag{15.2}$$

Fermat noted that if we set $y = x^2 + bx + c$ and plug this into Equation 15.2, then the x^4 terms cancel, leaving us with a cubic. Thus, a quadratic of this form and the curve C intersect in exactly three points (Why does this not contradict Bezout's theorem?). Thus, we can define a $*$ operation on this curve. That is, given two points P and Q on C, find b and c so that the quadratic goes through P and Q, and let this quadratic intersect C again at R. Define $P * Q = R$. As with the elliptic curve, choose a distinguished point O on C and define $P + Q = (P * Q) * O$. Show that the points of the curve C and this operation $+$ form a group.

15.4 A Mixture of Cevians

This section is inspired by a very nice article "My Favorite Elliptic Curve: A Tale of Two Types of Triangles", by Richard Guy [Gu2].

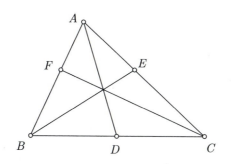

Figure 15.3

In this article, he asks the question: What are the integer triples (a, b, c) such that the triangle $\triangle ABC$ with sides a, b, and c has the property that the median from A, the angle bisector at B, and the altitude from C are coincident, as in Figure 15.3?

Let D be the midpoint of BC, let E be the point where the angle bisector of B intersects AC, and let F be the base of the altitude at C. By Ceva's theorem, this median, angle bisector, and altitude are concurrent if and only if

$$\frac{|BD|}{|DC|}\frac{|CE|}{|EA|}\frac{|AF|}{|FB|} = 1.$$

We know $|BD| = |DC| = a/2$, $\dfrac{|CE|}{|EA|} = \dfrac{a}{c}$ (by the angle bisector theorem), $|AF| = b\cos A$, and $|FB| = a\cos B$. Thus, we must have

$$1 = \frac{a/2}{a/2}\frac{a}{c}\frac{b\cos A}{a\cos B}$$

$$b\cos A = c\cos B.$$

Using the Law of Cosines, we get

$$\frac{b^2 + c^2 - a^2}{2c} = \frac{a^2 + c^2 - b^2}{2a}.$$

If we set

$$y = \frac{2b}{a+c}, \qquad x = \frac{2c}{a+c}$$

and

$$y^2 = x^3 - 4x + 4, \tag{15.3}$$

then we get the same equation. Thus, every triangle with this peculiar property and integer sides (a, b, c) generates a rational solution to the elliptic curve in Equation 15.3. The opposite, though, is not true. It is clear that the smallest triangle with this property is the equilateral triangle with sides $(1, 1, 1)$. If we further demand that the sides have coprime lengths, then the next smallest triangles with this property are the triangles with sides $(a, b, c) = (12, 13, 15)$, $(35, 277, 308)$, and $(26598, 26447, 3193)$. The solutions grow very rapidly. Guy discovered these by investigating the rational solutions on the elliptic curve.

Exercise 15.18. For the elliptic curve in Equation 15.3, find a formula for adding two distinct points P and Q if we choose the zero O to be the point at infinity. Find a formula to find $P + P$.

Exercise 15.19*. Let $P = (0, -2)$. We write nP for the point generated by adding P to itself n times. (So, for example, $3P = P + P + P$.) Let O be the point at infinity for the elliptic curve in Equation 15.3. Find the smallest $n > 2$ such that nP generates a triangle with this peculiar property. [A]

15.5 A Challenge for Fermat

The use of somewhat obscure geometric problems as a source for number theoretic problems (such as the one in the last section) is not new. Frenicle de Bessy, a contemporary of Fermat, was strangely obsessed with finding Pythagorean triples with interesting properties. One such problem which was posed of Fermat was to find a right angle triangle such that both the hypotenuse and sum of the other two sides are perfect squares.

The problem can be solved using a variation of the chord and tangent method. Let $\triangle ABC$ be a right angle triangle with sides $a < b$ and both $a + b$ and c perfect squares. Let us divide through by $a + b$ to get a triangle with rational sides whose hypotenuse is a square and whose other two sides sum to 1. Let

$$\frac{c}{a+b} = x^2,$$

and choose y so that the other two sides are

$$\frac{1-y}{2} = \frac{a}{a+b} \quad \text{and} \quad \frac{1+y}{2} = \frac{b}{a+b}.$$

Plugging these into the Pythagorean theorem, we get the curve C defined by the equation

$$C: \qquad y^2 = 2x^4 - 1. \tag{15.4}$$

This has the obvious solution $(1, 1)$, which gives a degenerate triangle. However, we can use this solution to find more solutions. Note that if we plug the equation

$$C': \qquad y = ax^2 + bx + c$$

into Equation 15.4, then we get a quartic in x which has four solutions. For each x, there is one value of y, so we get four points of intersection.[3] If three of these points of intersection are rational, then the fourth point must also be rational. Since the quadratic C' has three degrees of freedom, we can make it intersect C at the point $(1, 1)$ with multiplicity three, so that the fourth point of intersection is rational. We do this by setting the equations and their first and second derivatives equal at $x = 1$. Using implicit differentiation, we get

$$2yy' = 8x^3$$

$$y' = \frac{4x^3}{y}$$

$$y'' = \frac{12x^2y - 4x^3y'}{y^2}.$$

---

[3]If we homogenize these equations, we get a degree 2 curve and a degree 4 curve, which by Bezout's theorem, should intersect in eight points. The other four points of intersection are at infinity, where C has a singularity. In general, we should not be able to develop a 'chord and tangent' method on a degree four curve. We can do it in this case because C has a singularity.

Thus, at $x = 1$, we get $y(1) = 1$, $y'(1) = 4$, and $y''(1) = -4$, so we set

$$C' : \qquad y = 1 + 4(x-1) - 2(x-1)^2 = -2x^2 + 8x - 5.$$

The intersection of this curve with C gives

$$(-2x^2 + 8x - 5)^2 = 2x^4 - 1$$
$$0 = -x^4 + 16x^3 - 42x^2 + 40x - 13$$
$$= -(x-1)^3(x-13).$$

So, we get the solution $(13, -239)$. This point does not generate a triangle, so we look for more solutions. We can do the same trick again. This time, we choose C' so that it has a root at $x = 1$ with multiplicity two and goes through another rational point. If we choose the third root to be 13, then the fourth root must be 1 again, so instead, we note that Equation 15.4 is symmetric in x, so we choose the third root to be -13 and discover that the fourth root is $x = -\frac{1525}{1343}$. We repeat the process, and the next solution we find has $x = \frac{2165017}{2372159}$. This gives the triangle $\triangle ABC$ with sides of

$$a = 1061652293520 \qquad b = 4565486027761$$
$$c = 4687298610289.$$

It is truly remarkable that Fermat discovered this solution. He even claimed (in a less famous marginal note) that it is the smallest such solution. This was later verified by Lagrange in 1777.

Exercise 15.20**. In the same letter in which Frenicle posed the above question, he also asked for a right angle triangle with integer sides, $a > b$, and such that $(a-b)^2 - 2b^2$ is a perfect square. Find one such triangle. [A]

15.6   The Euler Characteristic in Algebraic Geometry

Using stereographic projection, we can view the real plane as being topologically equivalent to a punctured sphere. The complex projective line $\mathbb{P}^1(\mathbb{C})$ can be thought of as the real plane together with a point at infinity and so is topologically equivalent to a sphere. In fact, every curve $C \in \mathbb{P}^2(\mathbb{C})$ defined by a homogeneous polynomial is topologically equivalent to a (possibly punctured) compact surface, such as a sphere (a zero-holed surface), a torus (a one-holed surface), or a g-holed surface. The Euler characteristic χ_C of the curve C is given by

$$\chi_C = 2(1 - g).$$

There are many properties about curves in $\mathbb{P}^2(\mathbb{Q})$ which depend on the Euler characteristic of the curve when thought of as being in $\mathbb{P}^2(\mathbb{C})$. In this context, we usually refer to the value g, called the *genus* of the curve, rather than the Euler characteristic of the curve.

For example, in Section 15.1, we saw that there exists a map $(x(t), y(t))$ from the rationals $\mathbb{Q}$ to the set of rational points on the circle Γ described by $x^2 + y^2 = 1$, not including the point $(0, 1)$. Furthermore, the functions $x(t)$ and $y(t)$ are rational functions (that is, quotients of polynomials). We say that $\mathbb{Q}$ and Γ are *birational*. In general, we say a curve $\mathcal{C}$ in $\mathbb{Q}^2$ is birational to $\mathbb{Q}$ if there exists a map $(x(t), y(t))$ from all but finitely many points in $\mathbb{Q}$ to all but finitely many points in $\mathcal{C}$ such that $x(t)$ and $y(t)$ are rational functions. The surprising fact is that a curve $\mathcal{C}$ is birational to $\mathbb{Q}$ if and only if the genus of $\mathcal{C}$ is zero. Think about this for a moment. We have described a property of $\mathcal{C}$ when defined over the rationals in terms of the geometry of the curve when described over $\mathbb{C}$. This is a fundamental tenet of algebraic geometry – the idea that we can infer algebraic information from the geometry.

Let us describe another such property. Let P be a point in $\mathbb{P}^2(\mathbb{Q})$. By clearing denominators, we can write

$$P = (P_0, P_1, P_2)$$

with $P_i \in \mathbb{Z}$ and relatively prime P_i. Furthermore, this expression is unique, up to sign. We can therefore define

$$H(P) = \max\{|P_0|, |P_1|, |P_2|\},$$

which we call the *height* of P.

Given a curve $\mathcal{C}$ described by a polynomial f with rational coefficients, let us consider the quantity

$$N_\mathcal{C}(T) = \#\{P \in \mathcal{C} : H(P) < T\}.$$

It turns out that the behavior of this quantity $N_\mathcal{C}(T)$ can be characterized by the genus of $\mathcal{C}$. If $g = 0$ and $\mathcal{C}$ contains at least one rational point, then there exists an integer d and a constant $c > 0$ such that

$$\lim_{T \to \infty} \frac{N_\mathcal{C}(T)}{T^{2/d}} = c.$$

We say that $N_\mathcal{C}(T)$ grows *asymptotically* like $cT^{2/d}$. The integer d is the degree of the rational functions in the rational bijection with $\mathbb{Q}$. If the genus $g = 1$ and $\mathcal{C}$ contains infinitely many points, then $N_\mathcal{C}(T)$ grows asymptotically like $\ln(T)^{r/2}$, where r is an integer. If $g \geq 2$, then $\mathcal{C}$ contains only finitely many rational points. This last result, formerly Mordell's conjecture, was proved by Faltings in 1983.

15.7 Lattice Point Problems

Question: Given a disc D of radius r, how many integer points are inside D? For large r, we can make a fairly accurate estimate.

Let
$$N(r) = \#\{(a,b) \in \mathbb{Z}^2 : (a,b) \in D\}.$$

We can think of each integer point as representing a 1×1 tile. Since the area of the disc is πr^2, the circle touches at least πr^2 tiles. Suppose the circle intersects a particular tile, but that the integer point represented by that tile is not in the circle. Then, the integer point is at most a distance of $\sqrt{2}/2$ away from the circle. Similarly, if an integer point is inside D but the tile it represents is not entirely within D, then that point is at most $\sqrt{2}/2$ units away from the boundary of D. Thus,

$$\pi(r - \sqrt{2}/2)^2 \leq N(r) \leq \pi(r + \sqrt{2}/2)^2.$$

Said another way,

$$N(r) = \pi r^2 + R(r),$$

where $R(r)$ is a function such that

$$|R(r)| < 2\pi\sqrt{2}r + 1.$$

Let us now ask a similar question in hyperbolic geometry. Consider, for example, the tiling in Figure 8.8, which is a tiling with six 'squares' to a vertex. Let Λ be the lattice which consists of the set of points at the center of each 'square.' Let D be a disc of (hyperbolic) radius r. Let

$$N(r) = \#\{P \in \Lambda : P \in D\}.$$

What is the behavior of $N(r)$? The area of the disc is $4\pi \sinh^2(r/2)$ (see Exercise 7.88), and the area of each tile is $8\pi/3$ (Exercise 8.7), so we expect an answer of

$$N(r) = \frac{3e^r}{8} + R(r),$$

where the remainder $R(r)$ is not too large. If we try the same argument as we used in the Euclidean case, we find a bound on $R(r)$ that is of the same order of magnitude as the circumference, which is $\pi \sinh(r/2)\cosh(r/2)$. But for large r, this is approximately $4\pi e^r$, which is the same order of magnitude as the main term. No geometric argument has been discovered which gives a better bound on the remainder term $R(r)$. However, there are analytic methods. Patterson [P] has shown

$$|R(r)| < ce^{3r/4},$$

for some constant c.

Exercise 15.21**. The pseudosphere $\mathcal{V}^+$ given by $x^2 + y^2 - z^2 = -1$ and $z > 0$ contains a lot of integer points. How many integer points are there with $z < T$? More precisely, if we define

$$N(T) = \#\{(x,y,z) \in \mathbb{Z}^3 : x^2 + y^2 - z^2 = -1, \ 0 < z < T\},$$

then what is the asymptotic behavior of $N(T)$?   [H]

Exercise 15.22**. Let

$$N_{\vec{a}}(T) = \#\{(x, y, z) \in \mathbb{Z}^3 : x^2 + y^2 - z^2 = -1,\ 0 < \vec{a} \circ \vec{x} < T\}.$$

What conditions must be satisfied by $\vec{a}$ to make $N_{\vec{a}}(T)$ finite for all T? What is the asymptotic behavior of $N_{\vec{a}}(T)$?

15.8 Fractals and the Apollonian Packing Problem

In this section, we consider a couple of very odd looking regions. The first region is called Serpinski's carpet and appears in Figure 15.4. We construct this region by first taking a square of side s and dividing it into nine smaller squares. We remove the center square, and repeat this process for the remaining eight squares. We continue this process indefinitely.

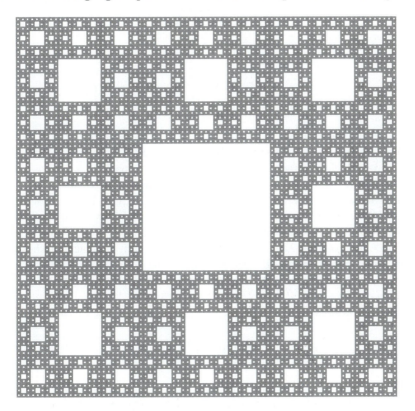

Figure 15.4. Serpinski's carpet.

Let us suppose that the area of this figure is given by $M(s)$. Every area formula has the property that

$$M(\lambda s) = \lambda^2 M(s).$$

Thus, if we triple the length of s, then we get a figure with 9 times the area of the original. On the other hand, we can construct the same figure with side of length $3s$ by reproducing the original 8 times. We therefore get

$$9M(s) = 8M(s),$$

from which we conclude that $M(s) = 0$. Serpinski's carpet has no area. That is, it is not two dimensional. We therefore feel compelled to conclude that it must be one dimensional, but then $M(\lambda s) = \lambda M(s)$, so we get

$$3M(s) = 8M(s).$$

This raises the interesting question: *Are there dimensions between 1 and 2?* Suppose there are, and that this carpet is d dimensional. Then we must have

$$M(\lambda s) = \lambda^d M(s),$$

where $M(s) \neq 0$ if $s \neq 0$. By setting $\lambda = 3$, we get

$$3^d M(s) = 8M(s)$$

$$d = \frac{\ln 8}{\ln 3} \approx 1.893.$$

We say that Serpinski's Carpet has a *fractal dimension* of $d = \ln 8/ \ln 3$.

Not all fractals are so nicely self-similar. For example, consider the Apollonian packing (see Figure 15.5). This is a packing of a disc using smaller discs. We begin with a disc and in it we remove two discs which are mutually tangent and tangent to the original circle. In this new figure, there are two *curvilinear triangles* – triangles whose sides are arcs of circles. In each of these two curvilinear triangular regions, we remove the unique incircle. We now have six curvilinear triangles left. In each of these, we again remove the incircles. We continue this process indefinitely. The region which is left is a fractal region which unfortunately has no nice self-similarity to exploit.

Let the radii of these circles be $\{r_1, r_2, ...\}$, and consider the function

$$f(s) = \sum_{i=1}^{\infty} r_i^s. \tag{15.5}$$

It is clear that this function converges for $s = 2$, diverges for $s = 0$, and where it is defined, it is a decreasing function. We therefore know there exists an α such that $f(s)$ converges for all $s > \alpha$ and diverges for all $s < \alpha$. Boyd [Bo] has shown that α is the fractal dimension of this set, and that

$$1.300197 < \alpha < 1.314534.$$

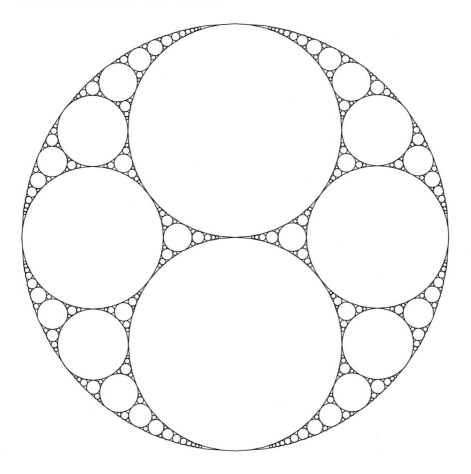

Figure 15.5. The Apollonian packing.

Though we have no intention of showing how α is calculated, we note that to do so, we must be able to calculate the radii r_i. This can be done using a very nice geometrical result due to Descartes. Recall first that the curvature of a circle with radius r is the quantity $t = 1/r$. Let us further adopt the convention that, in a configuration of three mutually tangent circles (as in Figure 15.6), if one circle bounds the other two, then the curvature of that circle is given a negative sign. With this convention, we have:

Theorem 15.8.1 (Descartes' Circle Theorem). *Suppose four circles with curvature t_1, t_2 and t_3 and t_4 are mutually tangent. Then*

$$2\left(t_1^2 + t_2^2 + t_3^2 + t_4^2\right) = \left(t_1 + t_2 + t_3 + t_4\right)^2.$$

We begin with a lemma:

Lemma 15.8.2. *Let A, B, and C be the centers of three mutually tangent circles with curvature t_a, t_b, and t_c, respectively. If the three circles are externally tangent, then the points of tangency are on the sides of $\triangle ABC$ and are where the incircle touches the sides. If the circles centered at B and C are inside the circle centered at A, then we take the curvature t_a to be negative; the points of tangency are on the sides of $\triangle ABC$ and are where the excircle centered at I_a intersect the sides. Let τ be the curvature of the incircle or excircle, depending on the case. Then, in both cases, we have*

$$\tau^2 = t_b t_c + t_a t_c + t_a t_b.$$

Proof. We leave as an exercise the proof that the incircle touches the sides of $\triangle ABC$ at the points of tangency of the circles.

Let us first consider the externally tangent case, as in Figure 15.6. By Exercise 1.81,

$$t_a = \frac{1}{s-a}, \qquad t_b = \frac{1}{s-b} \qquad \text{and} \qquad t_c = \frac{1}{s-c}.$$

Thus,

$$\tau^2 = \frac{1}{r^2} = \frac{s^2}{|\triangle ABC|^2} = \frac{s^2}{s(s-a)(s-b)(s-c)}$$

$$= t_a t_b t_c s = t_a t_b t_c \left(\frac{1}{t_a} + \frac{1}{t_b} + \frac{1}{t_c} \right)$$

$$= t_b t_c + t_a t_c + t_a t_b.$$

In the case when the circles centered at B and C are inside the circle centered at A (that is, when $t_a < 0$), we get

$$t_a = -\frac{1}{s}, \qquad t_b = \frac{1}{s-c} \qquad \text{and} \qquad t_c = \frac{1}{s-b},$$

so

$$\tau^2 = \frac{1}{r_a}^2 = \frac{(s-a)^2}{|\triangle ABC|^2} = \frac{(s-a)^2}{s(s-a)(s-b)(s-c)}$$

$$= -t_a t_b t_c (s-a) = t_a t_b t_c \left(\frac{1}{t_a} + \frac{1}{t_b} + \frac{1}{t_c} \right)$$

$$= t_b t_c + t_a t_c + t_a t_b. \qquad \square$$

Proof of Theorem 15.8.1. The four circles give four combinations of three mutually tangent circles and therefore generate four new circles with curvatures τ_1, τ_2, τ_3, and τ_4, where the circle with curvature τ_1 is generated by the circles with curvature t_2, t_3, and t_4, etc. (see Figure 15.7).

Note that the four new circles all intersect the original circles perpendicularly and at the points of tangency, so we get a new configuration of four mutually tangent circles. By Lemma 15.8.2, we get

$$\tau_1^2 = t_3 t_4 + t_2 t_4 + t_2 t_3$$

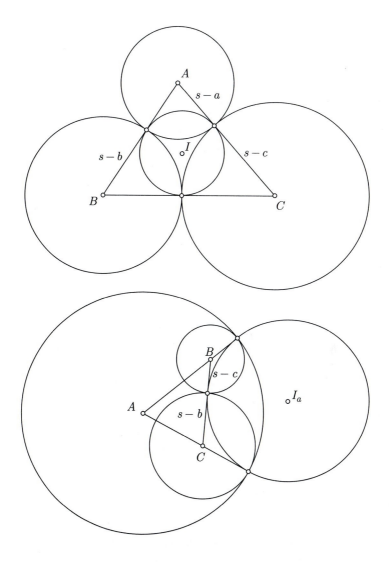

Figure 15.6

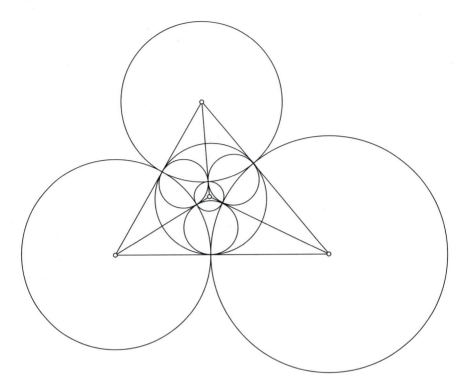

Figure 15.7

$$\tau_2^2 = t_3 t_4 + t_1 t_4 + t_1 t_3$$
$$\tau_3^2 = t_2 t_4 + t_1 t_4 + t_1 t_2$$
$$\tau_4^2 = t_2 t_3 + t_1 t_3 + t_1 t_2,$$

and summing, we get

$$\tau_1^2 + \tau_2^2 + \tau_3^2 + \tau_4^2 = 2t_1 t_2 + 2t_1 t_3 + 2t_1 t_4 + 2t_2 t_3 + 2t_2 t_4 + 2t_3 t_4$$

$$\sum_{k=1}^{4} \tau_k^2 + \sum_{k=1}^{4} t_k^2 = \left(\sum_{k=1}^{4} t_k \right)^2. \tag{15.6}$$

Thus, all we need to show is that $\sum t_k^2 = \sum \tau_k^2$. Reversing the roles of the four original circles and the four generated circles, we get

$$\sum_{k=1}^{4} (\tau_k^2 + t_k^2) = \left(\sum_{k=1}^{4} \tau_k \right)^2.$$

Hence

$$\left(\sum_{k=1}^{4} \tau_k \right)^2 = \left(\sum_{k=1}^{4} t_k \right)^2$$

$$\sum_{k=1}^{4} \tau_k = \pm \sum_{k=1}^{4} t_k.$$

Note that if $t_1 < 0$, then t_2, t_3, and t_4 are each positive and larger than $|t_1|$. Thus, $\sum t_k > 0$, and similarly, $\sum \tau_k > 0$, so

$$\sum_{k=1}^{4} \tau_k = \sum_{k=1}^{4} t_k.$$

Now, consider

$$
\begin{aligned}
(t_1 + t_2 + t_3 + t_4)(t_1 + t_2 + t_3 - t_4) &= (t_1 + t_2 + t_3)^2 - t_4^2 \\
&= t_1^2 + t_2^2 + t_3^2 + 2t_1t_2 + 2t_1t_3 + 2t_2t_3 - t_4^2 \\
&= t_1^2 + t_2^2 + t_3^2 + 2\tau_4^2 - t_4^2.
\end{aligned}
$$

Let us also substitute expressions in terms of the τ_k's for the t_k^2's to get

$$
\begin{aligned}
(t_1 + t_2 + t_3 + t_4)(t_1 + t_2 + t_3 - t_4) &= (\tau_2\tau_3 + \tau_1\tau_3 + \tau_1\tau_2) \\
&\quad + (\tau_3\tau_4 + \tau_1\tau_4 + \tau_1\tau_3) \\
&\quad + (\tau_2\tau_4 + \tau_1\tau_4 + \tau_1\tau_2) \\
&\quad - (\tau_2\tau_3 + \tau_1\tau_3 + \tau_1\tau_2) + 2\tau_4^2 \\
&= 2\tau_4(\tau_1 + \tau_2 + \tau_3 + \tau_4) \\
t_1 + t_2 + t_3 - t_4 &= 2\tau_4.
\end{aligned}
$$

In particular,

$$(t_1 + t_2 + t_3 - t_4)^2 = 4\tau_4^2,$$

and similarly,

$$
\begin{aligned}
(t_1 + t_2 - t_3 + t_4)^2 &= 4\tau_3^2 \\
(t_1 - t_2 + t_3 + t_4)^2 &= 4\tau_2^2 \\
(-t_1 + t_2 + t_3 + t_4)^2 &= 4\tau_1^2.
\end{aligned}
$$

Summing, we get

$$4(t_1^2 + t_2^2 + t_3^2 + t_4^2) = 4(\tau_1^2 + \tau_2^2 + \tau_3^2 + \tau_4^2),$$

and substituting into Equation 15.7, we get the desired result.   □

Exercise 15.23. Suppose three mutually externally tangent circles are centered at A, B, and C. Prove that the incircle of $\triangle ABC$ touches the sides of $\triangle ABC$ at the points of tangency of the original three circles.

Exercise 15.24. Prove that $f(s)$ in Equation 15.5 converges at $s = 2$, and that where it converges, $f(s)$ is a decreasing function.

Exercise 15.25*. Describe how to construct (using a straightedge and compass) the incircle of a curvilinear triangle formed by three mutually tangent circles. [H]

Exercise 15.26**. Suppose four mutually tangent circles in the complex plane have curvatures t_1, t_2, t_3, and t_4, and centers P_1, P_2, P_3, and P_4, where the centers are expressed as complex numbers. Prove that

$$2\left((t_1 P_1)^2 + (t_2 P_2)^2 + (t_3 P_3)^2 + (t_4 P_4)^2\right) = (t_1 P_1 + t_2 P_2 + t_3 P_3 + t_4 P_4)^2.$$

Exercise 15.27 (The Serpinski Gasket). The Serpinski gasket is created in a fashion similar to the construction of the Serpinski carpet. We begin with an equilateral triangle and remove the triangle whose vertices are the midpoints of the sides of the original triangle. In the remaining three triangles, we again remove a triangle, and we continue this process indefinitely. The first few steps are shown in Figure 15.8. What is the fractal dimension of the Serpinski gasket?

Figure 15.8. The Serpinski gasket and the Koch snowflake.

Exercise 15.28 (The Koch Snowflake). The Koch snowflake is produced by beginning with an equilateral triangle. We remove the middle third of each side, and replace it with two sides of an equilateral triangle (facing out). We repeat the procedure on the twelve new segments and continue the process indefinitely. The first few steps are shown in Figure 15.8. What is the dimension of the Koch snowflake?

Exercise 15.29 (The Cantor Set). To construct the Cantor set, we remove the middle third $(1/3, 2/3)$ of the segment $[0, 1]$ on the real line. We remove the middle thirds of the remaining two segments and continue this process indefinitely. What is the dimension of the Cantor set?

Figure 15.9. The Koch snowflake was featured in the logo for the '91 International Mathematical Olympiad.

Exercise 15.30 (The Menger Sponge). To construct the Menger sponge, we take a cube, and on each face we bore out the middle square. We repeat this process indefinitely. The first few steps are shown in Figure 15.10. What is the dimension of the Menger sponge?

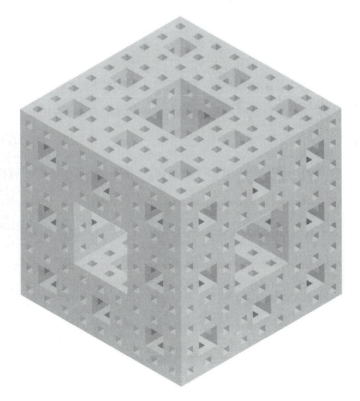

Figure 15.10. The Menger sponge.

15.9   Sphere Packing

Let us begin with a two-dimensional problem. How should we place unit discs in the plane so that they do not overlap and so that the greatest proportion of the plane is covered? One might think of trying an experiment: Take a handful of soda straws and wrap a rubber band around them. What we get is the *beehive* packing shown in Figure 15.11.

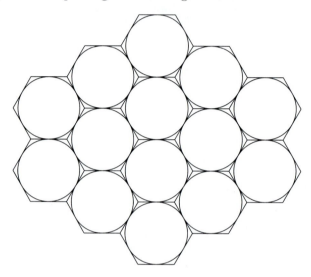

Figure 15.11

The *density* δ of a packing of a finite region R is the ratio of the area covered by the discs to the area of the region. The density of a packing of the entire plane is found by first taking a circle of radius r, finding the density $\delta(r)$ of the packing given by the intersection of the circle with the packing of the plane, and then finding the limit of $\delta(r)$ as r goes to infinity. If a packing of the plane has the structure of a tiling (like the beehive packing), then its density is just the ratio of the area of the disc in each tile to the area of each tile.

For the beehive packing where each disc has radius one, the the sides of the hexagonal tiles have length $s = \frac{2}{\sqrt{3}}$. Hence, the density δ of the beehive packing of the plane is

$$\delta = \frac{\pi}{2\sqrt{3}}.$$

Let us now ask, *Is this the best packing?* That is, is it possible to find a packing of the Euclidean plane with density $\delta > \frac{\pi}{2\sqrt{3}}$? The answer is no – the optimal packing is the beehive packing. This was shown by Axel Thue in 1890. The proof is not too difficult (see [Ha]).

In contrast, the three-dimensional version of this question is very difficult. Let us begin with an obvious packing, the *cannonball* packing, shown

Figure 15.12. The cannonball packing with $n = 3$.

in Figure 15.12. We begin with a triangular arrangement of cannonballs, n balls to a side, and on top of that, stack another triangular arrangement with $n - 1$ balls on each side. We continue until we have a tetrahedral arrangement. If the radius of each cannonball is one, then this cannonball stack gives us a packing of a tetrahedron with side

$$s = 2n + 3\sqrt{2} - 2.$$

The volume of a tetrahedron with side s is

$$V = \frac{s^3}{6\sqrt{2}}.$$

The number of cannon balls on the bottom layer is

$$n + (n - 1) + \ldots + 1 = \frac{(n + 1)n}{2},$$

so the total number of cannonballs is

$$\sum_{k=1}^{n} \frac{k^2 + k}{2} = \frac{1}{2}\left(\frac{(n + 1)(2n + 1)n}{6} + \frac{(n + 1)n}{2}\right) = \frac{n^3 + 3n^2 + 2n}{6}.$$

Thus, the density of this packing is

$$\delta = \frac{(n^3 + 3n^2 + 2n)4\pi}{18} \frac{6\sqrt{2}}{(2n + 3\sqrt{2} - 2)^3}.$$

As n approaches infinity, we get a packing of Euclidean three-space with density

$$\delta = \frac{\pi\sqrt{2}}{6}.$$

In 1611, Kepler asserted that no other packing has a greater density. His assertion has since become known as the *Kepler conjecture* and has remained unsolved for almost four centuries. A partial result was found by Gauss, who showed that $\delta = \frac{\pi\sqrt{2}}{6}$ is best possible if the packing has the structure of a tiling (like the cannonball packing). A general solution was recently proposed by Thomas Hales but is still under scrutiny by the mathematical community.

The question of finding optimal sphere packings in finite geometries might sound like just another academic exercise, but is, in fact, of relevance to modern telecommunications. When we transmit information, whether through a digital fax or phone, between ground control and robots on Mars, or between a memory chip and a computer processor, there is the possibility that the message will be garbled. We can guard against this by transmitting information using an *error correcting code*, a code which will detect and correct flipped digits in small blocks. Though the subject of error correcting codes is usually thought of as a topic in modern algebra, it has a geometric interpretation. Such a code can be thought of as a sphere packing in a finite geometry.

Consider a geometry with 2^n points, represented by $(\mathbb{Z}/2\mathbb{Z})^n$. In this geometry, we will say two points P and Q are a distance k apart if their representations in $(\mathbb{Z}/2\mathbb{Z})^n$ differ in k places. This distance is called the *Hamming* distance. Consider a packing of this space with spheres of radius $r > 1$ where r is not an integer. For example, suppose we have a packing with radius $r = 3/2$. Let $\{P_1, ..., P_N\}$ be the set of centers of this packing. These points are the allowable messages and might be thought of as *letters*. Suppose that a message Q is received and that Q is not one of these letters. We immediately know that an error has occurred during transmission, but do we know what the original letter was? If we assume that at most one digit was flipped, then this Q differs from the intended letter P_k by only one digit. That is, Q still lies within the sphere centered at P_k, so all we have to do is figure out which sphere it lies in, and we will know the intended letter. By doing so, we have corrected the error.

In a geometry with 2^n points, there are n points a Hamming distance 1 away from any fixed point. Thus, each sphere contains $n + 1$ points, so the number of spheres (or letters) N in any packing is at most

$$\frac{2^n}{n+1}.$$

If this packing has density one, then $n + 1$ divides 2^n, so

$$n + 1 = 2^k$$

for some integer k.

For example, with $k = 2$, we get $n = 3$ and there are at most two letters. It is in fact possible for N to equal 2, since the choice of 000 and 111 for letters works. Hence, when we send a message using this code, we repeat each digit three times. If we receive a message which includes a block of three digits which are not all equal, then we correct the message by choosing the digit which appears twice.

For $k = 3$, we get $n = 7$ and at most $2^4 = 16$ letters. It is again possible to pack in 16 spheres, and a choice of centers appears in Table 15.1. This code was developed in 1950 by Richard Hamming of Bell Labs.

0000000	1111111	0001111	1110000
1000110	0111001	1100011	0011100
0100101	1011010	1010101	0101010
0010011	1101100	1001001	0110110

Table 15.1. The sixteen letters of a seven-digit error correcting code.

The first four digits of these sixteen points are distinct, so uniquely identify the letter. We therefore often think of the first four digits as the information, while the last three digits are thought of as check digits. Note that the code corrects single-digit errors in the check digits too, and not just in the first four digits.

Let us compare these two codes. In order to communicate four digits of information, we must send twelve digits with the first code, but only seven digits are required with the second code. Both codes will properly correct a single-digit error. The first code is capable of correcting up to four flipped digits (in a twelve-digit block), but if we are unlucky, as few as two flipped digits can be corrected improperly. The second code will never properly correct two flipped digits. Thus, the first code has a slight error correcting advantage, but the second code is much more efficient. Extrapolating, it is not hard to imagine that longer codes are also of interest, particularly for applications where errors can happen, but are very rare.

Exercise 15.31. Derive the formula for the volume of a tetrahedron with side s.

Exercise 15.32. Prove that the smallest tetrahedron which contains a ball of radius 1 has sides of length $3\sqrt{2}$. Conclude that the smallest tetrahedron which can hold a cannonball stack with n balls of radius 1 per side is a tetrahedron with sides of length $s = 2n - 2 + 3\sqrt{2}$.

Exercise 15.33. Suppose we stack cannonballs begining with a square layer with n^2 balls. If we let n go to infinity, we get a packing of three-space. What is the density of this packing? [A]

Exercise 15.34. Suppose we tile the Euclidean plane with squares, and in each square, we inscribe a circle. What is the density of this packing?

Exercise 15.35. Prove that the density of a packing of the plane (or of three-space) which has the structure of a tiling is the same as the density of each tile.

Exercise 15.36. In our calculation of the density of the cannonball packing, we used tetrahedra instead of spheres. Justify our calculation.

Exercise 15.37. Suppose we have a tiling of hyperbolic geometry with n-gons, m per vertex. Inside each polygon, we can inscribe a circle, thereby giving a packing of the hyperbolic plane. What is the density of this packing?   [A]

Exercise 15.38. Show that the Hamming distance between any two points in Table 15.1 is either three, four, or seven. Conclude that the spheres of radius $3/2$ about each point do not intersect.   [H]

Exercise 15.39. What is the maximum possible density of packings with spheres of radius $3/2$ in geometries with 2^n points for every $n \leq 10$? For $n = 3$ and 7, the density is 1. In these two cases, the rate of information per transmission is $1/3 \approx 33\%$ and $4/7 \approx 57\%$, respectively. What is this rate for each $n \leq 10$, assuming the maximum possible density is obtainable?

Exercise 15.40*. We have been working in base two because it is the natural base for computers. What is the smallest $n \geq 2$ such that there exists a packing of $(\mathbb{Z}/3\mathbb{Z})^n$ with spheres of radius $3/2$ and density one. Find a set of centers which gives such a packing.   [A]

Exercise 15.41**. The DNA code is in base four. There is evidence that there are many natural strategies to prevent genetic mutations. Research the literature for evidence that using error correcting codes is one such strategy.

Chapter 16

A Selective Time Line of Mathematics

This is a selective time line of mathematicians who have been featured in this text or whose work is well known and is included for the sake of reference.

16.1 The Ancient Greeks

Pythagoras (ca. 572 – 500 B.C.) is usually credited with proving the Pythagorean theorem in Euclidean geometry (page 8). There are versions of the Pythagorean theorem in both spherical (page 168) and hyperbolic (page 211) geometry.

Plato (ca. 427 – 347 B.C.) is best known for his work the *Republic*. In this work, he asserts that the value of mathematics is how it trains the mind, and that its practical utility is of minor importance. Though the Platonic solids are named after him, this is because of his assertion that the particles of the four elements of fire, air, Earth, and water are, respectively, tetrahedrons, octahedrons, cubes, and icosahedrons. The dodecahedron represents the universe. The geometry of these solids was already known to the Pythagoreans and was fully developed by Theaetetus. Plato is also attributed with the compass and straightedge rules of construction.

Euclid (ca. 350 B.C.) wrote the *Elements*, a thirteen volume set on mathematics. Euclid's influence on modern mathematics is astounding. Even today, there are many professors who require that their students read at least part of the *Elements*.

Archimedes (287 – 212 B.C.) wrote *Psammites* (Sand-reckoner) as mentioned in Chapter 2. He devised a way of trisecting an arbitrary triangle using a compass and notched straightedge (page 84). Archimedes also found bounds on π; summed geometric series (both finite and infinite);

found formulas for the sum of integers and squares; and in his work, there is a strong hint of integral calculus (and to a lesser extent, differential calculus). He is probably best known for his water pump (the Archimedean screw); his machines of war, which were used to defend Syracuse (in Sicily) in the Second Punic War; and his discovery that a floating body displaces its weight in water. He is said to have made this last discovery while floating in a bath, after which he dashed through the streets naked, exclaiming "*Eureka, eureka*" ("I have found it, I have found it"). Archimedes was a contemporary of Eratosthenes of Cyrene (on the Red Sea) and often communicated with him. Archimedes died at the hands of a Roman soldier during the fall of Syracuse. It is said that Archimedes requested that his tombstone include a diagram of a sphere inscribed in a cylinder, in reference to his proof that the ratio of these two objects is $2/3$.

Apollonius of Perga (ca. 262 − 190 B.C.) is featured in this text in Exercise 1.84, and in the Apollonian packing problem in Section 15.8. Apollonius is best known for his work on conic sections.

Hipparchus (ca. 161 − 126 B.C.) developed a table of chords (essentially a table of sines).

Menelaus (ca. 100 A.D.) proved Menelaus' theorem in both the plane (page 53) and the sphere (page 217). He could also find the distance between cities, given (in modern language) their longitude and latitude. That is, he could solve a type of spherical triangle.

Ptolemy (ca. 100 − 170 A.D.) proved Ptolemy's theorem (page 51), which gives the angle sum formulas for (in modern language) sines and cosines. He used these to refine the table of chords. He used his tables to solve triangle problems in the plane using essentially the Law of Sines or Law of Cosines. He also used Menelaus' theorem on the sphere to prove the beginnings of spherical trigonometry, our Theorems 10.2.1 and 10.2.2.

Heron of Alexandria (3rd century A.D.?) is best known for Heron's Formula (page 39), though the result dates back to Archimedes. He also invented a number of mechanical contrivances, including a rudimentary steam turbine. It is not clear when Heron lived. Some place him after Ptolemy and before Pappus, but others place him as early as 100 B.C.

Pappus (ca. 320 A.D.) proved Pappus' theorem, which was introduced in Chapter 4 and proved in Chapter 11 (page 229). His work is also featured in Exercises 1.7 and 7.20. Pappus also developed the foci-focus-directrix and eccentricity definitions of conics.

Diophantus of Alexandria (325 − 410) wrote *Arithmetica*, a thirteen book set on algebra. Fermat wrote his famous marginal note in his copy of *Arithmetica*. Diophantine equations and Diophantine geometry are named after Diophantus.

Proclus (410 – 485 A.D.) wrote detailed commentaries on many of the works of the ancient Greeks, referring to and often quoting from works of antiquity which have since been lost. He is responsible for the preservation of much of the ancient Greek history of mathematics.

16.2 The Fifth Century A.D. to the Fifteenth Century A.D.

Brahmagupta (ca. 598 – 665) proved a version of Heron's formula for cyclic quadrilaterals (page 52).

Arabic Influence. We owe much to the developments of the Arabs, particularly in algebra. Our number system is Arabic, and the words 'algebra' and 'algorithm' have Arabic roots. Much of our knowledge of the works of the ancient Greeks is through Arabic translations. For example, Ptolemy's work, simply called *The Mathematical Collection*, is usually referred to as the *Almagest* (from *'al magest'* meaning 'the majestic'), which is the title the Arabs gave it.

Al-Biruni (973 – 1055) proved the Law of Sines in spherical geometry. By repeatedly using this result, Al-Biruni was able to solve more problems then we might at first think possible. His work was valued in the Muslim religion because of the importance of knowing the direction of Mecca.

Omar Khayyám (1048 – 1131), a poet, took up the challenge of trying to prove Euclid's fifth axiom from the first four. His work, through the writings of either Nasir Eddin al-Tusi (1201 – 1274) or his son, is probably the source of Saccheri's work. He also showed how to solve cubic equations in the fashion of Exercise 3.39.

Leonardo de Pisa, (1170 – 1240), better known as Fibonacci, wrote in Latin but was trained by and frequently visited Muslim scholars. He is best known for the Fibonacci sequence $\{1, 1, 2, 3, 5, 8, 13, ...\}$, where each term is the sum of the previous two.

16.3 The Renaissance to the Present

René Descartes (1596 – 1650) is credited with the invention of the Cartesian coordinate system. His circle theorem is featured in Theorem 15.8.1 on page 317. He is perhaps best known as a philosopher, and for the phrase *"Cogito ergo sum"* ("I think, therefore I am").

Gérard Desargues (1591 – 1661), influenced by the study of perspective by the painters of the Renaissance, began the development of projective geometry. He is best known for Desargues' theorem (pages 95 and 230).

Pierre de Fermat (1601 – 1665) is best known for 'Fermat's Last Theorem' (page 301), a conjecture only recently proved by Andrew Wiles. A taste of Fermat's extraordinary work is featured in Section 15.5.

Blaise Pascal (1623 – 1662) proved Pascal's theorem (page 231) when he was sixteen. He is also well known for Pascal's triangle, the triangular array of binomial coefficients. Pascal's triangle already appeared in a text by Jordanus de Nemore, ca. 1220, and much earlier in both Chinese and Arabic works.

Isaac Newton (1642 – 1727) and Leibniz are credited with independently inventing calculus. Newton is also credited with the concept of gravity.

Gottfried Wilhelm Leibniz (1646 – 1716) and Newton are credited with independently inventing calculus. Leibniz' notations for derivatives and integrals are still in use today.

Giovanni Ceva (1647 – 1736) proved Ceva's theorem (page 55).

Girolamo Saccheri (1667 – 1733) investigated the Saccheri quadrilateral (Figure 6.3, page 118), and by assuming the angles A and D are acute, he derived many 'strange' results but was unable to come to a contradiction.

Abraham DeMoivre (1667 – 1754) proved DeMoivre's theorem (pages 77, 80):
$$\cos n\theta + i \sin n\theta = (\cos \theta + i \sin \theta)^n.$$
In this text, we express this result in the form
$$e^{i\theta} = \cos \theta + i \sin \theta, \tag{16.1}$$
which is a formulation due to Euler.

Robert Simson (1687 – 1768) wrote a popular updated version of Euclid's elements which includes an attempt to fix Euclid's axioms. He is honored with the name of the Simson line (page 50), which was proved by William Wallace almost thirty years after his death.

Leonhard Euler (1707 – 1783) discovered the Euler line (page 44), and the Euler characteristic for polyhedra and surfaces (featured in Sections 5.3 and 15.6). He formulated Euler's formula, the identity in Equation 16.1, by noting that both sides satisfy the same differential equation (see Exercise 3.22). Euler was very prolific and published more than a 1000 papers in mathematics (only Paul Erdös has published more).

Jean-Robert Argand (1768 – 1822) and Caspar Wessel (1745 – 1818) independently developed the geometric interpretation of complex numbers.

Karl Friedrich Gauss (1777 – 1855) was one of the most influential mathematicians of all time. He claimed in a letter to János Bolyai that he had already discovered hyperbolic geometry. Though Gauss had a habit of not publishing his 'unpolished' results, of his work on this subject he wrote that his intention was "not to allow it to become known during my lifetime." Gauss was apparently afraid that the scientific community was not ready to understand this work. Despite Gauss' many achievements, he seemed to be most proud of his discovery (at a young age) that the regular 17-gon is constructible (see page 90 and Section 14.5) and requested that the figure be inscribed on his tombstone. He also proved that it is impossible to trisect an arbitrary angle and claimed that it is impossible to construct a regular n-gon for n not a Fermat prime (see page 83).

Charles-Julien Brianchon (1785 – 1864) proved Brianchon's theorem (page 239).

Jean-Victor Poncelet (1788 – 1867) discovered duality in projective geometry.

Augustin Louis Cauchy (1789 – 1857) refined the definition of a limit. In particular, his definition of a Cauchy sequence (page 195) allows one to define the convergence of a limit without *a priori* knowledge of the existence of the limit. We use it to axiomatically define the real numbers. This can also be done using *Dedekind cuts*, an idea which has its roots in Euclid's theory of proportions.

Nikolai Ivanovich Lobachevsky (1792 – 1856) is also sometimes credited with discovering hyperbolic geometry. He developed hyperbolic trigonometry.

Michel Chasles (1793-1880) discovered the cross ratio *in projective geometry.*

Franz Taurinus (1794 – 1874) experimented with the consequences of substituting ik for the radius r in formulas in spherical geometry. He called this *log-spherical geometry.* His work influenced Lobachevsky.

Karl Georg Christian von Staudt (1798 – 1867) developed a geometrically defined addition and multiplication on lines in the affine plane. Assuming Desargues' theorem (or using a set of axioms which imply Desargues' theorem), Hilbert showed that the algebra defined by von Staudt is a division ring (see Chapter 13).

Karl Feuerbach (1800 – 1834) proved Feuerbach's theorem (page 47).

Niels Henrik Abel (1802 – 1829) proved that it is impossible to solve an arbitrary quintic with radicals. Abelian groups are so named in his honor.

János Bolyai (1802 – 1860) wrote to Gauss about his discovery of hyperbolic geometry.

Évariste Galois (1811 – 1832) invented Galois theory, with which one can prove that it is impossible to trisect an arbitrary angle, square the circle (given Lindemann's result), double the cube, or solve an arbitrary quintic with radicals. Galois died in a duel at the age of 21.

Georg Bernhard Riemann (1826 – 1866) proved that all complex analytic functions preserve angles (except at a finite number of points). We saw this for fractional linear transformations. Riemann expanded on the idea of a geometry both in dimension and in type. The simplest idea is to consider surfaces in three dimensions and restrict the arclength differential to this surface. We did this for the sphere in Section 12.1. The study of such geometries is called *Riemann geometry.*

Eugenio Beltrami (1835 – 1900) developed the Beltrami-Klein model of hyperbolic geometry (page 259). Klein made this model more acceptable to the mathematical community by developing a distance function on it.

Moritz Pasch (1843 – 1930) renewed the investigation of the axiomatic foundations of geometry. He recognized the need for axioms of betweeness. Pasch's theorem appears on page 199.

Georg Cantor (1845 – 1918) invented the Cantor set (page 322). This set is a fractal which has been used to develop numerous intuition defying examples in analysis.

Felix Klein (1849 – 1925) developed the log-cross ratio definition of distance in the Poincaré models (Section 7.11), and a notion of distance in the Beltrami-Klein model (page 259).

Ferdinand Lindemann (1852 – 1939) proved that π is transcendental, thereby completing the proof that it is impossible to square the circle.

Hendrik Antoon Lorentz (1853 – 1928) is most famous for his work on the theory of the electron. He developed Lorentz space (see Chapter 12), a version of $\mathbb{R}^3$ in which hyperbolic geometry is naturally imbedded.

Jules Henri Poincaré (1854 – 1912) developed the Poincaré disc and upper half plane models of hyperbolic geometry (Chapter 7). Poincaré is often thought of as the father of modern topology.

David Hilbert (1862 – 1943) developed an axiomatic foundation for geometry, fixing the holes in Euclid's set of axioms. Some of his remarkable work in this area is featured in Chapter 13. Hilbert made significant contributions to many other areas of mathematics and is considered by some to be the most influential mathematician of the twentieth century.

Hermann Minkowski (1864 – 1909) developed Minkowski space-time, a four-dimensional model of Lorentz space in which he unified space and time (see Chapter 12). His work on this model influenced the work of his student Albert Einstein.

Appendix A

Quick Reviews

A.1   2 × 2 Matrices

A two by two matrix has the general form

$$\gamma = \begin{bmatrix} a & b \\ c & d \end{bmatrix},$$

where the entries a, b, c, and d are usually numbers, real or complex, but may also be functions. The determinant of γ is

$$\det \gamma = \det \begin{bmatrix} a & b \\ c & d \end{bmatrix} = ad - bc.$$

We multiply matrices as follows:

$$\begin{bmatrix} a & b \\ c & d \end{bmatrix} \begin{bmatrix} r & s \\ t & u \end{bmatrix} = \begin{bmatrix} ar + bt & as + bu \\ cr + dt & cs + du \end{bmatrix}.$$

Matrix multiplication is associative (that is, $(\alpha\beta)\gamma = \alpha(\beta\gamma)$), but it is not commutative (that is, except for unusual cases, $\alpha\beta \neq \beta\alpha$). Note that for two matrices α and β, $\det(\alpha\beta) = \det(\alpha)\det(\beta)$.

The *identity* is

$$I = \begin{bmatrix} 1 & 0 \\ 0 & 1 \end{bmatrix},$$

and for any matrix γ, $I\gamma = \gamma I = \gamma$.

We say a matrix γ is *invertible* if there exists a matrix γ^{-1} such that

$$\gamma\gamma^{-1} = \gamma^{-1}\gamma = I.$$

The matrix γ^{-1} is called the *inverse* of γ. A matrix γ is invertible if and only if $\det \gamma \neq 0$. For 2×2 matrices,

$$\gamma^{-1} = \begin{bmatrix} a & b \\ c & d \end{bmatrix}^{-1} = \frac{1}{ad - bc} \begin{bmatrix} d & -b \\ -c & a \end{bmatrix}.$$

A.2　Vector Geometry

Let $\vec{u} = (u_1, u_2, u_3)$ and $\vec{v} = (v_1, v_2, v_3)$ be vectors in $\mathbb{R}^3$. The length $||\vec{u}||$ of $\vec{u}$ is the distance from the origin $(0, 0, 0)$ to the point (u_1, u_2, u_3), which is

$$||\vec{u}|| = \sqrt{u_1^2 + u_2^2 + u_3^2}.$$

The dot product of $\vec{u}$ and $\vec{v}$ is

$$\begin{aligned} \vec{u} \cdot \vec{v} &= u_1 v_1 + u_2 v_2 + u_3 v_3 \\ &= ||\vec{u}|| ||\vec{v}|| \cos \theta, \end{aligned}$$

where θ is the angle between $\vec{u}$ and $\vec{v}$.

The cross product of $\vec{u}$ and $\vec{v}$ is

$$\vec{u} \times \vec{v} = (u_2 v_3 - u_3 v_2, u_3 v_1 - u_1 v_3, u_1 v_2 - u_2 v_1).$$

This vector $\vec{u} \times \vec{v}$ is perpendicular to both $\vec{u}$ and $\vec{v}$, and points in a direction according to the right-hand rule. The length of $\vec{u} \times \vec{v}$ is

$$||\vec{u} \times \vec{v}|| = ||\vec{u}|| ||\vec{v}|| \sin \theta.$$

Note also that

$$\vec{v} \times \vec{u} = -\vec{u} \times \vec{v}.$$

A.3　Groups

A group G is a set of objects together with an operation $*$, and with the following properties: For all elements a, b, and c in G, we have

1. Associativity:
$$(a * b) * c = a * (b * c).$$

2. The existence of an identity: There exists an element 1 such that for any $a \in G$,
$$a * 1 = 1 * a = a.$$

3. The existence of inverses: There exists an element a^{-1} such that
$$a * a^{-1} = a^{-1} * a = 1.$$

In the above, we have represented the operation as a multiplication. We also often represent the operation as an addition, in which case the properties are

1. Associativity:
$$(a + b) + c = a + (b + c).$$

2. The existence of an identity: There exists an element 0 such that for all $a \in G$,
$$a + 0 = 0 + a = a.$$

3. The existence of inverses: There exists an element $-a$ such that
$$a + (-a) = (-a) + a = 0.$$

However, we use the addition notation only if the group is commutative. That is, only if the group satisfies the following additional property:

4. Commutativity:
$$a + b = b + a.$$

Examples of groups include the integers under addition; the positive rationals under multiplication; the complex numbers without 0 under multiplication; $GL_2(\mathbb{R})$; and $SL_2(\mathbb{Z})$. The set of polynomials with coefficients in the rationals or reals under addition forms a group. The set of rational polynomials (that is, quotients of polynomials), not including the zero polynomial, together with multiplication, form a group. And of course, the set of isometries in Euclidean geometry under composition forms a (non-commutative) group.

A.4   Modular Arithmetic

Let $n \geq 2$ be an integer. We say that two integers a and b are *equivalent modulo n* if their difference $a - b$ is divisible by n. We write
$$a \equiv b \pmod{n}.$$

For any integer a, we can use the division algorithm to find integers q and r (the quotient and remainder) such that
$$a = qn + r$$

and $0 \leq r < n$. Thus, $a \equiv r \pmod{n}$. Hence, modulo n, every integer can be represented by an element of $\{0, ..., n-1\}$. We define addition and multiplication on this set via the addition and multiplication in the integers. That is, we say
$$a + b \equiv c \pmod{n}$$
if n divides $c - a - b$, and we similarly say
$$ab \equiv c \pmod{n}$$
if n divides $c - ab$.

We call the set $\{0, ..., n-1\}$ together with its algebra the *integers modulo n*, and we denote it with $\mathbb{Z}/n\mathbb{Z}$.

The operations of addition and multiplication modulo n satisfy several familiar properties. They are both closed (i.e., for any a and $b \in \{0, ..., n - 1\}$, there exist both c and $d \in \{0, ..., n - 1\}$ such that $a + b \equiv c \pmod{n}$ and $ab \equiv d \pmod{n}$); they are both associative and commutative; multiplication distributes over addition; there exist additive and multiplicative identities (0 and 1, respectively); and for any $a \in \{0, ..., n - 1\}$, there exists an additive inverse, namely the element $n - a$ which satisfies $a + (n - a) \equiv 0 \pmod{n}$. Multiplicative inverses do not always exist. For example, in $\mathbb{Z}/6\mathbb{Z}$, the element 2 does not have a multiplicative inverse, since there does not exist a b such that $2b \equiv 1 \pmod{6}$. Arithmetic modulo a prime number p is of particular interest because for any $a \neq 0$ in $\{0, ..., p - 1\}$, there exists an element b such that $ab \equiv 1 \pmod{p}$. That is, modulo a prime p, multiplicative inverses exist for all nonzero elements. Thus, the integers modulo a prime p form a field.

Appendix B

Hints, Answers, and Solutions

B.1 Hints to Selected Problems

Exercise 1.6. Show that triangle $\triangle ABB'$ is congruent to $\triangle C'BC$. What is the length $|C'D|$ in relation to $\triangle CBC'$?

Exercise 1.20. The proof requires Axiom 5.

Exercise 1.48. Let P' be another point on Γ and let $P'B$ intersect Γ' and Q'. Show $\triangle PQA \sim \triangle P'Q'A$. Now choose P' in a convenient location.

Exercise 1.49. Show that $\triangle PBA \sim \triangle PCD$. Let P' be another point, which generates the points C' and D'. Show that $\triangle PAD \sim \triangle P'AD'$.

Exercise 1.54. Start with the statement of the theorem and then decide where to place the circle.

Exercise 1.59. Not all exercises use the subject of the current section.

Exercise 1.78. Draw a triangle for which the quantity $a - b\cos C$ is negative.

Exercise 1.82. Use Exercise 1.80

Exercise 1.89. Consider the altitudes.

Exercise 1.100. Where is the centroid G?

Exercise 1.105. Look for a cyclic quadrilateral.

Exercise 1.116. What does the extended Law of Sines say for $\triangle AYZ$?

Exercise 1.124. Use Exercise 1.43.

Exercise 1.129. Use Exercise 1.128.

Exercise 3.39. Draw a circle centered at (a, b) which goes through $(0, 0)$. Where does this circle intersect the parabola?

Exercise 3.43. See the section titled "Nicomedes" in Chapter VII of [H3].

Exercise 4.37. Look at $\triangle BB'T$ and $\triangle CC'S$.

Exercise 5.25. The surface area of a sphere is the derivative of the volume with respect to r. Why?

Exercise 6.3. Show that at least one of the altitudes intersects the opposite side. Suppose AD is an altitude with D on BC. Show that the sum of angles in $\triangle ADB$ is $180°$.

Exercise 6.5. Stack congruent rectangles to show that arbitrarily tall rectangles exist. Given a point P and a line l, let Q be the point on l so that PQ is perpendicular to l. Align a congruent rectangle so that l is one of its sides, Q is one of its vertices, and P is on one of its sides. Use Exercise 6.2 to construct a rectangle with P as a vertex and two vertices on l. Use Exercise 6.4.

Exercise 6.6. Suppose they do. Find a triangle whose angles sum to more than $180°$.

Exercise 7.44. Use Exercise 7.38.

Exercise 7.48. Exercises 7.39, 7.46, and 7.47 are useful.

Exercise 7.81. Write $x + iy = \phi^{-1}(re^{i\theta})$. Conjugate to get $x - iy$ and subtract to get $2iy$. Do not bother expanding the denominator. Differentiate both expressions to get $dx \pm idy$. Multiply them together to get $dx^2 + dy^2$. Divide by y^2. The denominators of both expressions should cancel.

Exercise 7.83. Reflections are improper isometries and are their own inverses.

Exercise 7.89. In Euclidean geometry, the circumference of a circle is the derivative of the area. Why?

Exercise 7.105. Skip ahead to Section 10.5 on Heron's formula on the sphere.

Exercise 9.11. First, define *same orientation* for two triangles $\triangle ABC$ and $\triangle ABC'$.

Exercise 9.16. This is more of a sketch of the proof than a hint, but there are still a lot of details to be filled in. (1) Prove that there exist points R and S such that the intersection of $\mathcal{C}_P(r)$ with the ray PR is inside $\mathcal{C}_Q(s)$, and the intersection with the ray PS is outside $\mathcal{C}_Q(s)$. (2) Define a sequence of points T_n (in the fashion of the proof of Exercise 9.15) on the line segment RS. For each T_n, the line PT_n intersects the circle $\mathcal{C}_P(r)$ at, say, U_n. Decide whether $R_n = T_n$ or $S_n = T_n$ based on whether U_n is inside or outside $\mathcal{C}_Q(s)$. (3) Use completeness of the reals to argue that the sequence $\{T_n\}$ converges to a point T on RS. (4) Let PT intersect $\mathcal{C}_P(r)$ at U. Show $|QU| = s$.

Exercise 9.25. If l_1 and l_3 intersect, then let P be that point of intersection. What is the line through P which is parallel to l_2?

Exercise 9.26. Draw the diagonal AC and use ASA.

Exercise 10.2. Think of a circle on the sphere as a surface of revolution. Dig up that dusty calculus text.

Exercise 10.22. Write $\cos \Delta = \cos(\Delta_1 + \Delta_2)$ where Δ_1 and Δ_2 are the areas of right angle triangles.

Exercise 12.14. Suppose T is a matrix with positive determinant, and suppose vectors $\vec{u}$, $\vec{v}$, and $\vec{w}$ obey the right-hand rule. Then the vectors $T\vec{u}$, $T\vec{v}$, and $T\vec{w}$ also obey the right-hand rule.

Exercise 12.19. Use Exercises 12.10 and 12.11.

Exercise 12.22. Let $C = (0, 0, 1)$, $A = (\sinh b, 0, \cosh b)$, and $B = (0, \sinh a, \cosh b)$.

Exercise 12.24. Begin with $\vec{a} = (0, 0, 1)$. What are the images of planes and lines under linear transformations?

Exercise 13.4. Do the equations $x + y = 0$ and $2x + 2y = 0$ define the same line?

Exercise 13.14. See Theorem 11.3.1.

Exercise 13.17. There are two.

Exercise 13.22. Since $n > 2$, one can show that there are at least four parallel directions. Pick two intersecting lines to create a coordinate system. Use these to define the rows and columns. Pick two different intersecting lines, neither of which are parallel to the original two, to define a different coordinate system. Label points in this system with regiment and rank. Prove the desired result.

Exercise 13.27. To show Axiom 1, one must find a line through P_i and P_j. Consider the number $i - j$. For Axiom 5h, pick a point P not in $\mathfrak{p} = \mathfrak{p}(A, B, C)$. For any point $Q \in \mathfrak{p}$, show that there exists a point Q' which is not in $\mathfrak{p}$ and is on the line PQ. Conclude that $\mathcal{G}$ has at least twice as many points as $\mathfrak{p}$. Use Exercise 13.24.

Exercise 15.21. Show that the set of integer points on $\mathcal{V}^+$ is the orbit of the point $(0, 0, 1)$ under the action of the subgroup $O_J(\mathbb{Z})$. Show that the integer points on $\mathcal{V}^+$ form a lattice. Project onto the Poincaré disc $\mathcal{D}$ to find the fundamental domain.

Exercise 15.25. Find the 'altitudes' of the triangle. To find altitudes, we must find a circle which is perpendicular to two other circles. Note that such a circle has its center on the radical axis (see Exercise 1.65).

Exercise 15.38. Note that for every point P on the list, there is a complementary point P' a Hamming distance of seven away. That is, for every point P on the list, there is a point P' on the list which differs from P in every digit. Show that if a point Q is a distance k away from P, then Q is $7 - k$ away from P'. This observation significantly reduces the amount of checking required.

B.2 Answers to Selected Problems

Exercise 1.30. The angle $\angle ADC$ is equal to $60°$.

Exercise 1.31. The diameter is 15.

Exercise 1.45. The length $|AD|$ is 5.

Exercise 1.50. The length $|PD|$ is $\frac{8}{3}$.

Exercise 1.52. The length of $|BD|$ is $\frac{45}{13}$.

Exercise 1.55. The radius of the circle is 4.

Exercise 1.56. The area $|\triangle ABD|$ is $\frac{24}{5}$.

Exercise 1.67. The length $|BG|$ is $2\sqrt{2}$.

Exercise 1.69. The side c has length $\sqrt{5}$.

Exercise 1.73. The area of the quadrilateral is $6 + 4\sqrt{21}$.

Exercise 1.75. The area of the incircle is $\frac{8\pi}{3}$.

Exercise 1.76. The area of $\triangle ABC$ is $\frac{15}{4}\sqrt{7}$.

Exercise 1.77. The length $|AC|$ is $\sqrt{13 - 6\sqrt{2}}$.

Exercise 1.100. The length $|BC|$ is equal to 28.

Exercise 1.110. The third side has length $c = \frac{5\sqrt{7}+3\sqrt{39}}{4}$.

Exercise 3.26. The polynomial $f(x) = x^3 + x^2 - 2x - 1$ satisfies the conditions.

Exercise 5.26. The hyper-volume of the interior of a hyper-sphere of radius r in five dimensions is $\frac{8\pi^2 r^5}{15}$.

Exercise 7.49. The triangle $\triangle ABC$ is an equilateral triangle, and γ is rotation by $120°$ about the center of $\triangle ABC$.

Exercise 7.64. The isometry is $\gamma = \begin{bmatrix} 1 & -2 \\ 1 & 0 \end{bmatrix}$. This map is a rotation about the point $z = \dfrac{1 + i\sqrt{7}}{2}$.

Exercise 7.83. The set of reflections in $\mathcal{D}$ is the set of all maps $\gamma(-\bar{z})$ with $\gamma = \begin{bmatrix} a & b \\ b & a \end{bmatrix}$ and $b \in \mathbb{R}$.

Exercise 7.84. The area element in $\mathcal{D}$ is $dA = \dfrac{4r\,dr\,d\theta}{(1 - r^2)^2}$.

Exercise 7.85. The distance formula in $\mathcal{D}$ is the same as in $\mathcal{H}$. That is, $|PQ| = |\ln(P, Q; M, N)|$.

Exercise 7.105. Let triangle $\triangle ABC$ have sides a, b, and c, semiperimeter $s = \frac{a+b+c}{2}$, and area $\Delta = |\triangle ABC|$. Then

$$1 - \cosh \Delta = \frac{4 \sinh s \sinh(s - a) \sinh(s - b) \sinh(s - c)}{(1 + \cosh a)(1 + \cosh b)(1 + \cosh c)}.$$

Exercise 8.10. The value of a is $\sqrt{2} + \sqrt{3}$.

Exercise 8.14. One of the vertices is at $A = i(\sqrt{6} - \sqrt{2})/2$.

Exercise 9.4. We say P is *inside* $\angle BAC$ if P and C are on the same side of AB, and P and B are on the same side of AC.

Exercise 9.11. In the following, we define when two nondegenerate triangles $\triangle ABC$ and $\triangle A'B'C'$ have the *same orientation*. If two triangles have the same orientation, let us write $\triangle ABC \overset{o}{\sim} \triangle A'B'C'$. We define the relation $\overset{o}{\sim}$ to have the following properties:

1. $\triangle ABC \overset{o}{\sim} \triangle BCA$ and $\triangle ABC \overset{o}{\sim} \triangle CAB$

2. $\triangle ABC \overset{o}{\sim} \triangle ABC'$ if C and C' are on the same side of AB. If C and C' are on opposite sides, then $\triangle ABC \overset{o}{\sim} \triangle BAC'$.

3. The relation $\overset{o}{\sim}$ is transitive. That is, if $\triangle ABC \overset{o}{\sim} \triangle A'B'C'$ and $\triangle A'B'C' \overset{o}{\sim} \triangle A''B''C''$, then $\triangle ABC \overset{o}{\sim} \triangle A''B''C''$.

Note that Properties 1 and 2 are both transitive, so Property 3 does not contradict them. Hence, $\overset{o}{\sim}$ is well defined. Note that $\overset{o}{\sim}$ is reflexive. That is, if $\triangle ABC \overset{o}{\sim} \triangle A'B'C'$, then $\triangle A'B'C' \overset{o}{\sim} \triangle ABC$. Thus, by transitivity, $\triangle ABC \overset{o}{\sim} \triangle ABC$, so $\overset{o}{\sim}$ is symmetric too. Hence, $\overset{o}{\sim}$ is an equivalence relation.

Given any two nondegenerate triangles $\triangle ABC$ and $\triangle A'B'C'$, either $\triangle ABC \overset{o}{\sim} \triangle A'B'C'$ or $\triangle ABC \overset{o}{\sim} \triangle A'C'B'$. To see this, note that we can replace the vertices one at a time. That is, if C' is not on AB, then we can conclude that either $\triangle ABC \overset{o}{\sim} \triangle ABC'$ or $\triangle ABC \overset{o}{\sim} \triangle BAC'$. If C' is on AB, then one of A' or B' is not on AB, since $\triangle A'B'C'$ is not degenerate. Say that A' is not on AB. Then either $\triangle ABC \overset{o}{\sim} \triangle ABA'$ or $\triangle ABC \overset{o}{\sim} \triangle BAA'$. We repeat this step, replacing another of the vertices A or B with one of the vertices A', B', or C'. After two more steps, and using Property 1, we have either $\triangle ABC \overset{o}{\sim} \triangle A'B'C'$ or $\triangle ABC \overset{o}{\sim} \triangle B'A'C'$.

Thus, given some reference triangle $\triangle PQR$, every triangle $\triangle ABC$ has the same orientation as $\triangle PQR$ or $\triangle QPR$, so there are at most two equivalence classes. The problem now is to show there are at least two equivalence classes ....

Exercise 10.5. The corresponding result in Euclidean geometry is the statement $\cos A = \sin(\pi/2 - A)$.

Exercise 10.22. For small triangles, this formula approximates $\Delta = \frac{1}{2}hc$ ('the area is one half base times height').

Exercise 10.26. Each edge has length $\pi/5$.

Exercise 11.11. A *rotation* is a proper isometry which fixes a point. A *translation* is a proper isometry which fixes a line. Note that the identity is both a translation and a rotation. What we usually think of as rotations and translations in Euclidean geometry are still rotations and translations using this new definition. In hyperbolic geometry, rotations are still rotations, and hyperbolic translations are still translations, but parabolic translations fit neither definition. This is consistent with our discussion in Section 7.15, where we noticed that parabolic translations behave a little like rotations too. In elliptic geometry, every triangle is oriented the same way, so all isometries are proper. A rotation is also a translation and vice versa, since rotations about a point P fix the polar of P, and translations which fix a line p also fix the pole of p.

Exercise 11.18. The excenter has areal coordinates
$$I_a = \left(\frac{-a}{b+c-a}, \frac{b}{b+c-a}, \frac{c}{b+c-a} \right).$$

Exercise 12.10. The matrix representation of R_θ is
$$R_\theta = \begin{bmatrix} \cos\theta & \sin\theta & 0 \\ -\sin\theta & \cos\theta & 0 \\ 0 & 0 & 1 \end{bmatrix}.$$

Exercise 12.23. Polar coordinates. Given a point $Q = (0,0,1)$, the point $P(\phi, \theta)$ is a distance ϕ away from Q and makes an angle θ with a chosen axis. In Euclidean geometry, we usually denote ϕ with r. The model of Euclidean geometry which fits best as an analogue is the plane $z = 1$ parameterized by
$$P(\phi, \theta) = (\phi\cos\theta, \phi\sin\theta, 1).$$

Note that the linear approximation of $\sin\phi$ and $\sinh\phi$ at zero is just ϕ, and the linear approximation of $\cos\phi$ and $\cosh\phi$ at zero is just 1.

Exercise 15.19. The point $5P$ generates the triangle $(35, 277, 308)$.

Exercise 15.20. The smallest such triangle has $a = 1517$, $b = 156$, and $c = 1525$.

Exercise 15.33. The density is $\frac{\pi\sqrt{2}}{6}$, the same as for the tetrahedral cannonball packing.

Exercise 15.37. The density is

$$\delta = \frac{2}{\pi} \frac{\cos(\pi/m) - \sin(\pi/n))}{1 - \frac{2}{n} - \frac{2}{m}} \frac{(\pi/n)}{\sin(\pi/n)}.$$

Exercise 15.40. The smallest such n is $n = 4$. See Table B.1 for a list of centers.

0000	1011	2022
0112	0221	1120
1202	2101	2210

Table B.1

B.3 Solutions to Selected Problems

Exercise 1.25. Let us break this into several parts.
Part I: Show that a diameter which bisects a proper chord is perpendicular to it. Let the chord be AB with midpoint C. Let O be the center of the circle. Draw the radii OA and OB, as in Figure B.1. Then, $|OA| = |OB|$ since they are both radii, and $|AC| = |BC|$, since C is the midpoint. Thus, by SSS,

$$\triangle OCA \equiv \triangle OCB.$$

In particular,

$$\angle OCA = \angle OCB.$$

But these two angles sum to $180°$, since $\angle ACB = 180°$. Thus, OC is perpendicular to AB. Since all diameters go through the center O, the diameter through C includes the segment OC, and so is perpendicular to AB.
Part II: The perpendicular bisector of a chord goes through the center O.

Let C be the midpoint of AB. Draw the diameter through C. By Part I, we know OC is perpendicular to AB. Thus, OC is a perpendicular bisector of AB, and since perpendicular bisectors are unique, it is *the* perpendicular bisector. Thus, the perpendicular bisector goes through O.

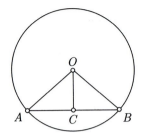

Figure B.1

Part III: A diameter which is perpendicular to a chord must bisect the chord.

Let this diameter intersect the chord at D. Let C be the midpoint of AB. Draw the diameter through C. Consider the triangle $\triangle OCD$. By Part I, $\angle OCD = 90°$. We are given $\angle ODC = 90°$. Thus, $\angle COD = 0°$. That is, O, C, and D are collinear. But C and D both lie on a line perpendicular to OD. Thus, since C and D both lie on two distinct lines, they must be the same point. Thus, this diameter OD bisects the chord. $\square$

Exercise 1.28. By the Star Trek lemma, $\angle ABC$ is half of the angular measure of the arc it subtends. Thus, the arc AB subtends is $180°$. That is, it is half a circle. Hence, the segment AB is a diameter. $\square$

Exercise 1.35. Draw the chord AB' in Figure B.2. By the Star Trek

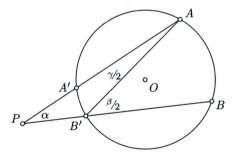

Figure B.2

lemma, $\angle PAB' = \dfrac{\gamma}{2}$, and $\angle AB'B = \dfrac{\beta}{2}$. Thus, since the exterior angle of a triangle is the sum of the other two interior angles, we get

$$\alpha + \frac{\gamma}{2} = \frac{\beta}{2}$$

$$\alpha = \frac{\beta - \gamma}{2}.$$ $\square$

Exercise 1.37. See Lemma 11.2.2.   □

Exercise 1.54. In this problem, we are asked to use a particular result to prove a theorem. Our solution should therefore start with a statement of the theorem we are proving.

Theorem (The Pythagorean Theorem). *Suppose* $\triangle ABC$ *is a right angle triangle with* $\angle ACB = 90°$. *Then*

$$a^2 + b^2 = c^2.$$

To use the tangential version of power of the point, we first wonder where there might be a natural place to put a right angle. We recall that the tangent to a circle is perpendicular to the radius, so let us draw the triangle centered at B and through C (see Figure B.3). This is the way our proof begins.

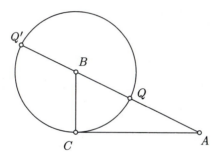

Figure B.3

Proof. In the right angle triangle $\triangle ABC$, draw the circle centered at B and passing through C. Since the angle at C is a right angle and BC is a radius, we know AC is tangent to the circle. Hence, we can apply the tangential case of power of the point to the point A, which gives

$$|AC|^2 = |AQ||AQ'|,$$

where Q and Q' are the points of intersection of the line AB with the circle. But $|AC| = b$, $|AQ| = c - a$, and $|AQ'| = c + a$. Thus, we get

$$b^2 = (c - a)(c + a)$$
$$= c^2 - a^2$$
$$a^2 + b^2 = c^2.$$ □

Exercise 1.79. Let O be the center of Γ. Note that the figure is symmetric about the line AO. Thus, the angle bisector of $\angle BAC$ goes through O (see

Figure B.4). Let it intersect Γ at the points I and I_a (which we will show are the incenter and excenter). Note that $\angle CBA$ subtends the arc BC, so the arc BC measures $2\angle CBA$. Since I is the midpoint of this arc, $\angle IBA = \frac{1}{2}\angle CBA$. That is, I is at the intersection of two angle bisectors, so is the incenter.

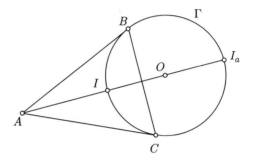

Figure B.4

Similarly, the exterior angle at C subtends the other arc BC, and I_a bisects that arc, so the ray BI_a bisects the exterior angle at B. Hence, since I_a is also on the angle bisector of A, we know it is one of the excenters, as desired. □

Exercise 1.82. Recall,

$$|\triangle ABC| = rs = r_a(s-a) = r_b(s-b) = r_c(s-c).$$

For simplicity, let us use Δ for the area of $\triangle ABC$. Then,

$$r = \frac{\Delta}{s}, \qquad r_a = \frac{\Delta}{s-a}, \qquad \text{etc.}$$

Hence,

$$rr_ar_br_c = \frac{\Delta^4}{s(s-a)(s-b)(s-c)}.$$

But, from Heron's formula,

$$\Delta^2 = s(s-a)(s-b)(s-c),$$

so

$$rr_ar_br_c = \frac{\Delta^4}{\Delta^2} = \Delta^2. \qquad □$$

Exercise 1.90. Consider the two triangles $\triangle ABD$ and $\triangle AEC$, as suggested. Note that $\angle ABD = \angle AEC$, since they subtend the same arc. Also, $\angle ADB = \angle ACE = 90°$. This is because AD is the altitude of $\triangle ABC$ and AE is a diameter. Thus,

$$\triangle ABD \sim \triangle AEC.$$

Hence,

$$\frac{|AD|}{|AC|} = \frac{|AB|}{|AE|}$$

$$|AD| = \frac{bc}{2R}.$$

Thus,

$$|\triangle ABC| = \frac{1}{2}a\left(\frac{bc}{2R}\right) = \frac{abc}{4R}.$$

Using the other hint, we note that

$$|AD| = c\sin B = c\left(\frac{b}{2R}\right) = \frac{bc}{2R},$$

and proceed as before. $\square$

Exercise 1.114. Consider Figure B.5. Let $\angle AOP = \alpha$ and $\angle BOA = \beta$,

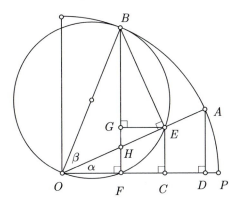

Figure B.5

so that $\angle BOP = \alpha + \beta$. Let BF be perpendicular to OP, and let BE be perpendicular to OA. Let EC be perpendicular to OP and let EG be perpendicular to BF. Note that $\angle EBF = \alpha$. To see this, consider the circle with diameter OB. Since the angles at both C and E are right angles, we know they are both on the circle. In particular, $\angle EBF = \angle EOF = \alpha$, since they both subtend the same arc.

We could also have noted that $\angle OHF = \angle BHE$, and since $\angle BEH = \angle HFO = 90°$, we get $\triangle OHF \sim \triangle BHE$. In particular, $\angle EBH = \angle HOF = \alpha$.

Now,

$$|FC| = |GE| = |BE|\sin\alpha = \sin\beta\sin\alpha$$

and

$$|OC| = |OE|\cos\alpha = \cos\beta\cos\alpha.$$

Thus,

$$\cos(\alpha + \beta) = |OC| = |OF| - |CF| = \cos\beta\cos\alpha - \sin\beta\sin\alpha.$$

Similarly,
$$|BG| = |BE|\cos\alpha = \sin\beta\cos\alpha$$

and
$$|GF| = |EC| = |OE|\sin\alpha = \cos\beta\sin\alpha.$$

Hence

$$\sin(\alpha + \beta) = |BC| = |BG| + |GC| = \sin\beta\cos\alpha + \cos\beta\sin\alpha. \qquad \square$$

Exercise 2.6. The triangle formed by Syene, Alexandria, and the sun is a long narrow one. We exaggerate that triangle in Figure B.6.

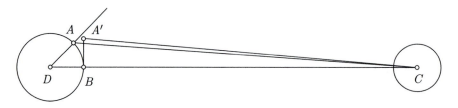

Figure B.6

In this diagram, Alexandria is labeled A, Syene is B, the center of the sun is C, and the center of the Earth is D. The exterior angle at A of $\triangle ADC$ is $7.2°$. Therefore,

$$\angle ADC + \angle ACD = 7.2°.$$

Rather than find $\angle ADC$, let us find an upper bound on it. To do this, we create a right angle triangle $\triangle A'BC$ with $|A'B| = 787$ km. Then

$$0 < \angle ACD < \angle A'CB,$$

since $|A'B|$ equals the arclength of the arc AB. But

$$\angle A'CB = \arctan\left(\frac{|A'B|}{|BC|}\right) = \arctan\left(\frac{787}{150{,}000{,}000}\right) = .0003°.$$

We now know the assumption is safe, since the difference between the measured $7.2°$ and the angle $\angle ADC = 7.1997$ is less than the implied accuracy of $7.2°$. Using this figure, we get the circumference of the Earth is 39,352 km, instead of 39,350 km.

Note that we used $|BC| = 150{,}000{,}000$ km, when we should have used $|DC| = 150{,}000{,}000$ km. The length $|DB|$ is approximately 6000 km, which is insignificant compared to $|DC|$, so this won't change our numbers. But I hear you thinking – isn't this a circular argument? How do we know what the radius of the Earth is, if that's what we're trying to calculate? Note that the question only asked us to evaluate Eratosthenes' assumption. It did not ask whether one can come up with a calculation which avoids that assumption.   □

Exercise 2.7. If the Earth is flat, then we get a triangle $\triangle ABC$ with Alexandria at A, Syene at B, and the sun at C (see Figure B.7). The angle $\angle ACB = 7.2°$, and $|AB| = 787$ km, so

$$\tan(7.2°) = \frac{|AB|}{|BC|} = \frac{787 \text{ km}}{|BC|}.$$

Thus, $|BC| = 6230$ km.   □

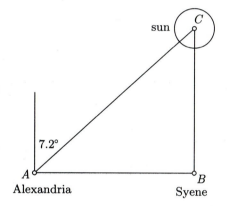

Figure B.7

Exercise 3.30. The regular 9-gon is constructible if and only if we can construct the length $\cos(2\pi/9)$. So let us set $\omega = e^{2\pi i/9} = \cos(2\pi/9) + i\sin(2\pi/9)$. Then, $\omega^9 = 1$, so

$$0 = \omega^9 - 1$$
$$= (\omega^3 - 1)(\omega^6 + \omega^3 + 1).$$

We note that the roots of $x^3 - 1$ are 1 and $e^{\pm 2\pi/3}$, so $\omega^3 - 1 \neq 0$, and hence ω is a root of

$$f(x) = x^6 + x^3 + 1.$$

Let $c = \omega + \omega^{-1} = 2\cos(2\pi/9)$. Then

$$c^3 = \omega^3 + 3\omega + 3\omega^{-1} + \omega^{-3},$$

so
$$c^3 - 3c + 1 = \omega^3 + 1 + \omega^{-3} = \omega^{-3}f(\omega) = 0.$$

Thus, c is the largest positive root of

$$g(x) = x^3 - 3x + 1.$$

To show that c is not constructible using Theorem 3.6.4, we must show that $g(x)$ is irreducible over the rationals. If $g(x)$ is reducible, then it factors into a linear and a quadratic rational polynomial, or into three linear factors. In either case, $g(x)$ has a linear factor, so has a rational root. By the rational root theorem, the only possible rational roots of $g(x)$ are ± 1. Since neither is a root, $g(x)$ cannot have any linear factors, so is irreducible. Since its degree is three, which is not a power of two, we know c is not constructible.   □

Exercise 3.39. Following the hint, the circle centered at (a, b) that goes through $(0, 0)$ has the equation

$$(x - a)^2 + (y - b)^2 = a^2 + b^2.$$

It is clear that we can construct such a circle if the lengths a and b are constructible. We find where this circle intersects the given parabola by substituting x^2 for y, to get

$$x^4 - (2b - 1)x^2 - 2ax = 0.$$

We knew this would have the solution $x = 0$ since we chose the circle so that it would go through the point $(0, 0)$, which is a point on the parabola. After we factor out x, we get the cubic

$$x^3 - (2b - 1)x - 2a = 0.$$

To construct the regular 7-gon, we must be able to construct a length w where w is a root of

$$w^3 + w^2 - 2w - 1 = 0$$

(see Exercise 3.26). We can get rid of the w^2 term by using a trick similar to completing the square: We make the substitution $w = u - 1/3$. We therefore get

$$(u - 1/3)^3 + (u - 1/3)^2 - 2(u - 1/3) - 1 = 0,$$

and after simplifying, we have

$$u^3 - \frac{7}{3}u - \frac{7}{27} = 0.$$

We therefore choose $a = \frac{7}{54}$ and $b = \frac{5}{3}$. These are both constructible numbers, so we can construct the circle centered at (a, b) which goes through $(0, 0)$. We drop a perpendicular from one of the other points of intersection to the x-axis (the line through the two given points) to find the point a distance u away from the origin. We subtract $1/3$ from u to get w, from which we can construct the regular 7-gon.   □

Exercise 6.1. Let the quadrilateral be $ABCD$. It looks as if all we have to do is point out that the sum of the angles in $\triangle ABC$ and $\triangle BCD$ are both less than $180°$, and the sum of the angles in the quadrilateral $ABCD$ is the sum of these, but it is possible that the points A and D are on the same side of the diagonal BC. However, in this case, either A is inside $\triangle BCD$ or D is inside $\triangle ABC$. In either case, the points B and C are on opposite sides of the diagonal AD, and the above argument works. $\square$

Exercise 6.4. Without loss of generality, we may assume l enters the quadrilateral $ABCD$. It therefore must intersect either BC or CD, since if it exits through either other side, we would have a contradiction to the first axiom. If it intersects BC, then we are finished, so suppose it intersects DC at E. Note that the sum of the angles in $\triangle ADE$ is $180°$, for if it were smaller, then the sum of the angles in the quadrilateral $ABCE$ would be larger than $360°$, which contradicts Exercise 6.1. Now reproduce a congruent quadrilateral $A_1B_1C_1D_1$ with the same orientation and with $A_1 = D$ and $B_1 = C$, as in Figure B.8. Let EF_1 be perpendicular to CD

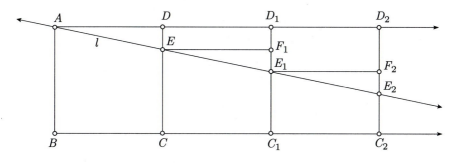

Figure B.8

with F_1 on C_1D_1. If l intersects CC_1 then we are finished, so suppose l intersects C_1D_1 at E_1. Then $|EF_1| = |DD_1| = |AD|$ (by Exercise 6.2), $\angle E_1EC_1 = 180° - 90° - \angle DEA = \angle DAE$, and $\angle EF_1E_1 = 90° = \angle ADE$ (by Exercise 6.2). Thus, $\triangle ADE \equiv \triangle EF_1E_1$, so $|D_1E_1| = 2|DE|$. We can continue this indefinitely, to get $|D_kE_k| = (k+1)|DE|$. For some k, we have $(k+1)|DE| > |DC|$, so l must eventually intersect the line BC. $\square$

Exercise 7.35. Let us first find the images of these points:

$$\begin{bmatrix} 2 & 1 \\ 1 & 1 \end{bmatrix} i = \frac{2i+1}{i+1} = \frac{(2i+1)(1-i)}{(1+i)(1-i)} = \frac{2i+1+2-i}{2} = \frac{3+i}{2}.$$

$$\begin{bmatrix} 2 & 1 \\ 1 & 1 \end{bmatrix}(-1+i) = \frac{-2+2i+1}{-1+i+1} = 2+i$$

$$\begin{bmatrix} 2 & 1 \\ 1 & 1 \end{bmatrix}(1+i) = \frac{2+2i+1}{1+i+1} = \frac{(3+2i)(2-i)}{5} = \frac{8+i}{5}.$$

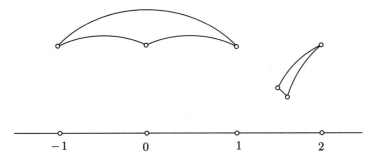

Figure B.9

The rest of the solution is Figure B.9. □

Exercise 7.37. The images of P and Q are

$$\begin{bmatrix} 1 & 2 \\ -1 & 2 \end{bmatrix} (2 + 4i) = \frac{2 + 4i + 2}{-2 - 4i + 2} = \frac{4 + 4i}{-4i} = i - 1.$$

$$\begin{bmatrix} 1 & 2 \\ -1 & 2 \end{bmatrix} \left(\frac{6 + 4i}{3} \right) = \frac{6 + 4i + 6}{-6 - 4i + 6} = \frac{12 + 4i}{-4i} = 3i - 1.$$

To find the distance between P and Q, we can either note that $-1 + i$ and $-1 + 3i$ lie on a vertical line and hence

$$|PQ| = |\gamma P \gamma Q| = |\ln(3/1)| = \ln 3,$$

or we can note that both $P = 2 + 4i$ and $Q = 2 + (4/3)i$ lie on a vertical line, so we can calculate $|PQ|$ directly:

$$|PQ| = \left| \ln \left(\frac{4}{4/3} \right) \right| = \ln 3.$$ □

Exercise 7.46. We solve

$$(z, 1; 0, \infty) = (w, 1; -i, i)$$

$$z = \frac{w + i}{w - i} \Big/ \frac{i + 1}{1 - i}$$

$$= \left(\frac{(1 - i)}{(1 - i)} \frac{(1 + i)}{(1 - i)} \right) \left(\frac{w + i}{w - 1} \right)$$

$$= \frac{-2i}{2} \left(\frac{w + i}{w - i} \right)$$

$$= \frac{w + i}{iw + 1}$$

$$= \begin{bmatrix} 1 & i \\ i & 1 \end{bmatrix} w$$

$$\begin{bmatrix} 1 & -i \\ -i & 1 \end{bmatrix} z = w.$$

Thus, the fractional linear transformation is given by $\begin{bmatrix} 1 & -i \\ -i & 1 \end{bmatrix}$. Since the line through 0, 1, and ∞ is the real line, we know that the real line goes through the line or circle that goes through the points 1, i, and $-i$. This is the unit circle. Note that i is sent to 0, so the upper half plane $\mathcal{H}$ is sent to the interior of the unit disk. Since fractional linear transformations preserve angles, a line or half circle perpendicular to the real axis is sent to a diameter or arc of a circle which is perpendicular to the unit circle and is inside the unit disk. This is the 'crutch' introduced in Chapter 6.   □

Exercise 7.60. We know the image of only two points, so we must first find the image of a third. The (Poincaré) line through $2i$ and ∞ goes through 0 too, so we seek the image of 0. This will be the other end of the Poincaré line through -1 and $3i$, which is a half circle centered at x, where the Euclidean distance from x to -1 is the same as from x to $3i$. That is,

$$(x+1)^2 = 3^2 + x^2$$
$$x^2 + 2x + 1 = x^2 + 9$$
$$x = 4.$$

Thus the radius of the half circle is 5, and the other end of the Poincaré line through -1 and $3i$ is at $4 + 5 = 9$. We can now solve for the map γ:

$$(z, 2i; \infty, 0) = (w, 3i; -1, 9)$$
$$\frac{z - \infty}{z - 0} \Big/ \frac{2i - \infty}{2i - 0} = \frac{w + 1}{w - 9} \Big/ \frac{3i + 1}{3i - 9}$$
$$\frac{2i}{z} = \left(\frac{w + 1}{w - 9}\right) \left(\frac{3i(1 + 3i)}{1 + 3i}\right)$$
$$\frac{2(w - 9)}{3(w + 1)} = z$$
$$\begin{bmatrix} 2 & -18 \\ 3 & 3 \end{bmatrix} w = z$$
$$w = \begin{bmatrix} 3 & 18 \\ -3 & 2 \end{bmatrix} z.$$

Thus, $\gamma = \begin{bmatrix} 3 & 18 \\ -3 & 2 \end{bmatrix}$. It has real entries, and positive determinant, so it is an isometry and is in $\mathrm{GL}_2(\mathbb{R})$. It sends $2i$ to $3i$ and ∞ to -1, as desired.   □

Exercise 7.63. We must first find the image of a third point under this isometry. The line through $1 + i$ and 1 also goes through ∞. The image of this line must be a line. So, we seek the other endpoint of the line through

2 and $1 + i$. This is the half circle centered at 1, and so the other endpoint is 0. Thus, we must solve

$$(z, 1 + i; 1, \infty) = (w, 1 + i; 2, 0)$$

$$\frac{z - 1}{i} = \frac{w - 2}{w} \Big/ \frac{-1 + i}{1 + i}$$

$$\frac{z - 1}{i} = -i \left(\frac{w - 2}{w} \right)$$

$$\begin{bmatrix} 1 & -1 \\ 0 & 1 \end{bmatrix} z = \begin{bmatrix} 1 & -2 \\ 1 & 0 \end{bmatrix} w$$

$$w = \begin{bmatrix} 0 & 2 \\ -1 & 1 \end{bmatrix} \begin{bmatrix} 1 & -1 \\ 0 & 1 \end{bmatrix} z$$

$$= \begin{bmatrix} 0 & 2 \\ -1 & 2 \end{bmatrix} z.$$

Since the determinant of this matrix is positive, it is an isometry of $\mathcal{H}$. By construction, it fixes $1 + i$ and sends 1 to 2. □

Exercise 7.77. Let $P = 12 + 5i$ and $Q = 5 + 12i$. To find $|PQ|$, we must fist find the endpoints of the Poincaré line through P and Q. In particular, we must find the real value x so that the Euclidean distance from P to x is the same as from Q to x. Thus, we solve

$$(12 - x)^2 + 5^2 = (5 - x)^2 + 12^2.$$

This is satisfied by $x = 0$. Then, the Euclidean distance from 0 to P is 13. Thus,

$$|PQ| = |\ln(12 + 5i, 5 + 12i; 13, -13)|$$

$$= \left| \ln \left(\frac{12 + 5i - 13}{12 + 5i + 13} \Big/ \frac{5 + 12i - 13}{5 + 12i + 13} \right) \right|$$

$$= \left| \ln \left(\frac{(5i - 1)}{(25 + 5i)} \frac{(18 + 12i)}{(-8 + 12i)} \right) \right|$$

$$= \left| \ln \left(\frac{(5i - 1)6i(-3i + 2)}{-5i(5i - 1)(-4)(2 - 3i)} \right) \right|$$

$$= |\ln(3/10)| = \ln(10/3). \qquad \square$$

Exercise 7.78. Let us solve

$$\gamma A = \frac{16 + 2i - 15}{-8 - i + 10} = i$$

$$\gamma M = \frac{10 - 9}{-5 + 6} = 1$$

$$\gamma N = \frac{4 - 3}{-2 + 2} = \infty.$$

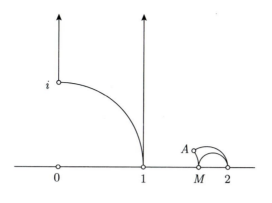

Figure B.10

Let us now draw both $\triangle AMN$ and its image (see Figure B.10). In the original, the angle at A looks like a right angle, but it may not be so easy to show. After applying γ, it is obvious that the angle at i is $\pi/2$. Thus, the area is

$$|\triangle AMN| = \pi - \pi/2 = \pi/2. \qquad \square$$

Exercise 7.81. Let

$$x + iy = \phi^{-1}(re^{i\theta}) = \frac{re^{i\theta} + i}{ire^{i\theta} + 1}.$$

Then

$$x - iy = \frac{re^{-i\theta} - i}{-ire^{-i\theta} + 1}.$$

Subtracting, we get

$$2iy = \frac{(re^{i\theta} + i)(-ire^{-i\theta} + 1) - (re^{-i\theta} - i)(ire^{i\theta} + 1)}{q\bar{q}}$$

$$= \frac{2i(1 - r^2)}{q\bar{q}},$$

where $q = ire^{i\theta} + 1$. Also,

$$dx + idy = \frac{e^{i\theta}(ire^{i\theta} + 1) - (re^{i\theta} + i)ie^{i\theta})dr}{q^2}$$

$$+ \frac{(ire^{i\theta}(ire^{i\theta} + 1) - (re^{i\theta} + i)(-re^{i\theta}))d\theta}{q^2}$$

$$= \frac{2e^{i\theta}dr + 2ire^{i\theta}d\theta}{q^2}.$$

Conjugating (which commutes with differentiation), we get

$$dx - idy = \frac{2e^{-i\theta}dr - 2ire^{-i\theta}d\theta}{\bar{q}^2}.$$

Thus,

$$ds^2 = \frac{dx^2 + dy^2}{y^2} = \frac{4(e^{i\theta}dr + ie^{i\theta}d\theta)(e^{-i\theta}dr - ire^{-i\theta}d\theta)(q\bar{q})^2}{q^2\bar{q}^2(1-r^2)^2}$$

$$= \frac{4(dr^2 + irdrd\theta - irdrd\theta + r^2d\theta^2)}{(1-r^2)^2}$$

$$= \frac{4(dr^2 + r^2d\theta^2)}{(1-r^2)^2}. \qquad \square$$

Exercise 7.82. We first note that

$$\begin{bmatrix} a & b \\ \bar{b} & \bar{a} \end{bmatrix}\begin{bmatrix} c & d \\ \bar{d} & \bar{c} \end{bmatrix} = \begin{bmatrix} ac + b\bar{d} & ad + b\bar{c} \\ \bar{b}c + \bar{a}\bar{d} & \bar{b}d + \bar{a}\bar{c} \end{bmatrix},$$

which is in Γ, so Γ is indeed a group. We next show that every element of Γ is an isometry of $\mathcal{D}$. Note that if an element in $SL_2(\mathbb{C})$ sends $\mathbb{R}$ to $\mathbb{R}$, then as a fractional linear transformation, it can be written so that it is in $SL_2(\mathbb{R})$, so it is an isometry of $\mathcal{H}$. Thus, any element of $SL_2(\mathbb{C})$ which sends $\mathcal{D}$ to $\mathcal{D}$ is an isometry of $\mathcal{D}$. We therefore check the boundary: Suppose $z\bar{z} = 1$ and $\gamma = \begin{bmatrix} a & b \\ \bar{b} & \bar{a} \end{bmatrix} \in \Gamma$. Then

$$\gamma z \overline{\gamma z} = \left(\frac{az+b}{\bar{b}z+\bar{a}}\right)\left(\frac{\bar{a}\bar{z}+\bar{b}}{b\bar{z}+a}\right)$$

$$= \frac{a\bar{a} + b\bar{a}\bar{z} + \bar{b}az + b\bar{b}}{\bar{b}b + \bar{a}b\bar{z} + \bar{b}az + a\bar{a}}$$

$$= 1.$$

That is, points a distance one away from zero are sent to points a distance one away from zero. We must also check that the interior is sent to the interior. To see this, note that

$$0 \leq \gamma(0)\overline{\gamma(0)} = \frac{b}{\bar{a}}\frac{\bar{b}}{a} = \frac{a\bar{a}-1}{a\bar{a}} = 1 - \frac{1}{a\bar{a}} < 1.$$

That is, the point 0 is sent to a point inside $\mathcal{D}$. Note that $a\bar{a} - b\bar{b} = 1$, so $a\bar{a} \neq 0$. Finally, we must check that every isometry of $\mathcal{D}$ is in Γ. Recall that σ and τ_a as a ranges over $\mathbb{R}$ generate the group of isometries of $\mathcal{H}$. Thus, we need only check that $\phi\sigma\phi^{-1}$ and $\phi\tau_a\phi^{-1}$ are in Γ.

$$\phi\sigma\phi^{-1} = -\frac{1}{2}\begin{bmatrix} -i & 1 \\ 1 & -i \end{bmatrix}\begin{bmatrix} 0 & -1 \\ 1 & 0 \end{bmatrix}\begin{bmatrix} i & 1 \\ 1 & i \end{bmatrix}$$

$$= \begin{bmatrix} -i & 0 \\ 0 & i \end{bmatrix} \in \Gamma.$$

$$\phi\tau_a\phi^{-1} = -\frac{1}{2}\begin{bmatrix} -i & 1 \\ 1 & -i \end{bmatrix}\begin{bmatrix} 1 & a \\ 0 & 1 \end{bmatrix}\begin{bmatrix} i & 1 \\ 1 & i \end{bmatrix}$$

$$= -\frac{1}{2} \begin{bmatrix} 2 - ia & a \\ a & 2 + ia \end{bmatrix} \in \Gamma. \qquad \square$$

Exercise 7.90. First, we note that every triangle $\triangle ABC$ can be imbedded in a triply asymptotic triangle (see Figure 7.18). Thus, the incircle of $\triangle ABC$ is smaller than the incircle of some triply asymptotic triangle. But all triply asymptotic triangles are congruent, so the largest circle that can be inscribed in a triangle is the incircle of a triply asymptotic triangle. Let us therefore consider the triangle $\triangle MNP$ with $M = -1$, $N = 1$, and $P = \infty$, as in Figure B.11. Since circles in $\mathcal{H}$ are Euclidean circles which lie in $\mathcal{H}$, it is clear that the incircle of $\triangle MNP$ is the one depicted. By symmetry with respect to the y-axis, the segment AB is a diameter, where $A = i$ and $B = 3i$. Finally, the length of AB is $\ln 3$. $\qquad \square$

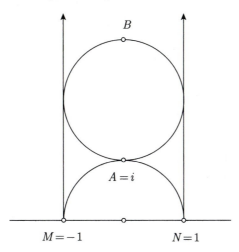

Figure B.11. See Exercise 7.90.

Exercise 8.9. Let the angles in the equilateral triangular tiles all be θ. Since we can tile with these with seven at each vertex, we know that

$$7\theta = 2\pi.$$

Thus, $\theta = 2\pi/7$, and the area of each tile is

$$\pi - 3(2\pi/7) = \pi/7. \qquad \square$$

Exercise 8.15. Let $a = e^r$, so that the circumcircle of the n-gon has a radius of hyperbolic length r. Let B be an adjacent vertex and let C be the circumcenter. Then $\triangle ABP$ is an isosceles triangle with $A = B = \pi/m$, $P = 2\pi/n$, and with equal sides $|PA|$ and $|PB|$ of length r. Let us apply the Law of Cosines for angles:

$$\cos(\pi/m) = -\cos(\pi/m)\cos(2\pi/n) + \sin(\pi/m)\sin(2\pi/n)\cosh r$$

$$\cosh r = \frac{\cos(\pi/m)(1 + \cos(2\pi/n))}{\sin(\pi/m)\sin(2\pi/n)}$$

$$= \frac{\cos(\pi/m)(2\cos^2(\pi/n))}{\sin(\pi/m)2\sin(\pi/n)\cos(\pi/n)}$$

$$= \cot(\pi/m)\cot(\pi/n).$$

Thus,

$$a = e^r$$

$$= \cosh r + \sinh r$$

$$= \cot(\pi/m)\cot(\pi/n) + \sqrt{1 + \cot(\pi/m)\cot(\pi/n)}$$

$$= \frac{\cos(\pi/m)\cos(\pi/n) + \sqrt{\sin(\pi/m)\sin(\pi/n) + \cos(\pi/m)\cos(\pi/n)}}{\sin(\pi/m)\sin(\pi/n)}. \qquad \square$$

Exercise 8.21. Let A be the center of the disc. Let B and C be adjacent vertices of the square centered at A. Let D be the center of the triangle with edge BC. Then the area of each fish is the same as the area of $ABDC$. Clearly, $A = \pi/2$ and $D = 2\pi/3$. We need six quadrilaterals congruent to $ABDC$ to tile around the point B, so $B = C = \pi/3$. Thus, the area of the quadrilateral $ABDC$ is

$$2\pi - \pi/2 - 2\pi/3 - 2(\pi/3) = \pi/6. \qquad \square$$

Exercise 9.3. Since l intersects $\triangle ABC$, it must intersect one of the sides, say AB. Thus, A and B are on opposite sides of l. The point C is on one of these sides, say the same side as B. Then l does not intersect BC, yet because 'same side' is an equivalence, A and C cannot be on the same side, so l intersects AC too. Thus, l intersects exactly two of the sides of $\triangle ABC$. $\qquad \square$

Exercise 9.10. Choose points B' and C' on the lines AB and AC, respectively, and such that A is between B and B', and also between C and C' (see Figure B.12). Suppose the segment BC' and the line l intersect at Q.

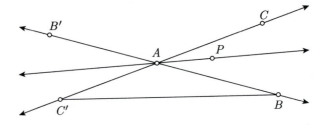

Figure B.12

Since $\triangle ABC$ is not degenerate, we know $Q \neq A$. By Exercise 9.7, Q is inside $\angle BAC'$, and hence, by Exercise 9.8, not inside $\angle BAC$ or $\angle B'AC'$. But, by Exercises 9.6 and 9.9, every point on the line $l = AP$ is either inside $\angle BAC$, inside $\angle B'AC'$, or is A. Thus, the point Q cannot exist. That is, B and C' are on the same side of the line l. Since C' and C are on opposite sides of l, we get that B and C are on opposite sides of l. Hence, the line l intersects the segment BC. □

Exercise 10.26. Since there are two pentagons and two triangles at each vertex and the sum of the angles at each vertex is 2π, one angle in the triangle and one angle in the pentagon sum to π. That is, opposite edges at a vertex form a straight line. Thus, there exists a sequence of edges which form a great circle. We count how many edges there are along a great circle (see page 105), and find that there are ten. Thus, each edge has length

$$s = \frac{2\pi}{10} = \frac{\pi}{5}.$$ □

Exercise 10.27. By Exercise 10.26, the sides of each triangle have length $\pi/5$. Let each angle be θ. Then, by the Law of Cosines for sides, we get

$$\cos(\pi/5) = \cos^2(\pi/5) + \sin^2(\pi/5)\cos\theta.$$

Note that

$$\cos(\pi/5) = \frac{1 + \sqrt{5}}{4},$$

so we get

$$\frac{1 + \sqrt{5}}{4} = \frac{3 + \sqrt{5}}{8} + \frac{5 - \sqrt{5}}{8}\cos\theta$$

$$\cos\theta = \frac{2 + 2\sqrt{5} - 3 - \sqrt{5}}{5 - \sqrt{5}}$$

$$= \frac{-1 + \sqrt{5}}{\sqrt{5}(\sqrt{5} - 1)} = \frac{1}{\sqrt{5}}.$$

Thus, the area A of each of the 20 triangles is

$$A = 3\theta - \pi = 3\arccos(1/\sqrt{5}) - \pi,$$

and the percentage of the sphere covered by triangles is

$$\frac{20A}{4\pi} = \frac{15\arccos(1/\sqrt{5})}{\pi} - 5 \approx 28.62\%.$$ □

Exercise 12.8. Since

$$x^2 + y^2 - z^2 = -1,$$

we have

$$2x\,dx + 2y\,dy - 2z\,dz = 0$$

$$dz = \frac{1}{z}(xdx + ydy).$$

We plug this into the formula for ds^2 to get

$$
\begin{aligned}
ds^2 &= \frac{1}{z^2}\left(z^2 dx^2 + z^2 dy^2 - (xdx + ydy)^2\right) \\
&= \frac{1}{z^2}\left((z^2 - x^2)dx^2 + (z^2 - y^2)dy^2 - 2xydxdy\right) \\
&= \frac{1}{z^2}\left((y^2 + 1)dx^2 + (x^2 + 1)dy^2 - 2xydxdy\right) \\
&= \frac{1}{z^2}\left(dx^2 + dy^2 + (y^2 dx^2 - 2xydxdy + x^2 dy^2)\right) \\
&= \frac{1}{z^2}\left(dx^2 + dy^2 + (ydx - xdy)^2\right). \qquad \square
\end{aligned}
$$

Exercise 13.20. Pick a line l and a point P not on l. For any point $Q \neq P$, the line PQ intersects l (by Axiom 4p) at, say, Q'. Thus, every point $Q \neq P$ lies on some line which intersects l. Furthermore, this line is unique (by Axiom 1). There are n points on l, so there are n such lines. Each line contains n points, one of which is P. Hence, there are $n(n-1) + 1 = n^2 - n + 1$ points in $\mathcal{G}$. $\qquad \square$

Exercise 13.23. Suppose this geometry has m points. Pick a point P. Every line through P contains three points and no two points are on two different lines through P, so there must be exactly $\frac{m-1}{2}$ lines through P. Note that this means m is odd. Suppose a line l does not include P. Since l has three points, there are exactly three lines through P which intersect l. Since there are at least two lines through P which do not intersect l, there must be at least five lines through P. Thus,

$$\frac{m-1}{2} \geq 5.$$

Hence, $m \geq 11$. Finally, let us count the total number of lines. There are m points, and $\frac{m-1}{2}$ lines going through each point. Since every line contains 3 points, we have counted each line three times. Thus, there are $\frac{m(m-1)}{6}$ lines. In particular, $m \not\equiv 2 \pmod 3$, so $m \neq 11$. Hence, $m \geq 13$. $\qquad \square$

Bibliography

[Be] Beyer, W.H., editor, *CRC Standard Mathematical Tables*, 27th edition, CRC Press, Boca Raton, FL, 1984.

[Bo] Boyd, D.W., "The residual set dimension of the Apollonian packing," *Mathematika* **20** (1973), 170 – 174.

[C] Coxeter, H. S. M., *Introduction to Geometry*, 2nd edition, John Wiley & Sons, Inc., New York, 1969.

[Dem] Dembowski, P., *Finite Geometries*, Springer Verlag, New York, 1968.

[Des] Desharnes, Robert, Néret, Gilles, *Salvador Dalí*, Borders Press, Köln, Germany, 1998.

[Di] Dickson, Leonard E., *History of the Theory of Numbers, Volume II, Diophantine Analysis*, Chelsea Publishing Co., New York, 1952.

[E] Ernst, Bruno, *The Magic Mirror of M.C. Escher*, Ballantine Books, New York, 1976.

[G-L] Graham, L., Lagarias, J., Mallows, C., Wilks, A., Yan, C., "Integral Apollonian circle packings," to appear.

[Gre] Greenberg, Marvin J., *Euclidean and Non-Euclidean Geometries*, 3rd edition, W. H. Freeman and Co., New York, 1993.

[Gro1] Grossman, J.P., "Ye olde geometry shoppe – Part I, Triangular trivia," *Mathematical Mayhem* **6**, No. 2 (1994), 13 – 17.

[Gro2] Grossman, J.P., "Ye olde geometry shoppe – Part II," *Mathematical Mayhem* **6**, No. 3 (1994), 7 – 12.

[Gro3] Grossman, J.P., "Ye olde geometry shoppe – Part III," *Mathematical Mayhem* **6**, No. 4 (1994), 19 – 24.

[Gro4] Grossman, J.P., "Ye olde geometry shoppe – Part IV," *Mathematical Mayhem* **6**, No. 5 (1994), 15 – 19.

[Gu1]   Guy, Richard K., *Unsolved Problems in Number Theory*, 2nd edition, Springer Verlag, New York, 1994.

[Gu2]   Guy, Richard K., "My favorite elliptic curve: A tale of two types of triangles," *Math. Monthly* **102**, No. 9 (1995), 771 – 781.

[Ha]   Hales, Thomas C., "Cannonballs and Honeycombs," *Notices of the Amer. Math. Soc.* **47**, No. 4 (2000), 440 – 449.

[H-S]   Hall, H. S., Stevens, F. H., *Euclid, Books I to VI and XI*, MacMillan and Co., Ltd., New York, 1903.

[H1]   Heath, Sir Thomas L., *The Thirteen Books of Euclid's Elements*, Vol. 1 (Books I and II), translated from the text of Heiberg, Dover Publications, New York, 1956.

[H2]   Heath, Sir Thomas L., *Greek Astronomy*, AMS Press, New York, 1969, first published 1932 by J. M. Dent and Sons.

[H3]   Heath, Sir Thomas L., *A History of Greek Mathematics, Volume I*, Oxford University Press, London, 1960, first published in 1921.

[H4]   Heath, Sir Thomas L., *A History of Greek Mathematics, Volume II*, Oxford University Press, London, 1960, first published in 1921.

[Her]   Herstein, I. N., *Topics in Algebra*, 2nd Edition, John Wiley and Sons, Inc., New York, 1975.

[Hi]   Hilbert, David, *Foundations of Geometry*, Open Court Publishing Company, Peru, IL, 1994. (Translated by Leo Unger from the tenth revised and enlarged edition by Paul Barnays.)

[Kz]   Katz, V. J., *A History of Mathematics*, 2nd Edition, Addison-Wesley, Reading, MA, 1998.

[Ky]   Kay, David C., *College Geometry, A Discovery Approach*, Harper-Collins, New York, 1994.

[Ke]   Kedlaya, Kiran S., *Notes on Euclidean Geometry*, unpublished.

[Kl]   Kline, Morris, *Mathematical Thought from Ancient to Modern Times*, Volume 3, Oxford University Press, New York, 1972.

[M]   Manzel, Donald H., *Astronomy*, Random House, New York, 1975.

[P]   Patterson, S. J., "A lattice-point problem in hyperbolic space," *Mathematika* **22**, No. 1 (1975), 81–88.

[Ra]   Ratcliffe, John G., *Foundations of Hyperbolic Manifolds*, Springer Verlag, New York, 1994.

[Ro] Rosen, M., "The history of Fermat's Last Theorem," *Modular forms and Fermat's Last Theorem*, ed. by G. Cornell, J.H. Silverman, and G. Stevens, Springer Verlag, New York, 1997, 505 – 525.

[Sib] Sibley, Thomas Q., *The Geometric Viewpoint*, Addison-Wesley, Reading, MA, 1998.

[Sim] Simmons, George F., *Calculus Gems*, McGraw-Hill, Inc., New York, 1992.

[Sil] Silverman, J. H., *A Friendly Introduction to Number Theory*, Prentice Hall, Upper Saddle River, NJ, 1997.

[S-T] Silverman, J. H., Tate, J., *Rational Points on Elliptic Curves*, Undergraduate Texts in Mathematics, Springer Verlag, New York, 1992.

[Sm] Smart, James R., *Modern Geometries*, 4th edition, Brooks/Cole Publishing Co., Pacific Grove, CA, 1994.

[Sti] Stillwell, John, "Exceptional Objects," *Amer. Math. Monthly* **105**, No. 9 (1998), 850 – 858.

[Str] Struik, Dirk J., *A Concise History of Mathematics*, 4th revised edition, Dover Publications, Toronto, Canada, 1987.

[VdP] van der Poorten, Alf, *Notes on Fermat's Last Theorem*, John Wiley & Sons, Inc., New York, 1996.

[VY] van Yzren, Jan, "A simple proof of Pascal's hexagon theorem," *Amer. Math. Monthly* **100**, No. 10 (1993), 930 – 931.

[W] Weil, André, *Number Theory, An Approach Through History*, Birkhäuser, Boston, MA, 1984.

Index